Applied
Mathematics
in Engineering
Practice

FREDERICK S. MERRITT

Consulting Engineer, Syosset, New York

McGRAW-HILL BOOK COMPANY

*New York   St. Louis   San Francisco
Düsseldorf   London   Mexico
Panama   Sydney   Toronto*

Sponsoring Editor   Tyler G. Hicks
Director of Production   Stephen J. Boldish
Editing Supervisor   Frank Purcell
Designer   Naomi Auerbach
Editing and Production Staff   Gretlyn Blau,
   Teresa F. Leaden, George E. Oechsner

APPLIED MATHEMATICS IN ENGINEERING PRACTICE

07-041511-0

1234567890 MAMM 7543210

Applied
Mathematics
in Engineering
Practice

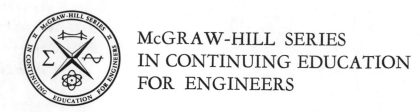

McGRAW-HILL SERIES IN CONTINUING EDUCATION FOR ENGINEERS

Preface

To use the newest high-speed computers profitably, engineers must be familiar with appropriate mathematical methods. Some of these methods may be new to you. Other methods, if you are to use them, may require review to refresh your memory. Thus, to advance with your profession, you have to continue your education in mathematics. The purpose of this book is to help you do this.

A major objective, therefore, is to make self-study as easy as possible. To achieve this, the author and editor have strived for simplicity, clarity, and conciseness. Nevertheless the book deals with advanced concepts, and despite all efforts, you may find progress difficult at times.

The book provides a quick review of essential undergraduate-level mathematics, covers applied calculus, presents the basics of functions of complex variables, and introduces you to probability and statistics. The text assumes that you are familiar with the mathematics generally regarded as prerequisite for study of differential and integral calculus in college. Hence, the book does not include these subjects. (For a quick review, refer to F. S. Merritt, "Mathematics Manual," McGraw-Hill Book Company, New York.)

In general, to conserve space and to insure continuity of thought, the book omits history, proofs, and derivations. Readers who would like this information or more details can find it in books listed in bibliographies at the end of the chapters. Some derivations are included, however, to clarify concepts.

The first two chapters review differential and integral calculus. They cover methods that will be needed later in the book as well as methods that you will find useful in engineering practice.

Chapters 3 through 7 deal with methods for solving differential equations likely to be encountered in engineering work. Solutions may be classified as open or closed. Open solutions can be expressed as formulas, which can be evaluated for many given conditions on substitution of appropriate values for the variables. Closed solutions provide numerical values for specific conditions, or one set of values for the variables. While high-speed computers are useful in obtaining both open and closed solutions, these machines are especially valuable for performing the tedious and numerous calculations often required for closed solutions of differential equations. Thus, computers have made practical the solution of engineering problems that previously could not be solved or had to be solved by approximate methods. In the chapters on differential equations, you will find many methods that lend themselves to computer solution.

Chapter 8 presents additional integrating techniques in which use is made of complex variables. It introduces the powerful technique of conformal mapping. This often simplifies the solution of differential equations by transforming complicated geometric boundaries into simpler ones, sometimes into shapes for which solutions are already known.

The last two chapters are intended to help you solve problems in which uncertainty is a factor. They provide an introduction to probability and statistics.

To help you get the most out of this book, the following suggestions are offered:

Read each article through for the main ideas. If you have any questions as you proceed, wait until you finish the article before you try to answer them. Often, you will find the answer in the text before the end of the article, perhaps in the next sentence or in the next paragraph. After the first reading of an article, go through it again. This time, be sure you understand the meaning of each sentence.

Problems and solutions are provided at the end of each chapter. (There are about 300 in total.) By trying to solve the problems, you will improve your grasp of the subject.

You will profit most from time spent on this book if, as you read it, you will relate what you are learning to your every-day activities.

If the author is successful in his aims, you should be inspired on completion of your studies with this book to continue your education in mathematics, perhaps with the aid of the specialized texts in the bibliographies. You also will find it worthwhile to become familiar with the techniques discussed in a supplemental volume to this, "Modern Mathematical Methods in Engineering," which includes matrix algebra and vector and tensor calculus.

The author is indebted to many sources, too numerous to list, for the information in this book. His treatment of the various topics, however, usually is different from that of the source material, which often was theoretical or academic. The objective throughout this volume is to show how to apply mathematics in engineering practice.

Frederick S. Merritt

Contents

ONE

Differential Calculus

Differential calculus permits the development of methods for determining the instantaneous rate of change of a variable. For example, suppose that an airplane flies 2,000 miles between two cities in 5 hours. Its average speed is $2,000/5 = 400$ mph. But suppose now it is important for you to know what its cruising speed was, or how fast it was going when it was halfway between the two cities. The average speed is not the correct answer, because the time of flight includes time for takeoff, climbing, and landing. You want to know the speed of the airplane at a specific instant, or when the plane was in specific locations. You can solve this problem by differential calculus if you know the relation between location of plane and time.

1-1 *Limits of Functions.* Let us examine the behavior of a function $f(x)$ of a real variable x as x approaches a specific real number a. Since $f(x)$ is an arbitrary function, it may be continuous or discontinuous at a; $f(x)$ may have any value we desire to assign to it at a, or it may not exist. Yet, $f(x)$ may have a definite value as x approaches a. We call such a value the limit of $f(x)$ if it satisfies the following definition:

1

f(x) converges to a limit L, a constant, as x approaches a if and only if for every positive real number ε there exists a positive number δ such that when $0 < |x - a| < δ$, then f(x) is defined and $|f(x) - L| < ε$.

We write "*x* approaches *a*" as "$x \to a$" and the limit as

$$\lim_{x \to a} f(x) = L$$

The latter is read, "The limit of $f(x)$, as x approaches a, equals L."

For example, let us find the limit of $(1 - x)/(1 - x^2)$ as $x \to 1$. The function has the indeterminate value $0/0$ at $x = 1$. Let us assume, therefore, that $x \neq 1$ and divide numerator and denominator by $x - 1$:

$$\frac{1 - x}{1 - x^2} = \frac{1 - x}{(1 - x)(1 + x)} = \frac{1}{1 + x}$$

Now, let x differ from 1 by only a small amount δ. Then $1/(1 + x)$ will differ from ½ by only a small amount ε. We can decrease this difference by decreasing the difference between x and 1. Hence,

$$\lim_{x \to 1} \frac{1 - x}{1 - x^2} = \frac{1}{2}$$

Some functions converge to a definite value even when x increases without bound ($x \to \infty$). We call such a value the limit of $f(x)$ if it satisfies the following definition:

f(x) converges to a limit L as $x \to \infty$ if and only if for every positive real number ε there exists a positive real number N such that when $x > N$, then f(x) is defined and $|f(x) - L| < ε$.

For example, let us find the limit of $(6x^3 - 2x^2)/(2x^3 - 3x + 2)$ as $x \to \infty$. The function becomes ∞/∞ if $x = \infty$. But let us divide numerator and denominator by x^3 and then observe the behavior of the function as $x \to \infty$.

$$\frac{6x^3 - 2x^2}{2x^3 - 3x + 2} = \frac{6 - 2/x}{2 - 3/x^2 + 2/x^3}$$

Notice that the larger x becomes, the smaller are $2/x$, $3/x^2$, and $2/x^3$. Hence, when x is a very large number N, the function differs from $6/2 = 3$ by only a small amount. We can decrease this difference by making x larger than N. Therefore,

$$\lim_{x \to \infty} \frac{6x^3 - 2x^2}{2x^3 - 3x + 2} = 3$$

The following theorems are useful in determining limits of sequences and functions of real variables.   We assume that the limits given below exist and that a may be finite or infinite.

$$\lim_{x \to a} [f(x) \pm g(x)] = \lim_{x \to a} f(x) \pm \lim_{x \to a} g(x) \tag{1-1}$$

$$\lim_{x \to a} f(x)g(x) = \lim_{x \to a} f(x) \lim_{x \to a} g(x) \tag{1-2a}$$

If we take $g(x)$ as a constant k, Eq. (1-2a) becomes

$$\lim_{x \to a} kf(x) = k \lim_{x \to a} f(x) \tag{1-2b}$$

$$\lim_{x \to a} \frac{f(x)}{g(x)} = \frac{\lim\limits_{x \to a} f(x)}{\lim\limits_{x \to a} g(x)} \qquad \text{if} \qquad \lim_{x \to a} g(x) \neq 0 \tag{1-3}$$

As an example, let us evaluate $e = \lim\limits_{x \to 0} (1 + x)^{1/x}$.   First, expand the function by the binomial theorem:

$$(1 + x)^{1/x} = 1 + 1 + \frac{1}{2!}(1 - x) + \frac{1}{3!}(1 - x)(1 - 2x) + \cdots$$

$$+ \frac{1}{(n-1)!}(1 - x)(1 - 2x)(1 - 3x) \cdots [1 - (n-2)x] + \cdots$$

Now, apply Eqs. (1-1) and (1-2):

$$\lim_{x \to 0} (1 + x)^{1/x} = 2 + \lim_{x \to 0} \frac{1}{2!}(1 - x) + \lim_{x \to 0} \frac{1}{3!}(1 - x)(1 - 2x) + \cdots$$

$$= 2 + \frac{1}{2!} + \frac{1}{3!} + \frac{1}{4!} + \cdots = 2.7183 = e$$

Another useful theorem is the following:

Suppose that $f(x) \geq g(x)$ and $f(x) \leq h(x)$ for all values of x in an interval containing $x = a$, except possibly at $x = a$.   If $\lim\limits_{x \to a} g(x) = \lim\limits_{x \to a} h(x) = L$, then $\lim\limits_{x \to a} f(x) = L$.

We can use this theorem, for example, to show that

$$\lim_{x \to 0} \frac{\sin x}{x} = 1$$

where x is an angle, in radians.   In the circle with unit radius in Fig. 1-1, draw a chord AB.   Let x be half the central angle AOB subtended by AB.   Then, $AB = 2 \sin x$ and arc $AB = 2x$, since the radius is unity. Now, draw tangents PA and PB from an external point P to points A

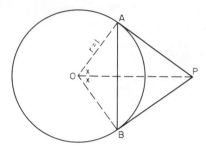

Fig. 1-1

and B on the circle.   Since $PA + PB = 2 \tan x$, and

$$\text{chord } AB < \text{arc } AB < PA + PB$$
$$2 \sin x < \quad 2x \quad < 2 \tan x$$

If we divide the inequality by $2 \sin x$ and invert, we obtain

$$1 > \frac{\sin x}{x} > \cos x$$

The limit of $\cos x$ as $x \to 0$ is 1.   Hence, the last theorem requires that $\lim\limits_{x \to 0} (\sin x / x) = 1$.

At the start of this article, your attention was called to the fact that $f(a)$ and $\lim\limits_{x \to a} f(x)$ may not be equal.   For continuous functions, however, these values are equal.

If defined throughout a neighborhood of $x = a$, $f(x)$ is continuous at $x = a$ if and only if for every positive real number ϵ there exists a positive number δ such that when $|x - a| < \delta$, then $|f(x) - f(a)| < \epsilon$.

Similarly, a function $f(x_1, x_2, \ldots , x_n)$ of several variables is continuous at the point $(a_1, a_2, \ldots , a_n)$ if and only if the function is defined throughout a neighborhood of the point and the limit of the function as $x_1 \to a_1$, $x_2 \to a_2$, $\ldots , x_n \to a_n$ equals $f(a_1, a_2, \ldots , a_n)$.   Also, we consider a function continuous on a set of points; for instance, an interval or region, if and only if the function is continuous at every point of the set.

Many methods and formulas of calculus are based on the assumption of continuous functions.   You will find the preceding definitions and the following theorems useful in verifying that the functions you are dealing with in solving engineering problems are continuous.

If $f(x)$ and $g(x)$ are continuous at $x = a$, their sum, difference, and product are each continuous at $x = a$.   Also, $f(x)/g(x)$ is continuous at that point unless $g(a) = 0$.

If $f(x)$ is a polynomial, $a_0 x^n + a_1 x^{n-1} + \cdots + a_n$, it is continuous for all values of x.   Also, if $f(x)$ is a rational function, the quotient of two polynomials $g(x)/h(x)$, then $f(x)$ is continuous for all values of x except the zeros of $h(x)$.

If f and g are two continuous functions of x, the composite function $h = g(f)$ is also continuous. For example,

$$\lim_{x \to 0} \sin\left(x + \frac{\pi}{2}\right) = \sin \lim_{x \to 0}\left(x + \frac{\pi}{2}\right) = \sin\frac{\pi}{2} = 1$$

If $f(x)$ is continuous on a closed interval $a \leq x \leq b$, the function has a maximum and minimum value in that interval. And $f(x)$ assumes every value between the maximum and minimum at least once in the interval. Furthermore, if $f(a) < 0$ and $f(b) > 0$, then somewhere in the interval $f(x) = 0$. Thus, if a continuous curve lies below the x axis in part of an interval and above the axis elsewhere, the curve must cross the axis in the interval.

We consider $1/x$ discontinuous at $x = 0$, because the function is not defined at that point. Similarly, $1/(x^2 - 9)$ is discontinuous at $x = \pm 3$. In contrast, -1^x is discontinuous at $x = 0$ even though the function is defined there as $(-1^0 = 1)$, because $\lim_{x \to 0}(-1)^x$ does not exist. The function shifts from 1 to -1 as x assumes different values close to 0; for example, if $x = \frac{2}{9}$, $-1^x = 1$ and if $x = \frac{1}{9}$, $-1^x = -1$.

1-2 First-order Derivatives. The concept of limit is basic in defining the instantaneous rate of change of a function. We call such a rate of change the derivative of the function. We call the process of determining the derivative differentiation.

Let $y = f(x)$ be a function of the real variable x throughout a neighborhood of a specific point. The first-order derivative of $f(x)$ with respect to x at the point is then defined as

$$\lim_{\Delta x \to 0} \frac{f(x + \Delta x) - f(x)}{\Delta x} = \lim_{\Delta x \to 0} \frac{\Delta y}{\Delta x} \tag{1-4}$$

A variety of symbols are used to denote this derivative. Included are $\frac{dy}{dx}$, $\frac{d}{dx}f(x)$, $D_x y$, $f'(x)$, y', and $\dot{y}$.

As an example of the use of the definition to determine the derivative of a function, let us differentiate $y = x^2$.

$$y' = \lim_{\Delta x \to 0} \frac{(x + \Delta x)^2 - x^2}{\Delta x} = \lim_{\Delta x \to 0}(2x + \Delta x) = 2x$$

As an example of the application of differentiation, consider the following problem: A dropped weight falls in accordance with the law $s = 16t^2$, where t is the time, sec, from start of motion, and s is the distance, feet, from point of release. What is the speed of the weight at the end of 4 sec?

The speed at any instant is the first derivative of s with respect to t at that time. Hence,

$$\frac{ds}{dt}\bigg]_{t=4} = \frac{d}{dt}[16t^2]_{t=4} = 16 \times 2t\bigg]_{t=4} = 128$$

Geometrically, the derivative $f'(x)$ represents the instantaneous direction, or slope of the tangent, to the curve representing $f(x)$. The equation of the tangent to the curve at a point $P_1(x_1,y_1)$ is

$$y - y_1 = (x - x_1)f'(x) \tag{1-5}$$

The equation of the normal to the curve at the same point is

$$x - x_1 = (y_1 - y)f'(x) \tag{1-6}$$

From the definition of first derivative, the differentiation formulas in Tables 1-1 and 1-2 may be derived. With the aid of these formulas, you can develop the derivatives of more complicated functions.

TABLE 1-1   Basic Differentiation Formulas for Algebraic Functions

$$\frac{dc}{dx} = 0 \qquad c = \text{constant}$$

$$\frac{dx}{dx} = 1$$

$$\frac{d}{dx}(u + v - w) = \frac{du}{dx} + \frac{dv}{dx} - \frac{dw}{dx}$$

$$\frac{d}{dx}cv = c\frac{dv}{dx} \qquad c = \text{constant}$$

$$\frac{d}{dx}uv = u\frac{dv}{dx} + v\frac{du}{dx}$$

$$\frac{d}{dx}(u_1,u_2, \ldots ,u_n) = (u_2,u_3, \ldots ,u_n)\frac{du_1}{dx} + (u_1,u_3, \ldots ,u_n)\frac{du_2}{dx}$$
$$+ \cdots + (u_1,u_2, \ldots ,u_{n-1})\frac{du_n}{dx}$$

$$\frac{dx^n}{dx} = nx^{n-1} \qquad n = \text{constant}$$

$$\frac{du^n}{dx} = nu^{n-1}\frac{du}{dx} \qquad n = \text{constant}$$

$$\frac{d}{dx}\frac{u}{c} = \frac{1}{c}\frac{du}{dx} \qquad c = \text{constant}$$

$$\frac{d}{dx}\frac{u}{v} = \left(v\frac{du}{dx} - u\frac{dv}{dx}\right)v^{-2}$$

TABLE 1-2   Basic Differentiation Formulas for Transcendental Functions

$$\frac{d}{dx} e^u = e^u \frac{du}{dx} \qquad e = 2.718281828459045$$

$$\frac{d}{dx} a^u = a^u \frac{du}{dx} \log_e a \qquad a = \text{constant}$$

$$\frac{d}{dx} u^v = vu^{v-1} \frac{du}{dx} + u^v \frac{dv}{dx} \log_e u$$

$$\frac{d}{dx} \log_e u = \frac{1}{u} \frac{du}{dx} \qquad \log_e u = \log_e 10 \log_{10} u = 2.302585092994046 \log_{10} u$$

$$\frac{d}{dx} \log_{10} u = \frac{\log_{10} e}{u} \frac{du}{dx} \qquad \log_{10} e = 0.434294481903252$$

$$\frac{d}{dx} \sin u = \cos u \frac{du}{dx} \qquad\qquad \frac{d}{dx} \sinh u = \cosh u \frac{du}{dx} = \frac{e^u + e^{-u}}{2} \frac{du}{dx}$$

$$\frac{d}{dx} \cos u = -\sin u \frac{du}{dx} \qquad\qquad \frac{d}{dx} \cosh u = \sinh u \frac{du}{dx} = \frac{e^u - e^{-u}}{2} \frac{du}{dx}$$

$$\frac{d}{dx} \tan u = \sec^2 u \frac{du}{dx} \qquad\qquad \frac{d}{dx} \tanh u = \operatorname{sech}^2 u \frac{du}{dx}$$

$$\frac{d}{dx} \cot u = -\csc^2 u \frac{du}{dx} \qquad\qquad \frac{d}{dx} \coth u = -\operatorname{csch}^2 u \frac{du}{dx}$$

$$\frac{d}{dx} \sec u = \sec u \tan u \frac{du}{dx} \qquad\qquad \frac{d}{dx} \operatorname{sech} u = -\operatorname{sech} u \tanh u \frac{du}{dx}$$

$$\frac{d}{dx} \csc u = -\csc u \cot u \frac{du}{dx} \qquad\qquad \frac{d}{dx} \operatorname{csch} u = \operatorname{csch} u \coth u \frac{du}{dx}$$

$$\frac{d}{dx} \arcsin u = \frac{1}{\sqrt{1 - u^2}} \frac{du}{dx} \qquad\qquad \frac{d}{dx} \operatorname{argsinh} u = \frac{1}{\sqrt{u^2 + 1}} \frac{du}{dx}$$

$$\frac{d}{dx} \arccos u = -\frac{1}{\sqrt{1 - u^2}} \frac{du}{dx} \qquad\qquad \frac{d}{dx} \operatorname{argcosh} u = \pm \frac{1}{\sqrt{u^2 - 1}} \frac{du}{dx} \qquad u > 1$$

$$\frac{d}{dx} \arctan u = \frac{1}{1 + u^2} \frac{du}{dx} \qquad\qquad \frac{d}{dx} \operatorname{argtanh} u = \frac{1}{1 - u^2} \frac{du}{dx} \qquad -1 < u < 1$$

$$\frac{d}{du} \operatorname{arccot} u = -\frac{1}{1 + u^2} \frac{du}{dx} \qquad\qquad \frac{d}{dx} \operatorname{argcoth} u = \frac{1}{1 - u^2} \frac{du}{dx} \qquad |u| > 1$$

For example, differentiate $y = x^3 - 3x^2 + 4 + 1/x^2 + \sqrt[3]{x}$. Examination of Table 1-1 indicates that the derivative can be obtained by application of the formulas, $d(u + v - w)/dx = du/dx + dv/dx - dw/dx$, $d(cv)/dx = c(dv/dx)$, and $du^n/dx = nu^{n-1}$, where n is a constant. Thus,

$$\frac{dy}{dx} = \frac{d}{dx} x^3 - 3 \frac{d}{dx} x^2 + \frac{d}{dx} 4 + \frac{d}{dx} x^{-2} + \frac{d}{dx} x^{1/3}$$

$$= 3x^{3-1} - 3 \times 2x^{2-1} + 0 - 2x^{-2-1} + \tfrac{1}{3} x^{1/3-1}$$

$$= 3x^2 - 6x - \frac{2}{x^3} + \frac{1}{3x^{2/3}}$$

As another example, differentiate $(\sin x)/x$. One way to do this is to use the formula $d(u/v)/dx = [v(du/dx) - u(dv/dx)]v^{-2}$, from Table 1-1, with $u = \sin x$ and $v = x$. The derivative of $\sin x$, from Table 1-2, is $\cos x$. Hence, $d[(\sin x)/x]/dx = (x \cos x - \sin x)x^{-2}$. Another way to obtain the derivative is to use the formula $d(uv)/dx = u(dv/dx) + v(du/dx)$, from Table 1-1, with $u = \sin x$ and $v = x^{-1}$. Then,

$$\frac{d[(\sin x)/x]}{dx} = -x^{-2} \sin x + x^{-1} \cos x = (x \cos x - \sin x)x^{-2}$$

as before.

Following are some additional rules to help you differentiate complicated functions:

If y is a function of a variable u, and u is a function of another variable x, then

$$\frac{dy}{dx} = \frac{dy}{du}\frac{du}{dx} \tag{1-7}$$

For example, differentiate $y = \log_e (x^2 + 1)$. Let $u = x^2 + 1$. Then, $y = \log_e u$ and $y' = (1/u)(du/dx)$. Since $du/dx = 2x$, $y' = 2x/(x^2 + 1)$.
    When dx/dy is obtained more easily than dy/dx, you can use

$$\frac{dy}{dx} = \frac{1}{(dx/dy)} \tag{1-8}$$

For example, differentiate arcsec x. Let $y = \text{arcsec } x$. Then, $x = \sec y$ and $dx/dy = \sec y \tan y = \sec y \sqrt{\sec^2 y - 1} = x \sqrt{x^2 - 1}$. Therefore, $d(\text{arcsec } x)/dx = dy/dx = 1/(x \sqrt{x^2 - 1})$.

If a function is given in parametric form, $x = f_1(t)$, $y = f_2(t)$, then

$$\frac{dy}{dx} = \frac{(dy/dt)}{(dx/dt)} \tag{1-9}$$

Find, for example, the slope of the cycloid

$$x = a(\theta - \sin \theta) \qquad y = a(1 - \cos \theta)$$

where θ is a parameter. In this case, we find the slope dy/dx by differentiating the given equations with respect to θ and using Eq. (1-9):

$$\frac{dx}{d\theta} = a(1 - \cos \theta)$$

$$\frac{dy}{d\theta} = a \sin \theta$$

$$\frac{dy}{dx} = \frac{(dy/d\theta)}{(dx/d\theta)} = \frac{a \sin \theta}{a(1 - \cos \theta)} = \frac{\sin \theta}{1 - \cos \theta}$$

When you are given a relation $f(x,y) = 0$, one way to obtain dy/dx is to differentiate f, treating y as a function of x, and solve for dy/dx. For instance, what is the slope of the ellipse $4x^2 + 9y^2 = 36$ at any point? Let $f = 4x^2 + 9y^2 - 36 = 0$. Then, $df/dx = 4(2x) + 9(2y)(dy/dx) = 0$. From this, we obtain $dy/dx = -4x/9y$.

Another method requires use of a formula containing partial derivatives (Art. 1-5):

$$\frac{dy}{dx} = -\frac{(\partial f/\partial x)}{(\partial f/\partial y)} \tag{1-10}$$

In brief, $\partial f/\partial x$ is the derivative of f with respect to x with y considered a constant, and $\partial f/\partial y$ is the derivative of f with respect to y with x considered a constant. Let us apply Eq. (1-10) to the preceding problem.

$$\frac{\partial f}{\partial x} = \frac{\partial}{\partial x}(4x^2 + 9y^2 - 36) = 8x$$

$$\frac{\partial f}{\partial y} = \frac{\partial}{\partial y}(4x^2 + 9y^2 - 36) = 18y$$

$$\frac{dy}{dx} = -\frac{(\partial f/\partial x)}{(\partial f/\partial y)} = -\frac{8x}{18y} = -\frac{4x}{9y}$$

1-3 Higher-order Derivatives of Functions. The derivative of the first derivative of a function is called the second derivative. If $y = f(x)$, the second derivative of y with respect to x may be denoted by d^2y/dx^2, D_x^2y, dy'/dx, $f''(x)$, y'', or $\ddot{y}$.

We observed in Art. 1-2 that velocity may be considered the first derivative of distance with respect to time. Similarly, we may consider acceleration as the first derivative of velocity with respect to time, since acceleration is the instantaneous rate of change of velocity. Hence, acceleration is the second derivative of distance with respect to time.

Suppose the distance, ft, of a body from a fixed point is given by $s = 1/(t + 1)$, where t is the time in minutes. What are the velocity and acceleration of the body after 2 min?

$$\text{Velocity} = v = \frac{ds}{dt} = \frac{d}{dt}(t + 1)^{-1} = -(t + 1)^{-2}$$

After 2 min,

$$v = -(2 + 1)^{-2} = -\frac{1}{9}\text{ fpm}$$

$$\text{Acceleration} = a = \frac{dv}{dt} = \frac{d}{dt} - (t + 1)^{-2} = 2(t + 1)^{-3}$$

After 2 min,

$$a = 2(2 + 1)^{-3} = \frac{2}{27} \text{ fpm}^2$$

The second derivative of a function is useful in determining the curvature of a plane curve representing the function. **Curvature** is the rate of change of direction of the curve. In Cartesian plane coordinates, it is given by

$$K = \frac{y''}{(1 + y'^2)^{\frac{3}{2}}} = \frac{-x''}{(1 + x'^2)^{\frac{3}{2}}} \tag{1-11}$$

where y', y'' = first and second derivatives of y with respect to x
$\quad\quad x'$, x'' = first and second derivatives of x with respect to y

For example, what is the curvature of the ellipse $4x^2 + 9y^2 = 36$ at $x = 0$? In Art. 1-2, we found $y' = -4x/9y$. Differentiating, we get

$$y'' = -\frac{4}{9} \frac{y - xy'}{y^2}$$

Then, the curvature at any point of the ellipse is, from Eq. (1-11),

$$K = \frac{-(\frac{4}{9})[(y - xy')/y^2]}{(1 + y'^2)^{\frac{3}{2}}} = -\frac{4(y - xy')}{9y^2(1 + y'^2)^{\frac{3}{2}}}$$

When $x = 0$, $y = \pm 2$, and $y' = 0$. Hence, the curvature at $x = 0$ is

$$K = -\frac{4}{9y} = \pm \frac{2}{9}$$

In structural analysis, the bending moment at any point of a loaded elastic beam is related to the curvature by

$$M = EIK \tag{1-12}$$

where E is the modulus of elasticity of the material and I the moment of inertia of the beam cross section. For practical purposes, beam deflections are kept small. Hence, it is reasonable to assume that y', the slope of the beam at any point, may be omitted from Eq. (1-11). Therefore, the bending moment is usually expressed in terms of the second derivative of the beam deflection:

$$M = EIy'' \tag{1-13}$$

Suppose, for example, that the deflection of a beam is determined to be $y = -(xL^3 - 2x^3L + x^4)w/24EI$, where w is the load, lb per in., and L is the span of the beam, in. What is the bending moment at a distance x from one support? Differentiating, we obtain the slope of the elastic curve (deflection curve), $y' = -(L^3 - 6x^2L + 4x^3)(w/24EI)$. Differentiating again, we get the bending moment $M = EIy'' = -(-12xL + 12x^2)(w/24) = (w/2)[x(L - x)]$.

The shear at any point in a beam is given by

$$V = \frac{dM}{dx} \tag{1-14}$$

Hence, for the conditions of the problem, $V = (w/2)(L - 2x) = wL/2 - wx$. This indicates that the shear equals half the load on the beam (the load carried by one support) minus the load between the support and the point for which the shear is being determined.

The load at any point in a beam is given by

$$-w = \frac{dV}{dx} = \frac{d^2M}{dx^2} \tag{1-15}$$

This is easily verified for the conditions of the problem.

Let us now return to the concept of curvature. Computed from Eq. (1-11), curvature may be positive or negative.

When curvature is positive, the curve is concave upward (tangents are below the curve). When curvature is negative, the curve is concave downward (tangents are above the curve).

Examination of Eq. (1-11) indicates that y', being squared, does not affect the sign of the curvature. Hence, the concavity of the curve is determined by y''. This observation is useful in determining whether a critical point of the curve is a maximum or a minimum or neither (see Art. 1-8).

It is sometimes desirable to express the change of direction of a plane curve as a **radius of curvature,** the reciprocal of the curvature.

$$R = \frac{1}{K} = \frac{(1 + y'^2)^{3/2}}{y''} \tag{1-16}$$

For example, let us compute the radius of curvature of the parabola $y = x^2$ at its vertex. At $(0,0)$, $y' = 2x = 0$ and $y'' = 2$. Therefore, the radius of curvature is

$$R = \frac{(1 + 0)^{3/2}}{2} = \frac{1}{2}$$

We can continue to find derivatives of derivatives as long as they exist. For instance, the derivative with respect to x of the second derivative is called the third-order derivative. It may be denoted by d^3y/dx^3, D_x^3y, $f^{(3)}(x)$, or $y^{(3)}$. In the beam problem, we observed that $V = dM/dx$

and $M = EIy''$.   Therefore,

$$V = EIy^{(3)} \tag{1-17}$$

Also, since the load $w = -dV/dx$, then

$$w = -EIy^{(4)} \tag{1-18}$$

1-4 *Differentials.* In Art. 1-2, we defined the first derivative of a function $f(x)$ as the limit of a ratio as a real variable x approaches zero. This concept requires the numerator and denominator of the representation of the derivative, dy/dx, to be inseparable.   But sometimes, for ease of algebraic manipulation, it is desirable to separate them.

For the purpose, we define dy as

$$dy = f'(x)\,dx \tag{1-19}$$

where $y = f(x)$ and $f'(x) = dy/dx$.   We call dy and dx differentials.

Geometrically, $f'(x)$ gives the slope of the tangent to the curve representing $f(x)$.   Hence, if we view dx as an infinitesimal increment to x, then the product of the slope of the tangent and dx gives the corresponding infinitesimal increment dy to $f(x)$.

As one example of the use of differentials, let us differentiate the relation $x^2 + y^2 = r^2$.   We apply Eq. (1-19) to both x^2 and y^2 and set the differential of r equal to zero because r is a constant.   Thus, we obtain

$$2x\,dx + 2y\,dy = 0$$

Dividing through by $2y\,dx$ and transposing terms, we get the desired derivative

$$\frac{dy}{dx} = -\frac{2x}{2y} = -\frac{x}{y}$$

As another example, let us determine what function is its own derivative.   Let y be the function.   Then, $dy/dx = y$.   If we multiply both sides of this equation by dx/y, we get $dy/y = dx$.   From Table 1-2, we see that $d(\log_e y)/dy = 1/y$.   Hence, $dy/y = d(\log_e y)$.   Also, we can write the right-hand side of the equation as $dx + 0 = d(x + C)$, where C is a constant.   We then have $d(\log_e y) = d(x + C)$.   Thus, $\log_e y = x + C$.   From the definition of logarithm, $y = e^{x+C} = e^C e^x$.   Therefore, the function that is its own derivative is ke^x, where k is any constant.

For still another application of differentials, let us consider again a plane curve representing $y = f(x)$.   If we are using Cartesian coordinates, then the increments dx to x and dy to y form two sides of a right triangle. The third side, the hypotenuse, is an infinitesimal length of the curve, to which we assign the value ds.   Then,

$$ds^2 = dx^2 + dy^2 \tag{1-20}$$

We can also rearrange the right-hand side of this equation to obtain

$$ds = \sqrt{1 + y'^2}\, dx = \sqrt{1 + x'^2}\, dy \tag{1-21}$$

In Chap. 2, methods are given for using these equations to compute the length of a curve.

1-5   *Partial Derivatives.* Sometimes, when we have a problem involving several variables, we want to know the rate of change of the function with respect to only one of the variables.   In such a case, we can use the partial derivative.

The partial derivative of a function is the derivative with respect to one variable, all other variables being treated as constants.

Various symbols are in use for denoting partial derivatives.   For example, if $z = f(x,y)$, the partial derivative of z with respect to x may be written $\partial z/\partial x$, z_x, or $f_x(x,y)$.   Similarly, the partial derivative of z with respect to y may be written $\partial z/\partial y$, z_y, or $f_y(x,y)$.

Geometrically, we may consider $z = f(x,y)$ as a surface.   If we take y as a constant, we determine a curve, the intersection of the surface and the plane y = a constant.   The partial derivative $\partial z/\partial x$ gives the slope of the tangent in that plane at each point of that curve.

For example, what is the slope of $xyz = a^3$ in the planes $z = k$?   Differentiating the equation with respect to x while holding z constant, we obtain

$$\frac{\partial}{\partial x} xyz = yz + xz\frac{\partial y}{\partial x} = 0$$

$$\frac{\partial y}{\partial x} = -\frac{y}{x}$$

If the function to be differentiated is in the form $f(x,y,z) - 0$, it some times is easier to use

$$\frac{\partial z}{\partial x} = -\frac{\partial f/\partial x}{\partial f/\partial z} \qquad \frac{\partial z}{\partial y} = -\frac{\partial f/\partial y}{\partial f/\partial z} \tag{1-22}$$

To solve the previous problem with Eqs. (1-22), we first write $f(x,y,z) = xyz - a^3 = 0$, then differentiate.

$$\frac{\partial f}{\partial x} = yz \qquad \frac{\partial f}{\partial y} = xz$$

$$\frac{\partial y}{\partial x} = -\frac{yz}{xz} = -\frac{y}{x}$$

Suppose that x and y in $z = f(x,y)$ are, in turn, functions of other variables, for instance, $x = f_1(u,v)$ and $y = f_2(u,v)$. In that case, you can obtain the partial derivatives of z with respect to the new variables from

$$\frac{\partial z}{\partial u} = \frac{\partial z}{\partial x}\frac{\partial x}{\partial u} + \frac{\partial z}{\partial y}\frac{\partial y}{\partial u} \qquad (1\text{-}23)$$

$$\frac{\partial z}{\partial v} = \frac{\partial z}{\partial x}\frac{\partial x}{\partial v} + \frac{\partial z}{\partial y}\frac{\partial y}{\partial v} \qquad (1\text{-}24)$$

By differentiating partial derivatives, we can find higher-order derivatives, as long as they exist:

$$\frac{\partial}{\partial x}\frac{\partial z}{\partial x} = \frac{\partial^2 z}{\partial x^2} = \frac{\partial^2 f}{\partial x^2} = f_{xx} = z_{xx}$$

$$\frac{\partial}{\partial y}\frac{\partial z}{\partial y} = \frac{\partial^2 z}{\partial y^2} = \frac{\partial^2 f}{\partial y^2} = f_{yy} = z_{yy}$$

$$\frac{\partial}{\partial x}\frac{\partial z}{\partial y} = \frac{\partial}{\partial y}\frac{\partial z}{\partial x} = \frac{\partial^2 z}{\partial x \, \partial y} = \frac{\partial^2 f}{\partial x \, \partial y} = f_{xy} = z_{xy}$$

Note that in obtaining the second derivative with respect to x and y, the order of differentiation is immaterial if the derivatives are continuous.

1-6   *Directional Derivatives.* In Art. 1-5, you saw that the partial derivative of a function $z = f(x,y)$ geometrically represents the slope at each point of the curve of intersection of the surface $f(x,y)$ and a plane normal to one of the coordinate axes. Suppose instead that you wished to find the slope for a plane at a different angle.

Consider the case in which $f(x,y)$ intersects a plane parallel to the z axis and at an angle θ with the xz plane. The slope of the tangent at each point of the curve of intersection is given by the directional derivative

$$\frac{\partial z}{\partial l} = \frac{\partial z}{\partial x}\cos\theta + \frac{\partial z}{\partial y}\sin\theta \qquad (1\text{-}25)$$

You can use directional derivatives to solve problems of the following type: The temperature at any point of a plate lying in the xy plane is given by $t = e^{-x^2-y^2}t_0$. What is the rate of change of temperature at a point $(1,1)$ in the direction making an angle of $60°$ with the x axis? To solve the problem, you must determine the directional derivative of t from Eq. (1-25) with $\theta = 60°$.

$$\left.\frac{\partial t}{\partial x}\right]_{1,1} = \left.-2xe^{-x^2-y^2}t_0\right]_{1,1} = -2e^{-2}t_0$$

$$\left.\frac{\partial t}{\partial y}\right]_{1,1} = \left.-2ye^{-x^2-y^2}t_0\right]_{1,1} = -2e^{-2}t_0$$

$$\left.\frac{\partial t}{\partial l}\right]_{1,1} = -2e^{-2}t_0\cos 60° - 2e^{-2}t_0\sin 60° = -e^{-2}t_0(1+\sqrt{3})$$

1-7 *Total Differential and Total Derivative.* Suppose that we are given a function of several variables $y = f(x_1, x_2, \ldots, x_n)$. Suppose also that x_1 is given an increment dx_1, x_2 an increment dx_2, and x_n an increment dx_n. The resulting increment in y is called the total differential of y.

$$dy = \frac{\partial y}{\partial x_i}\, dx_i = \frac{\partial y}{\partial x_1}\, dx_1 + \frac{\partial y}{\partial x_2}\, dx_2 + \cdots + \frac{\partial y}{\partial x_n}\, dx_n \qquad (1\text{-}26)$$

where the repeated index i indicates summation from $i = 1$ to n. For example, let us determine the total differential of u if $u = xyz$. From Eq. (1-26), we find $du = yz\, dx + xz\, dy + xy\, dz$.

If the variables $x_1, x_2, \ldots, x_n$ are all functions of a parameter t, then $y = f(x_1, x_2, \ldots, x_n)$ also is a function of t. The derivative dy/dt is called the total derivative of y with respect to t. You can evaluate this derivative by substituting functions of t for the variables $x_1, x_2, \ldots, x_n$ in $f(x_1, x_2, \ldots, x_n)$ and then differentiating with respect to t. But usually you will find it easier to determine the total derivative from

$$\frac{dy}{dt} = \frac{\partial y}{\partial x_i}\frac{dx_i}{dt} = \frac{\partial y}{\partial x_1}\frac{dx_1}{dt} + \frac{\partial y}{\partial x_2}\frac{dx_2}{dt} + \cdots + \frac{\partial y}{\partial x_n}\frac{dx_n}{dt} \qquad (1\text{-}27)$$

For example, find dz/dt for $z = x^2 + 4y^2$ when $x = \sin t$ and $y = \cos t$. From Eq. (1-27),

$$\frac{dz}{dt} = \frac{\partial z}{\partial x}\frac{dx}{dt} + \frac{\partial z}{\partial y}\frac{dy}{dt} = 2x \cos t - 8y \sin t.$$

1-8 *Maximum and Minimum Values of a Function.* Let us consider now the types of problems in which we wish to determine the maximum or minimum values of a function and the values of the variables that produce these maximums and minimums. As a first step, let us examine the case of a function of one variable.

Suppose that the curve representing the continuous function $y = f(x)$ has a maximum in a given interval at a point x_1. Then, for $x < x_1$ in the neighborhood of x_1, the tangent to the curve has a positive slope (Fig. 1-2a). And for $x > x_1$ in the neighborhood of x_1, the tangent has a negative slope. Therefore, at a maximum point, the slope of the tangent, or derivative y', changes sign. Furthermore, if y' is continuous at x_1, $y' = 0$ there. Similarly, at a minimum, y' changes from negative to positive as it passes through the critical point and is zero there if the derivative is continuous (Fig. 1-2b). These observations lead to the following rules:

To find a maximum or minimum value of a function of one variable, set the first derivative of the function equal to zero. Solve the equation to find the critical values of the variable.

Test the derivative first with a value a trifle less than each of the critical

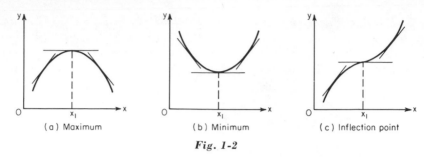

(a) Maximum   (b) Minimum   (c) Inflection point

Fig. 1-2

values, then with a value a trifle larger than each. If the sign of the deriva-
tive is first plus then minus, the function is at a maximum. But if the sign
is first minus then plus, the function is at a minimum.

An alternative test is to substitute the critical values in the second derivative of the function. If this derivative is negative, the function is at a maximum. If the second derivative is positive, the function is at a minimum. The test fails, however, if the second derivative is also zero. In such a case, test successive derivatives with the critical value until a derivative is found that is not zero. If the derivative is an even derivative, the function is at a maximum or minimum. If the critical value makes the nonvanishing even derivative negative, the function is at a maximum. If the derivative is positive, the function is at a minimum. But if the nonvanishing derivative is odd, the function is at an inflection point, where the curvature changes sign (Fig. 1-2c).

As an illustration of the mechanics of the computation, let us find the maximum value of $y = x^3 - 3x^2$. In accordance with the rules, we differentiate y, set the derivative equal to zero, and solve for the critical values of x:

$$y' = 3x^2 - 6x = 0$$

Thus, we find $x = 0, 2$. Substitution in the second derivative yields

$$y''\Big]_0 = [6x - 6]_0 = -6 \qquad y''\Big]_2 = [6x - 6]_2 = 12 - 6 = 6$$

Since y'' is negative at $x = 0$, the function is at a maximum there.

In practice, you may encounter problems in the following form: The bending moment in a beam is given by $M = wx(L - x)/2$, where w is the load, lb per ft; L the span, ft; and x the distance from one support, ft. Where does the maximum bending moment occur and what is its value? (The size of the beam will be determined by this moment.) Differentiating and recalling that $dM/dx = V$, the shear, we obtain $M' = V = wL/2 - wx = 0$. Hence, $x = L/2$. At this point, $M = wL^2/8$. This is a maximum, because the second derivative $M'' = -w$ is negative.

Sometimes, you will find that a constraint is placed on the function to be maximized. For example, what is the area of the largest rectangular field that can be enclosed with 900 lin ft of fencing? Let x be one side of the field, y the other, and A the area. The function to be maximized is $A = xy$. But it is constrained by $2x + 2y = 900$. Elimination of y from these equations yields

$$A = x \left(\frac{900 - 2x}{2} \right) = 450x - x^2$$

Differentiation yields $A' = 450 - 2x = 0$, from which $x = 225$ ft. And substitution in the original equations gives $y = x = 225$ ft and $A = 50,625$ sq ft.

Suppose now that we have two variables x and y. The function $z = f(x,y)$ may be represented by a surface. At its critical points, the tangent plane will be parallel to the xy plane. Thus,

$$\frac{\partial z}{\partial x} = 0 \qquad \frac{\partial z}{\partial y} = 0 \tag{1-28}$$

To test the critical points for a maximum or minimum, calculate Q from

$$Q = \left(\frac{\partial^2 z}{\partial x \, \partial y} \right)^2 - \frac{\partial^2 z}{\partial x^2} \frac{\partial^2 z}{\partial y^2} \tag{1-29}$$

If Q and $\partial^2 z/\partial x^2$ (or $\partial^2 z/\partial y^2$) are negative, z will be at a maximum.
If $Q < 0$ and $\partial^2 z/\partial x^2$ (or $\partial^2 z/\partial y^2$) is positive, z will be at a minimum.
If $Q > 0$, z will be neither at a maximum or a minimum.
If $Q = 0$, the test fails.

For example, let us determine the maximum and minimum values of $z = x^2 + xy + y^2 - 6x - 3y$. Differentiation yields

$$\frac{\partial z}{\partial x} = 2x + y - 6 = 0 \qquad \frac{\partial z}{\partial y} = x + 2y - 3 = 0$$

Simultaneous solution of these equations gives the critical point $x = 3$, $y = 0$. Hence, $z = -9$. To test for a maximum or minimum, we differentiate again:

$$\frac{\partial^2 z}{\partial x \, \partial y} \bigg]_{3,0} = 1 \qquad \frac{\partial^2 z}{\partial x^2} \bigg]_{3,0} = 2 \qquad \frac{\partial^2 z}{\partial y^2} \bigg]_{3,0} = 2$$

Substitution of these derivatives in Eq. (1-29) yields $Q = 1 - 4 = -3$. Since $Q < 0$ and the second derivatives are positive, the point $(3,0,-9)$ is a minimum.

1-9 *Indeterminate Forms.* In trying to evaluate some functions of a variable x for a specific value a of x, you may obtain an indeterminate form, such as $0/0$, ∞/∞, or $0 \cdot \infty$. Thus, the function may be undefined for $x = a$. But the function may have a limit as x approaches a.

L'Hôpital's rule gives a way to find the limit of functions of the form $f(x)/g(x)$ that approach $0/0$ or ∞/∞ as $x \to a$. The rule states that

If $f(x)$ and $g(x)$ have derivatives of all orders up to n, then the limit of $f(x)/g(x)$ equals the first of the following limits that is not indeterminate (if a limit exists):

$$\lim_{x \to a} \frac{f'(x)}{g'(x)}, \lim_{x \to a} \frac{f''(x)}{g''(x)}, \cdots, \lim_{x \to a} \frac{f^{(n)}(x)}{g^{(n)}(x)}$$

If the first of these limits becomes infinite, then $f(x)/g(x)$ also becomes infinite.

As an example, let us determine the limit of $(x^2 - 4)/(x^2 + x - 6)$ as $x \to 2$. If $x = 2$ is substituted in this function, it takes the form $0/0$. But we can find the limit as $x \to 2$ by applying L'Hôpital's rule.

$$\lim_{x \to 2} \frac{x^2 - 4}{x^2 + x - 6} = \lim_{x \to 2} \frac{(d/dx)(x^2 - 4)}{(d/dx)(x^2 + x - 6)} = \frac{2x}{2x + 1}\bigg]_2 = \frac{4}{5}$$

As another example, evaluate $\lim (x + \log_e x)/x \log_e x$ as $x \to \infty$. As x increases without bound, the function approaches ∞/∞. But from L'Hôpital's rule, we find that

$$\lim_{x \to \infty} \frac{x + \log x}{x \log x} = \lim_{x \to \infty} \frac{1 + 1/x}{\log x + 1} = 0$$

Suppose that as x approaches a specific value, $f(x) \cdot g(x) \to 0 \cdot \infty$. In that case, put the product in the form $f(x)/[1/g(x)]$ or $g(x)/[1/f(x)]$ so that it approaches $0/0$ or ∞/∞. You can then find the limit (if one exists) by L'Hôpital's rule.

For example, determine the limit of $x \log_e (1 + 1/x)$ as $x \to \infty$. If $x = \infty$ is substituted in this function, it takes the form $\infty \cdot 0$. So we rewrite it and apply L'Hôpital's rule:

$$\lim_{x \to \infty} x \log \left(1 + \frac{1}{x}\right) = \lim_{x \to \infty} \frac{\log (1 + 1/x)}{1/x} = \lim_{x \to \infty} \frac{-1/(1 + 1/x)x^2}{-1/x^2}$$

$$= \lim_{x \to \infty} \frac{1}{1 + 1/x} = 1$$

Another indeterminate form that can be transformed into one to which L'Hôpital's rule may be applied is $\infty - \infty$. For example, let us determine $\lim (\csc \theta - \cot \theta)$ as $\theta \to 0$ from the positive side. If $\theta = 0$ is substituted in this function, it takes the form $\infty - \infty$. Let us rewrite the function in terms of $\sin \theta$ and $\cos \theta$:

$$\csc \theta - \cot \theta = \frac{1}{\sin \theta} - \frac{\cos \theta}{\sin \theta} = \frac{1 - \cos \theta}{\sin \theta}$$

In its new form, the function approaches $0/0$ as $\theta \to 0$. Now, we can differentiate numerator and denominator and let $\theta \to 0^+$.

$$\lim_{\theta \to 0^+} (\csc \theta - \cot \theta) = \lim_{\theta \to 0^+} \frac{1 - \cos \theta}{\sin \theta} = \left. \frac{\sin \theta}{\cos \theta} \right]_{\theta \to 0^+} = 0$$

Indeterminate forms with exponents may sometimes be transformed with logarithms to permit use of L'Hôpital's rule. Suppose as $x \to a$, $y = f(x)^{g(x)} \to 0^0$, 1^∞, or ∞^0. Take logarithms of both sides of the equation:

$$\log_e y = g(x) \log_e f(x)$$

Now, the right-hand side of the equation approaches $0 \cdot \infty$ as $x \to a$. Therefore, let us rewrite the equation as

$$\log_e y = \frac{\log_e f(x)}{1/g(x)}$$

This permits us to apply L'Hôpital's rule to find the limit of $\log_e y$. If this limit is z, then $y = e^z$.

For example, let us find $\lim (\sin \theta)^{\sec \theta}$ as θ approaches $\pi/2$. If $\theta = \pi/2$ is substituted in this function, it takes the form 1^∞. Let $y = (\sin \theta)^{\sec \theta}$. Then, $\log_e y = \sec \theta \log_e \sin \theta$. This takes the form $\infty \cdot 0$ when $\theta \to \pi/2$. Let us rewrite it as

$$\log_e y = \frac{\log_e \sin \theta}{1/\sec \theta} = \frac{\log_e \sin \theta}{\cos \theta}$$

The right-hand side of this equation takes the form $0/0$ when $\theta \to \pi/2$. We can now apply L'Hôpital's rule.

$$\lim_{\theta \to \pi/2} \log_e y = \lim_{\theta \to \pi/2} \frac{\log_e \sin \theta}{\cos \theta} = \left. \frac{\cos \theta}{-\sin^2 \theta} \right]_{\pi/2} = 0$$

Therefore, $\lim (\sin \theta)^{\sec \theta}$ as $\theta \to \pi/2$ is $e^0 = 1$.

1-10 *Bibliography*

R. V. ANDREE, "Introduction to Calculus," McGraw-Hill Book Company, New York.
H. M. BACON, "Differential and Integral Calculus," McGraw-Hill Book Company, New York.

W. R. BLAKELEY, "Calculus for Engineering Technology," John Wiley & Sons, Inc., New York.

S. K. STEIN, "Calculus," McGraw-Hill Book Company, New York.

PROBLEMS

1. Evaluate the following limits:
   (a) $x^2 + 2x$ as $x \to 2$.
   (b) $(x + 4)/5x$ as $x \to \infty$.
   (c) $(3x^3 - 4x^2)/(5x^4 - 8x^2)$ as $x \to 0$.
   (d) $(1 - \sqrt{x - 4})/(x - 5)$ as $x \to 5$.
   (e) $x \sin (1/x)$ as $x \to 0$.
   (f) $(\cos x)/(x - \pi/2)$ as $x \to \pi/2$.
   (g) $(e^x + 1)/(e^x - 1)$ as $x \to \infty$.
   (h) $x^{1/(1-x)}$ as $x \to 1$.

2. (a) For what value of x is $f(x) = (x^3 - 8)/(x - 2)$ discontinuous?
   (b) What value assigned to $f(x)$ at the discontinuity will make it continuous?

3. (a) Give the equation of the tangent to $y = x^2 - x + 1$ at $x = 0$.
   (b) Give the equation of the normal to $y = x^2$ at $x = 10$.

4. Suppose that you heat a circular metal plate with radius r.
   (a) What is the rate of change of circumference with r?
   (b) What is the rate of change of area with r?
   (c) If the radius increases 0.001 in. per sec, how fast is the area changing when $r = 100$ in.?

5. A flexible cable is hung between two supports 100 ft apart. One support is 5 ft lower than the other. The cable weighs 10 lb per lin ft and is under 1,000-lb tension. How far does the cable sag below the upper support? How far from that support is the low point?

6. Given is the position of a projectile at any time t, sec, as

$$x = v_0 t \cos \theta$$
$$y = v_0 t \sin \theta - \frac{1}{2} gt^2$$

where v_0 = initial velocity, ft per sec
   θ = angle of projection above the horizontal
   g = 32 ft per sec²
   (a) Give the equation of motion in rectangular coordinates. What is the shape of the path?
   (b) Find the range (determined by $y = 0$ when $t \neq 0$).
   (c) What is the maximum height reached by the projectile?
   (d) Determine the angle of projection for maximum range.

7. A cylindrical vessel containing a liquid is rotated with angular velocity ω about a vertical axis through the center of mass. This imposes a centripetal force $C = M\omega^2 r$ on a small mass M of the liquid, where r is the distance of that mass from the axis of rotation. If the resultant of the liquid pressure F is normal to the free surface, what is the equation of that surface? (Take the intersection

of surface and axis of rotation as origin.)   What is the shape of the surface?

8. A tub contains a liquid to a depth H, ft.   It discharges the liquid from the bottom at the rate of $48\sqrt{h}$ cfm, where h is the depth of liquid at any time t, min, after discharge starts.   The area of the free surface of the liquid in the tub at time t is $4h$ sq ft.   How long will it take to empty the tub?

9. Flow of liquid, cu ft per ft of width, in a rectangular channel is given by $q = d\sqrt{2g(H-d)}$, where d is the depth of flow, ft, and H is total head, ft. What is the maximum flow and the depth at that stage (critical depth)?

10. The principle of least work, used in structural analysis, states that *the strain energy in a system is the minimum consistent with equilibrium.*   Suppose that a vertical load P is suspended from a flat ceiling on three bars in a vertical plane.   All bars have the same area A and modulus of elasticity E.   The bars are connected at their lower ends, where P is attached.   One bar is vertical and has a length L.   The other two bars are of equal length, $L/\cos\alpha$, where α is the angle they make with the vertical.   What is the stress in the vertical bar? (The strain energy in a tension bar is $W^2L/2AE$, where W is the stress, lb, in the bar.)

11. The rate of biochemical oxidation of organic matter in a body of water is proportional to the remaining concentration of unoxidized substance: $dL/dt = -KL$, where t is time, days, and K is a constant.   What is the oxygen demand L at the end of t days?

12. A body is moving so that its position at any time t is $y = t^3 + 3t^2 - 8t + 2$.   What is its acceleration when $t = 6$?

13. What is the slope of the ellipse $x^2 + 2y^2 = 27$ at the point where $x = y$? What is the curvature there?

14. Find $\partial z/\partial x$ and $\partial z/\partial y$ when
   (a) $z = \sin(ax + by)$.
   (b) $x^2 + y^2 + z^2 = 16$.
   (c) $z = u^2 + uv + v^2$, where $u = 2x + y$ and $v = x - 2y$.

15. Power, watts, consumed in an electrical resistor is given by $P = E^2/R$. Suppose that $E = 220$ volts and $R - 10$ ohms.   If you approximate the change in power by using dP, how much does the power change if E drops 10 volts and R rises by $\frac{1}{2}$ ohm?

16. The electric potential at any point in a flat plate is given for $\rho \geq 1$ by $E = \log_e \rho$, where ρ is the radius vector from the origin.   Determine the rate of change of E at $\rho = 5$, $\theta = \arcsin 0.8$ in the direction toward the point $\rho = 10$, $\theta = \pi/2$.

ANSWERS

1. (a)
$$\lim_{x\to2} x^2 = 4 \text{ and } \lim_{x\to2} 2x = 4$$

Hence,
$$\lim_{x\to2}(x^2 + 2x) = 4 + 4 = 8$$
See Art. 1-1.

(b) Division of numerator and denominator by x yields

$$\lim_{x\to\infty} \frac{x+4}{5x} = \lim_{x\to\infty} \frac{1+4/x}{5} = \frac{1}{5}$$

because

$$\lim_{x\to\infty} 4/x = 0$$

See Art. 1-1.

(c) Assume that $x \neq 0$. After division of numerator and denominator by x^2, $(3x^3 - 4x^2)/(5x^4 - 8x^2) = (3x - 4)/(5x^2 - 8)$. Then,

$$\lim_{x\to 0} \frac{3x-4}{5x^2-8} = \frac{4}{8} = \frac{1}{2}$$

See Art. 1-1.

(d) Assume that $x \neq 5$. Then,

$$\frac{1-\sqrt{x-4}}{x-5} \frac{1+\sqrt{x-4}}{1+\sqrt{x-4}} = \frac{1-(x-4)}{(x-5)(1+\sqrt{x-4})}$$

$$= \frac{-(x-5)}{(x-5)(1+\sqrt{x-4})} = -\frac{1}{1+\sqrt{x-4}}$$

Therefore,

$$\lim_{x\to 5} \frac{1-\sqrt{x-4}}{x-5} = \lim_{x\to 5} -\frac{1}{1+\sqrt{x-4}} = -\frac{1}{2}$$

See Art. 1-1.

(e) Let $y = 1/x$. Then, $x \sin(1/x) = (1/y) \sin y$. But

$$\lim_{y\to\infty} \frac{1}{y} \sin y = 0$$

Hence,

$$\lim_{x\to 0} x \sin \frac{1}{x} = 0$$

(f) By L'Hôpital's rule,

$$\lim_{x\to\pi/2} \frac{\cos x}{x-\pi/2} = \lim_{x\to\pi/2} \frac{d(\cos x)/dx}{d(x-\pi/2)/dx} = \frac{-\sin x}{1}\bigg]_{\pi/2} = -1$$

See Art. 1-9.

(g) Multiplication of numerator and denominator by e^{-x} yields

$$\lim_{x\to\infty} \frac{e^x+1}{e^x-1} = \lim_{x\to\infty} \frac{1+e^{-x}}{1-e^{-x}} = 1$$

See Art. 1-1.

(h) Let $y = x^{1/(1-x)}$ and $\log_e y = (\log_e x)/(1 - x)$.   Then, by L'Hôpital's rule,

$$\lim_{x \to 1} \log_e y = \lim_{x \to 1} \frac{d(\log_e x)/dx}{d(1 - x)/dx} = \lim_{x \to 1} \frac{1/x}{-1} = -1$$

Hence,

$$\lim_{x \to 1} y = e^{-1}$$

See Art. 1-9.

2. (a) $x = 2$, because division by zero is not defined.

   (b) Use of L'Hôpital's rule gives

$$\lim_{x \to 2} \frac{x^3 - 8}{x - 2} = \lim_{x \to 2} \frac{(d/dx)(x^3 - 8)}{(d/dx)(x - 2)} = \lim_{x \to 2} \frac{3x^2}{1} = 12$$

If you define $f(x) = 12$ at $x = 2$, $f(x)$ will be continuous there.   See Arts. 1-1 and 1-9.

3. (a) $y'\Big]_{x=0} = [2x - 1]_0 = -1$.   Hence, the equation of the tangent is $y - 1 = -x$.   See Art. 1-2.

   (b) $y'\Big]_{x=10} = 2x\Big]_{10} = 20$.   Therefore, the equation of the normal is $x - 10 = 20(100 - y)$.   See Art. 1-2.

4. (a) The circumference $C = 2\pi r$. $dC/dr = 2\pi$.   See Art. 1-2.

   (b) The area $A = \pi r^2$. $dA/dr = 2\pi r$.   See Art. 1-2.

   (c) $dA/dt = (dA/dr)(dr/dt) = 2\pi r(dr/dt) = 2\pi(100)(0.001) = 0.2\pi$ sq in. per sec.   See Eq. (1-7).

5. Let R be the vertical load on the upper support.   Then, by taking moments of the loads about the lower support, we find $100R - 5 \times 1{,}000 - 10 \times 100 \times 100/2 = 0$, from which $R = 55{,}000/100 = 550$ lb.   Let x be the horizontal distance measured from the upper support and y the vertical sag of the cable below the upper support.   The bending moment in the cable must be zero at all points, because it is flexible.   Hence, $M = 550x - 1{,}000y - (10/2)x^2 = 0$.   Thus, $y = (550x - 5x^2)/1{,}000$.   The sag will be a maximum for $y' = (550 - 10x)/1{,}000 = 0$.   Solving for x, we find that the low point will be at $x = 55$ ft.   The maximum sag below the upper support is

$$y_{\max} = \frac{550 \times 55 - 5(55)^2}{1{,}000} = 15.125 \text{ ft}$$

See Art. 1-8.

6. (a) $t = x/v_0 \cos \theta$.   Hence,

$$y = v_0 \left(\frac{x}{v_0 \cos \theta}\right) \sin \theta - \frac{1}{2} g \left(\frac{x^2}{v_0^2 \cos^2 \theta}\right) = x \tan \theta - \frac{g}{2v_0^2 \cos^2 \theta} x^2$$

The path is a parabola.

(b) From $y = v_0t \sin \theta - gt^2/2 = 0$, we find $t = (2v_0/g) \sin \theta$. Therefore, the range is $x = v_0t \cos \theta = v_0[(2v_0/g) \sin \theta] \cos \theta = (v_0^2/g) \sin 2\theta$.

(c) For maximum height, $y' = \tan \theta - (g/v_0^2 \cos^2 \theta)x = 0$. Solving for x, we find that the projectile will be highest when $x = (v_0^2/g) \sin \theta \cos \theta$. The maximum height is

$$y = \left(\frac{v_0^2}{g} \sin \theta \cos \theta\right) \tan \theta - \frac{g}{2v_0^2 \cos^2 \theta}\left(\frac{v_0^2}{g} \sin \theta \cos \theta\right)^2$$

$$= \frac{v_0^2 \sin^2 \theta}{g} - \frac{v_0^2 \sin^2 \theta}{2g} = \frac{v_0^2 \sin^2 \theta}{2g}$$

See Art. 1-8.

(d) For maximum range, $dx/d\theta = (2v_0^2/g) \cos 2\theta = 0$. Hence, $\theta = \pi/4 = 45°$. See Art. 1-8.

7. Let $y =$ height of free surface above its low point. The forces acting on the mass are C horizontally, the force of gravity $W = Mg$ downward, and F. The slope of tangents to the curve of intersection of the free surface and a vertical plane through the axis of rotation is $dy/dr = C/W = M\omega^2r/Mg = \omega^2r/g$. Thus, $dy = (\omega^2r/g)\, dr$. (See Art. 1-4.) From Table 1-1, $d(r^2)/dr = 2r$, from which $r\, dr = d(r^2)/2$. Also, $dy = d(y + C)$, where C is a constant. Consequently, $y + C = \omega^2r^2/2g$. Since $y = 0$ when $r = 0$, C must be zero. Therefore, the equation of the curve in the vertical plane is $y = \omega^2r^2/2g$. This also is the equation of the free surface with r the radius vector from the axis of rotation. The surface is a paraboloid of revolution.

8. During time dt, the volume of liquid, cu ft, discharged is $dV = 48 \sqrt{h}\, dt$. The drop in volume, cu ft, in the tub is $dV = -4h\, dh$, where dh is the decrease in height of liquid. Then, $-4h\, dh = 48 \sqrt{h}\, dt$, or $dt = -(h^{1/2}\, dh)/12$. From Table 1-1, $d(h^{3/2})/dh = \tfrac{3}{2}h^{1/2}$, from which $\tfrac{2}{3}d(h^{3/2}) = h^{1/2}\, dh$. Also, $dt = d(t + C)$, where C is a constant. Hence, $t + C = -(\tfrac{2}{3}h^{3/2})/12 = -h^{3/2}/18$. When $t = 0$, $h = H$. So $C = -H^{3/2}/18$. Therefore, $t = (H^{3/2} - h^{3/2})/18$. And the time to empty $(h = 0)$ is $H^{3/2}/18$. See Art. 1-4.

9. Setting the derivative of q with respect to d equal to zero and dividing by $\sqrt{2g}$ yields $\sqrt{H - d_c} - \tfrac{1}{2} \dfrac{d_c}{\sqrt{H - d_c}} = 0$, from which $d_c = \tfrac{2}{3}H$. The maximum flow, therefore, is $\sqrt{g}\, d_c^{3/2}$. See Art. 1-8.

10. Let X be the stress, lb, in the vertical bar. Then, $(P - X)/\cos \alpha$ is the sum of the stresses in the other two bars. So the total strain energy in the system is

$$U = \frac{X^2L}{2AE} + 2\left(\frac{P - X}{2 \cos \alpha}\right)^2 \frac{L/\cos \alpha}{AE}$$

By the principle of least work,

$$\frac{dU}{dx} = \frac{XL}{AE} + \frac{(P - X)L}{2AE \cos^3 \alpha} = 0$$

Therefore, $X = P/(1 + 2 \cos^3 \alpha)$. See Art. 1-8.

11. $dL/L = -K\,dt$.   From Table 1-2, $(d/dL)\log_{10}L = 0.434/L$, from which $dL/L = 2.303d(\log_{10}L)$.   Also, $dt = d(t + C_1)$, where C_1 is a constant.   Hence, $\log_{10}L = -K_1t + C_2$, where K_1 and C_2 are constants.   From the definition of logarithm, $L = C_3 10^{-K_1t}$, where C_3 is a constant.   If the initial oxygen demand $(t = 0)$ is L_0, $C_3 = L_0$.   Therefore, $L = L_0 10^{-K_1t}$.

12. Velocity $= 3t^2 + 6t - 8$.   Acceleration $= a = 6t + 6$.   When $t = 6$, $a = 42$.   See Art. 1-3.

13. $2x\,dx + 4y\,dy = 0$.   Hence, $dy/dx = y' = -2x/4y = -x/2y$.   When $x = y$, $y' = -\frac{1}{2}$. $y'' = dy'/dx = (-2y + 2xy')/4y^2$.   From the equation of the ellipse, $x = y = 3$.   So $y''\Big]_{3,3} = [-2 \times 3 + 2 \times 3(-\frac{1}{2})]/(4)(3)^2 = -\frac{1}{4}$. Therefore, curvature $= -\frac{1}{4}/\sqrt{1 + (-\frac{1}{2})^2} = -1/2\sqrt{5}$.   See Art. 1-3.

14. (a) $\partial z/\partial x = a\cos(ax + by)$. $\partial z/\partial y = b\cos(ax + by)$.   See Art. 1-5 and Table 1-2.

(b) From Eq. (1-22),

$$\frac{\partial z}{\partial x} = -\frac{2x}{2z} = -\frac{x}{z} \qquad \frac{\partial z}{\partial y} = -\frac{2y}{2z} = -\frac{y}{z}$$

(c) From Eqs. (1-23) and (1-24),

$$\frac{\partial z}{\partial x} = \frac{\partial z}{\partial u}\frac{\partial u}{\partial x} + \frac{\partial z}{\partial v}\frac{\partial v}{\partial x} = (2u + v)2 + (u + 2v) = 5u + 4v$$

$$\frac{\partial z}{\partial y} = \frac{\partial z}{\partial u}\frac{\partial u}{\partial y} + \frac{\partial z}{\partial v}\frac{\partial v}{\partial y} = (2u + v) + (u + 2v)(-2) = -3v$$

15. By Eq. (1-26),

$$dP = \frac{\partial P}{\partial E}\,dE + \frac{\partial P}{\partial R}\,dR = \frac{2E}{R}\,dE - \frac{E^2}{R^2}\,dR$$

$$= \frac{2 \times 220}{10}(-10) - \frac{(220)^2}{(10)^2}\left(\frac{1}{2}\right) = -682 \text{ watts}$$

16. In Cartesian coordinates, you need $\partial E/\partial l$ at $(3,4)$ in the direction toward $(0,10)$.   This direction is such that θ in Eq. (1-25) is $\arctan -(10 - 4)/3 = -2$. Hence, $\sin\theta = 2/\sqrt{5}$ and $\cos\theta = 1/\sqrt{5}$.   Then,

$$\frac{\partial E}{\partial l} = \frac{\partial E}{\partial x}\cos\theta + \frac{\partial E}{\partial y}\sin\theta = \cos\theta\frac{\partial}{\partial x}\log_e\sqrt{x^2 + y^2} + \sin\theta\frac{\partial}{\partial y}\log_e\sqrt{x^2 + y^2}$$

$$= \cos\theta\frac{2x}{2(x^2 + y^2)} + \sin\theta\frac{2y}{2(x^2 + y^2)}$$

At $(3,4)$, $\partial E/\partial l = (-1/\sqrt{5})(3/25) + (2/\sqrt{5})(4/25) = \sqrt{5}/25$.

TWO

Integral Calculus

Contemporary textbooks usually approach integral calculus from either of two viewpoints. One defines integration as the inverse of differentiation and then shows that integration may be interpreted in other ways. The other approach defines integration as the limit of a sum and then shows that integrals may be evaluated by using the inverse of differentiation. If the second viewpoint is adopted, you are offered the choice of two definitions. One leads to the Riemann integral, the second to the Lebesgue integral. Both integrals have the same form, and in general integration is carried out in the same manner. But some functions may be integrable under the Lebesgue definition but not under the Riemann definition.

In this book, we shall adopt the first viewpoint. We shall then interpret integrals as limits of sums in agreement with the Riemann definition. This approach provides you with sufficient information to solve the engineering problems that you are likely to encounter. Nevertheless, you will find it well worth while in continuing your study of mathematics to become familiar with the Lebesgue integral.

2-1 *Indefinite Integrals.* *Integration is the process of finding a function when given its differential.* Equation (1-19) defined the differential of $y = f(x)$ as $dy = df(x) = f'(x)\, dx$, where $f'(x) = dy/dx = y'$. Thus, integration transforms $f'(x)$ into $f(x)$. Or given any function $f(x)$, integration transforms it into its integral $F(x)$. The operation is indicated by a symbol resembling an elongated S.

$$\int f(x)\, dx = F(x) + C \qquad (2\text{-}1)$$

The left-hand side of Eq. (2-1) is called an indefinite integral. The sign before $f(x)$ is the integration sign; $f(x)$ is called the integrand. The expression is read "the integral of $f(x)\, dx$."

On the right-hand side of Eq. (2-1), $F(x)$ is any function whose derivative is $f(x)$. C is a constant. (Since the derivative of a constant is zero, all functions differing only by a constant have the same derivative.) $F(x)$ plus any constant is called an integral of $f(x)$.

You may recall from Chap. 1 that velocity is the derivative of distance with respect to time and that acceleration is the derivative of velocity with respect to time. In some of the problems in that chapter, you were given distance as a function of time and asked to compute velocity or acceleration. Now you will be given acceleration and asked to find the velocity or location of a moving body at a given time.

For example, if a body is dropped from a height of 1,000 ft and has a constant acceleration $g = 32$ ft per sec², what is its velocity after 3 sec? How high above the ground will it be?

If the downward direction is taken as negative, the acceleration $g = -dv/dt$, where v is the velocity of the body and t is time. Therefore,

$$\int dv = v = -\int g\, dt = -gt + C_1$$

At the start, both t and v are zero. Hence, $C_1 = 0$; so $v = -gt$. At the end of 3 sec, $v = -32 \times 3 = -96$ fps.

Since velocity $v = ds/dt$, the rate of change of distance with respect to time, then

$$\int ds = s = \int v\, dt = -\int gt\, dt = -gt^2/2 + C_2$$

At the start, $t = 0$ and $s = 1,000$ ft. Therefore, $C_2 = 1,000$. Then, $s = 1,000 - gt^2/2$. And at the end of 3 sec, the height of the body will be $s = 1,000 - \frac{1}{2} \times 32(3)^2 = 856$ ft.

2-2 *Integration Methods.* By the rules given in Chap. 1, differentiation is a straightforward process, and often simple. However, integration in general is neither straightforward nor simple. Integrands may not be in a form that you recognize as being a result of differentiation; for instance, see one of the derivatives in Tables 1-1 and 1-2. In nonelementary cases, you have to resort to transformations and other techniques to convert the integrands into recognizable derivatives.

As an alternative, you might seek an integral in a table of integrals. The table of standard elementary integrals on pages 29 and 30 is a small sample of a table of integrals. See the Bibliography at the end of this chapter for the titles and publishers of such tables.

As a first step in transforming an integrand into recognizable form, you might employ certain basic properties of integrals:

$$\int du = u + C \qquad \int \frac{du}{dx} dx = u + C \qquad \frac{d}{dx} \int u\, dx = u + C \quad (2\text{-}2)$$

where C is a constant and u is a function of x.

If k is a constant, then

$$\int k\, du = k\int du \qquad\qquad\qquad\qquad\qquad\qquad\qquad (2\text{-}3)$$

Thus, a constant factor may be moved outside the integral sign.

If $du, dv, dw, \ldots$ are the differentials of a finite number of functions, then

$$\int (du + dv + dw + \cdots) = \int du + \int dv + \int dw + \cdots$$
$$= u + v + w + \cdots \quad (2\text{-}4)$$

Thus, the integral of the algebraic sum of a finite number of functions is the algebraic sum of the integrals of the functions.

For example, find $\int (3x^2 - 4x + 5)\, dx$.

$$\begin{aligned}
\int (3x^2 - 4x + 5)\, dx &= \int 3x^2\, dx - \int 4x\, dx + \int 5\, dx \\
&= 3\int x^2\, dx - 4\int x\, dx + 5\int dx \\
&= 3(x^3/3) - 4(x^2/2) + 5x + C \\
&= x^3 - 2x^2 + 5x + C
\end{aligned}$$

As this example indicates, you can use the basic properties to simplify an integral. In such a case, the integrand is simplified by first writing it as the sum of integrals of simple functions. Sometimes other techniques must be adopted to obtain simple functions.

If the integrand is a rational fraction with the numerator of degree equal to or greater than that of the denominator, carry out the indicated division until the remainder is of lower degree than the denominator. If necessary, break the remainder into **partial fractions** for further simplification.

Often, **substitution** of another variable for a portion of the integrand may facilitate integration.

Standard Elementary Integrals

$$\int dx = x + C$$

$$\int u^n\, du = \frac{u^{n+1}}{n+1} + C \qquad n \neq -1$$

$$\int \frac{du}{u} = \log_e u + C$$

$$\int e^u\, du = e^u + C$$

$$\int a^u\, du = \frac{a^u}{\log_e a} + C$$

$$\int \sin u\, du = -\cos u + C$$

$$\int \cos u\, du = \sin u + C$$

$$\int \tan u\, du = \log_e \sec u + C$$

$$\int \cot u\, du = \log_e \sin u + C$$

$$\int \sec u\, du = \log_e (\sec u + \tan u)C$$

$$\int \csc u\, du = \log_e (\csc u - \cot u)C$$

$$\int \sec^2 u\, du = \tan u + C$$

$$\int \csc^2 u\, du = -\cot u + C$$

$$\int \sec u \tan u\, du = \sec u + C$$

$$\int \csc u \cot u\, du = -\csc u + C$$

$$\int \frac{du}{u^2 + a^2} = \frac{1}{a} \arctan \frac{u}{a} + C$$

$$\int \frac{du}{u^2 - a^2} = \frac{1}{2a} \log_e \frac{a - u}{a + u} + C$$

$$\int \frac{du}{a^2 - u^2} = \frac{1}{2a} \log_e \frac{a + u}{a - u} + C$$

$$\int \frac{du}{\sqrt{u^2 \pm a^2}} = \log_e (u + \sqrt{u^2 \pm a^2})C$$

$$\int \frac{du}{\sqrt{a^2 - u^2}} = \arcsin \frac{u}{a} + C$$

Standard Elementary Integrals (Continued)

$$\int \sqrt{a^2 - u^2}\, du = \frac{u}{2}\sqrt{a^2 - u^2} + \frac{a^2}{2}\arcsin\frac{u}{a} + C$$

$$\int \sqrt{u^2 \pm a^2}\, du = \frac{u}{2}\sqrt{u^2 \pm a^2} \pm \frac{a^2}{2}\log_e(u + \sqrt{u^2 \pm a^2}) + C$$

$$\int \sinh u\, du = \cosh u + C$$

$$\int \cosh u\, du = \sinh u + C$$

$$\int \tanh u\, du = \log_e \cosh u + C$$

$$\int \coth u\, du = \log_e |\sinh u| + C$$

Thus, to find $\int \sin\theta \cos\theta\, d\theta$, let $u = \sin\theta$. Then, $du = \cos\theta\, d\theta$. Substitution in the integral yields

$$\int \sin\theta \cos\theta\, d\theta = \int u\, du = u^2/2 + C = \tfrac{1}{2}\sin^2\theta + C$$

As another example, find $\int x(x^2 + 2)^9\, dx$. This could be integrated by expanding $(x^2 + 2)^9$ and integrating the result term by term. Simpler, however, is the substitution $u = x^2 + 2$. Then, $du = 2x\, dx$ and $x\, dx = du/2$. Hence,

$$\int x(x^2 + 2)^9\, dx = \int u^9 \left(\frac{1}{2}\, du\right) = \frac{1}{2}\int u^9\, du$$

$$= \frac{1}{2}\frac{u^{10}}{10} + C = \frac{(x^2 + 2)^{10}}{20} + C$$

In general, if an integral contains

$f(\sqrt{a^2 - u^2})$ try substitution of $u = a\sin\theta$
$f(\sqrt{a^2 + u^2})$ try substitution of $u = a\tan\theta$
$f(\sqrt{u^2 - a^2})$ try substitution of $u = a\sec\theta$
$f(x^{p/q})$ try substitution of $x = z^n$

where n is the least common denominator of the fractional exponents of x.

$f[x,(a + bx)^{p/q}]$ try substitution of $a + bx = z^n$

where n is the least common denominator of the fractional exponents of $a + bx$.

Another technique you may find useful is **integration by parts**. This is based on $d(uv) = u\, dv + v\, du$. If u and v are functions of a single variable, then

$$\int u \, dv = uv - \int v \, du \tag{2-5}$$

Hence, $u \, dv$ can be integrated if $v \, du$ can be integrated.

For example, to find $\int x \sin x \, dx$, let $u = x$ and $dv = \sin x \, dx$. Then

$$du = dx \qquad v = \int \sin x \, dx = - \cos x$$
$$\int x \sin x \, dx = -x \cos x - \int - \cos x \, dx = -x \cos x + \sin x + C$$

As a more complicated example, find $y = \int e^x \sin x \, dx$. Let $u = e^x$ and $dv = \sin x \, dx$. Then,

$$du = e^x \, dx \qquad v = \int \sin x \, dx = - \cos x$$
$$y = -e^x \cos x + \int e^x \cos x \, dx$$

Since the result contains an integral of the same form as the given integral, try integration by parts again. Let $u = e^x$ and $dv = \cos x \, dx$. Then,

$$du = e^x \, dx \qquad v = \int \cos x \, dx = \sin x$$
$$\int e^x \cos x \, dx = e^x \sin x - \int e^x \sin x \, dx = e^x \sin x - y$$

Substitution of this result in the previous expression for y gives

$$y = -e^x \cos x + e^x \sin x - y$$

Solving for y, we get

$$y = e^x(\sin x - \cos x)/2$$

2-3 *Definite Integrals.* From the observation that all functions differing only by a constant have the same derivative, we have concluded that the integral $F(x)$ of a function $f(x)$ contains an arbitrary constant C; that is, $\int f(x) \, dx = F(x) + C$. Suppose that $F(x)$ takes the value $F(b)$ at $x = b$ and $F(a)$ at $x = a$. We indicate $F(b) - F(a)$ by

$$F(b) - F(a) = \int_a^b f(x) \, dx \tag{2-6}$$

The expression on the right in Eq. (2-6) is called a definite integral. It is read "definite integral from a to b of $f(x)$." a is called the lower limit (though "bound" probably would be preferable to avoid confusion with limit defined in Art. 1-1). b is the upper limit.

A definite integral is evaluated by first integrating as for indefinite integrals. The limits are then substituted for the variable. Finally, the result obtained with the lower limit is subtracted from that obtained with the upper limit.

$$\int_a^b f(x) \, dx = F(x) \Big]_a^b = F(b) - F(a) \tag{2-7}$$

Note that a definite integral is a function of the limits.   It is independent of the variable of integration.   Note also that the constant of integration is eliminated in the subtraction.

If, to integrate, you substitute other variables, you must also change the limits in accordance with the equation of substitution.   For example, to evaluate $\int_1^4 dx/\sqrt{x}$, let $x = z^2$ and $dx = 2z\,dz$.   When the substitution is made, the limits must be changed: when $x = 1$, $z = 1$ and when $x = 4$, $z = 2$.   Therefore,

$$\int_1^4 \frac{dx}{\sqrt{x}} = \int_1^2 \frac{2z\,dz}{z} = 2z\Big]_1^2 = 2 \times 2 - 2 \times 1 = 2$$

Reversal of the limits of a definite integral changes its sign:

$$\int_a^b f(x)\,dx = -\int_b^a f(x)\,dx \tag{2-8}$$

If $a < b < c$, then

$$\int_a^c f(x)\,dx = \int_a^b f(x)\,dx + \int_b^c f(x)\,dx \tag{2-9}$$

Let us now consider the curve representing $y = F(x)$.   The slope at any point of the curve is given by $y' = f(x)$.   Then, $f(x)\,dx = dy$ may be considered as a differential increment in y corresponding to a differential increment dx in x.   If dy were a finite increment Δy, summation of the increments from $x = a$ to $x = b$ would give $F(b) - F(a)$ (Fig. 2-1). This suggests that the definite integral $\int_a^b f(x)\,dx = F(b) - F(a)$ is a summation.   Let us delve into this concept.

Suppose that $f(x)$ is continuous in the closed interval $a \leq x \leq b$.   Let us divide this interval into n parts such that the point x_1 is contained in the interval Δx_1, x_2 in Δx_2, . . . , x_n in Δx_n, and $\sum_1^n \Delta x_i = b - a$.   From Fig. 2-2, you can see that $\sum_1^n y_i\,\Delta x_i = \sum_1^n f(x_i)\,\Delta x_i$ approximates the area between the curve $y = f(x)$ and the x axis and between the ordinates at

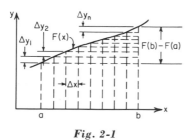

Fig. 2-1

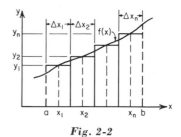

Fig. 2-2

$x = a$ and $x = b$. If we let n approach infinity, then the area is given exactly by

$$A = \lim_{n \to \infty} \sum_{1}^{n} f(x_i) \, \Delta x_i$$

Now, let ΔA be the increment of area under $f(x)$ between the ordinates at x_i and $x_i + \Delta x_i$. Then,

$$f(x_i) \, \Delta x_i \leq \Delta A \leq f(x_i + \Delta x_i) \, \Delta x_i$$

Dividing by Δx_i and letting $\Delta x_i \to 0$, we get

$$f(x) \leq \lim_{\Delta x \to 0} \frac{\Delta A}{\Delta x} \leq \lim_{\Delta x \to 0} f(x + \Delta x) = f(x)$$

But

$$\lim_{\Delta x \to 0} \frac{\Delta A}{\Delta x} = \frac{dA}{dx}$$

Hence, $dA/dx = f(x)$, $dA = f(x) \, dx$, and $A = \int f(x) \, dx = F(x) + C$. When $x = a$, $A = 0$; so $C = -F(a)$, and $A = F(x) - F(a)$. Furthermore, when $x = b$,

$$A = F(b) - F(a) = \int_a^b f(x) \, dx \qquad (2\text{-}10a)$$

Therefore,

$$\int_a^b f(x) \, dx = \lim_{n \to \infty} \sum_{1}^{n} f(x_i) \, \Delta x_i \qquad (2\text{-}10b)$$

As an example of the use of the summation concept, consider the following problem: If a fluid weighs w lb per cu ft, how much pressure does it exert on a 1-ft width of wall 10 ft high? Assume that the unit pressure, psf, varies directly with depth.

Let y be the unit pressure on a strip of wall dx high at a distance x from the top of the wall. The total pressure on the strip is $y \, dx$. With Eq. (2-10) as justification, we find the total pressure P on the 10-ft wall from

$$P = \int_0^{10} y \, dx$$

The unit pressure y equals the weight of a column of fluid x ft high and 1 ft square in cross section; that is, $y = wx$. Hence,

$$P = \int_0^{10} wx \, dx = \frac{1}{2} wx^2 \bigg]_0^{10} = \frac{1}{2} w(100 - 0) = 50w$$

2-4 *Improper Integrals.* If either or both of the limits of a definite integral is infinity, or if the integrand becomes infinite in the interval of integration, the definite integral is called an improper integral.

Suppose that an integral has infinity as one of its limits a (or b).   Let us treat a (or b) as if it were a variable approaching infinity.   Let us also define the given integral as the limit, if one exists, of the definite integral as a (or b) approaches infinity.   For example,

$$\int_2^\infty \frac{dx}{x^2} = \lim_{b \to \infty} \int_2^b \frac{dx}{x^2} = \lim_{b \to \infty} \left(-\frac{1}{b} + \frac{1}{2} \right) = \frac{1}{2}$$

Suppose that an integrand becomes infinite at one of the limits a (or b). Again, let us treat the limit as if it were a variable approaching a (or b). Let us also define the given integral as the limit, if one exists, of the definite integral as the variable approaches a (or b).   For instance,

$$\int_0^4 \frac{dx}{\sqrt{x}} = \lim_{a \to 0} \int_a^4 \frac{dx}{\sqrt{x}} = \lim_{a \to 0} (2\sqrt{4} - 2\sqrt{a}) = 4$$

Finally, suppose that an integrand becomes infinite within the interval of integration, say at $x = b$, where $a < b < c$.   If both $\int_a^b f(x)\,dx$ and $\int_b^c f(x)\,dx$ exist, let us define

$$\int_a^c f(x)\,dx = \int_a^b f(x)\,dx + \int_b^c f(x)\,dx$$

For example, in the following integral, the integrand becomes infinite at $x = 0$.

$$\int_{-1}^1 \frac{dx}{x^{2/3}} = \lim_{a \to 0+} \int_a^1 \frac{dx}{x^{2/3}} + \lim_{b \to 0-} \int_{-1}^b \frac{dx}{x^{2/3}}$$
$$= \lim_{a \to 0+} (3 - 3a^{1/3}) + \lim_{b \to 0-} (3b^{1/3} + 3) = 3 + 3 = 6$$

2-5 *Multiple Integrals.* The symbols

$$\int_a^b dx \int_m^n f(x,y)\,dy = \int_a^b \int_m^n f(x,y)\,dy\,dx$$

where m and n may be functions of x, represent an iterated or repeated integral.   The symbols indicate that integration is to be executed first with respect to y, holding x constant, then with respect to x.

Let $f(x,y)$ be continuous throughout a region S of the xy plane.   Now, consider S subdivided into n subregions so that (x_1,y_1) is contained in the subregion ΔS_1, (x_2, y_2) in ΔS_2, (x_n, y_n) in ΔS_n, and $\sum\limits_{i}^{n} \Delta S_i = S$.   Then,

$$\lim_{n \to \infty} [f(x_1, y_1) \, \Delta S_1 + f(x_2, y_2) \, \Delta S_2 + \cdots + f(x_n, y_n) \, \Delta S_n]$$

$$= \iint_{S} f(x, y) \, dS \quad (2\text{-}11)$$

The expression on the right-hand side of Eq. (2-11) is called a double integral. It requires $f(x,y)$ to be integrated over the entire region S. The left-hand side indicates that the double integral is the limit of a sum. It can be evaluated as an iterated integral.

$$\iint_{S} f(x, y) \, dS = \int_{a}^{b} \int_{m}^{n} f(x, y) \, dy \, dx = \int_{c}^{d} \int_{p}^{q} f(x, y) \, dx \, dy \quad (2\text{-}12)$$

where m and n may be functions of x, and p and q may be functions of y.

For example, let us evaluate the double integral of $x^2 y^3$ over the rectangle A whose vertices are $(0,0)$, $(3,0)$, $(3,2)$, and $(0,2)$.

$$\iint_{A} x^2 y^3 \, dA = \int_{0}^{3} \int_{0}^{2} x^2 y^3 \, dy \, dx = \int_{0}^{3} \left[\frac{1}{4} x^2 y^4 \right]_{0}^{2} dx = \int_{0}^{3} 4x^2 \, dx$$

$$= \left[\frac{4}{3} x^3 \right]_{0}^{3} = 36$$

As an example of the summation concept of the double integral, let us find the area of one quadrant of a circle. Start with a differential area $dy \, dx$ (Fig. 2-3a).   Now, holding x constant, integrate with respect to y between the circle and the x axis.

$$A = \iint_{A} dy \, dx = \iint_{0}^{\sqrt{r^2 - x^2}} dy \, dx = \int [y]_{0}^{\sqrt{r^2 - x^2}} dx = \int \sqrt{r^2 - x^2} \, dx$$

The result is the area of a strip with height $\sqrt{r^2 - x^2}$ and width dx (Fig. 2-3b).   Finally, integrate with respect to x from $x = 0$ to $x = r$, to sum

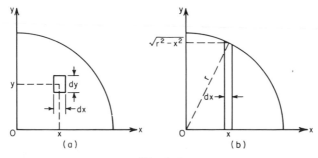

(a)   (b)

Fig. 2-3

all such strips in the quadrant.   The result is the area of the quadrant:

$$A = \int_0^r \sqrt{r^2 - x^2}\, dx = \left[\frac{x}{2}\sqrt{r^2 - x^2} + \frac{r^2}{2}\arcsin\frac{x}{r}\right]_0^r = \frac{\pi r^2}{4}$$

2-6   *Line Integrals.* If $M(x,y)$ and $N(x,y)$ are continuous functions that are single-valued at every point of a plane curve $C(x,y)$ between two points $A(x_1,y_1)$ and $B(x_2,y_2)$, then

$$I = \int_C [M(x,y)\, dx + N(x,y)\, dy]$$

integrated over C between A and B is called a line integral.

Suppose, for example, that a body moves along the parabola $y = x^2$ acted upon by a force F.   Suppose also that the component of F parallel to the x axis is $X = k$, a constant, and the component parallel to the y axis is $Y = 2kx$.   How much work is done when the body moves from $(0,0)$ to $(2,4)$?   Since work is the product of force and movement in the direction of the force, then

$$W = \int_C F\, ds = \int_C (X\, dx + Y\, dy) = \int_{0,0}^{2,4} (k\, dx + 2kx\, dy)$$
$$= k\int_0^2 dx + 2k\int_0^4 y^{1/2}\, dy = kx\Big]_0^2 + \tfrac{4}{3}ky^{3/2}\Big]_0^4 = 2k + \tfrac{32}{3}k = \tfrac{38}{3}k$$

If $\partial M/\partial y = \partial N/\partial x$, the line integral around any closed curve is zero. Also, the value of the integral between any two points is independent of the path between them and hence is a function of the end points only.

If $P(x,y,z)$, $Q(x,y,z)$, and $R(x,y,z)$ are single-valued continuous functions along a space curve C, then the line integral is

$$I = \int_C (P\, dx + Q\, dy + R\, dz)$$

If $\partial P/\partial y = \partial Q/\partial x$, $\partial Q/\partial z = \partial R/\partial y$, and $\partial R/\partial x = \partial P/\partial z$, the line integral around any closed curve in space is zero.   Also, the value of the integral between any two points is independent of the path between them.

Green's theorem relates a line integral over a closed curve in a plane to a double integral over the plane region bounded by the curve.   Let $M(x,y)$, $N(x,y)$, $\partial M/\partial y$, and $\partial N/\partial x$ be continuous single-valued functions over the closed, simply connected region R and its boundary curve C. Then, the theorem states

$$\iint_R \left(\frac{\partial M}{\partial y} - \frac{\partial N}{\partial x}\right) dx\, dy = -\int_C (M\, dx + N\, dy) \tag{2-13}$$

The double integral is taken over the given region and the line integral is taken counterclockwise along C.

From Eq. (2-13), we can derive a useful formula for the area enclosed by C. Set $M = -y$ and $N = x$ in Eq. (2-13). The result is

$$\iint_R 2dy\, dx = \int_C (-y\, dx + x\, dy)$$

The double integral on the left equals twice the area of the enclosed region. Hence, the area bounded by C is

$$A = \frac{1}{2} \int_C (-y\, dx + x\, dy) \tag{2-14}$$

For example, let us find the area between $x^2 = 4y$ and $y^2 = 4x$. The curves intersect at $(0,0)$ and $(4,4)$. The integration should be carried out first along $x^2 = 4y$ from $(0,0)$ to $(4,4)$, then along $y^2 = 4x$ from $(4,4)$ to $(0,0)$.

$$A = \frac{1}{2} \int_{0,0}^{4,4} (-y\, dx + x\, dy) + \frac{1}{2} \int_{4,4}^{0,0} (-y\, dx + x\, dy)$$

$$= \frac{1}{2} \int_0^4 \left(-\frac{x^2}{4}\, dx + x\, \frac{x}{2}\, dx \right) + \frac{1}{2} \int_4^0 \left(-y\, \frac{y}{2}\, dy + \frac{y^2}{4}\, dy \right)$$

$$= \frac{x^3}{24} \Big]_0^4 - \frac{y^3}{24} \Big]_4^0 = \frac{16}{3}$$

2-7 Differentiation under the Integral Sign. You may sometimes find it necessary to differentiate an integral. Several cases are possible. One simple case is given in Eq. (2-2):

$$\frac{d}{dx} \int u\, dx = u + C$$

where u is a function of x and C is a constant. Differentiation in this case removes the integral sign.

Suppose that the integrand is a function of a parameter as well as the variable of integration:

$$F(\alpha) = \int_a^b f(x,\alpha)\, dx \tag{2-15}$$

Suppose also that the partial derivative of the integrand with respect to the parameter exists and is continuous. Then, $f(x,\alpha)$ may be differentiated with respect to the parameter before integration.

$$\frac{dF}{d\alpha} = \frac{d}{d\alpha} \int_a^b f(x,\alpha)\, dx = \int_a^b \frac{\partial f}{\partial \alpha}\, dx \tag{2-16}$$

Equation (2-16) may be used to derive a formula for the deflection of a beam (dummy-unit-load method).   The strain energy due to bending is

$$U = \int_0^L \frac{M^2\,dx}{2EI}$$

where L is the span, M is the bending moment at a distance x from a support, E is the modulus of elasticity of the beam material, and I is the moment of inertia of the beam cross section.   Castigliano's first theorem states that the partial derivative of the strain energy with respect to a force gives the deformation corresponding to that force.   Hence, place a load P on the beam at the point for which the deflection is to be computed. Then, the deflection of the beam due to the loading causing M is

$$\delta = \frac{\partial U}{\partial P} = \frac{\partial}{\partial P}\int_0^L \frac{M^2\,dx}{2EI} = \int_0^L \frac{M}{EI}\frac{\partial M}{\partial P}\,dx$$

$\partial M/\partial P$ equals the bending moment m produced by a unit load at the point where the deformation is to be measured and in the direction of the deformation.   Hence,

$$\delta = \int_0^L \frac{Mm}{EI}\,dx$$

Equation (2-16) may also be useful in evaluating certain integrals. For example, let us determine $y = \int_{-\infty}^0 [(e^{\alpha x} - e^x)/x]\,dx$, where $\alpha \geq 1$. The first step is to differentiate with respect to α:

$$\frac{dy}{d\alpha} = \int_{-\infty}^0 e^{\alpha x}\,dx = \frac{1}{\alpha}e^{\alpha x}\Big]_{-\infty}^0 = \frac{1}{\alpha} - 0$$

The second step is to integrate $dy = d\alpha/\alpha$.   This yields $y = \log_e \alpha + C$, where C is a constant.   To evaluate C, return to the given integral and let $\alpha = 1$.

$$y(1) = \int_{-\infty}^0 \frac{e^x - e^x}{x}\,dx = 0 = \log_e 1 + C$$

Since $\log_e 1 = 0$, $C = 0$.   Hence, the required integral is $\log_e \alpha$.

If the limits of a definite integral, as well as the integrand, are functions of a parameter α, then

$$\frac{dF}{d\alpha} = \int_{a(\alpha)}^{b(\alpha)} \frac{\partial f(x,\alpha)}{\partial \alpha}\,dx + f(b,\alpha)\frac{db}{d\alpha} - f(a,\alpha)\frac{da}{d\alpha} \qquad (2\text{-}17)$$

For example, let us find $dy/d\alpha$ when $y = \int_{\pi/2\alpha}^{\pi/\alpha} [(\sin \alpha x)/x] \, dx$, where $\alpha \neq 0$.   Differentiation under the integral sign gives

$$\frac{dy}{d\alpha} = \int_{\pi/2\alpha}^{\pi/\alpha} \cos \alpha x \, dx + \frac{\sin \pi}{\pi/\alpha} \left(-\frac{\pi}{\alpha^2} \right) - \frac{\sin \pi/2}{\pi/2\alpha} \left(-\frac{\pi}{2\alpha^2} \right)$$

$$= \frac{1}{\alpha} \sin \alpha x \Big]_{\pi/2\alpha}^{\pi/\alpha} + 0 + \frac{1}{\alpha} = 0 - \frac{1}{\alpha} + \frac{1}{\alpha} = 0$$

2-8 Bibliography

R. V. Andree, "Introduction to Calculus," McGraw-Hill Book Company; New York.

H. M. Bacon, "Differential and Integral Calculus," McGraw-Hill Book Company, New York.

W. R. Blakeley, "Calculus for Engineering Technology," John Wiley & Sons, Inc., New York.

R. S. Burington, "Handbook of Mathematical Tables and Formulas," McGraw-Hill Book Company, New York.

H. B. Dwight, "Tables of Integrals and Other Mathematical Data," The Macmillan Company, New York.

B. O. Peirce, "A Short Table of Integrals," Ginn and Company, Boston.

S. K. Stein, "Calculus," McGraw-Hill Book Company, New York.

PROBLEMS

1. Find the area under the curve $y = \sin x$ between $x = 0$ and $x = \pi$.

2. Find the area of the ellipse $x = a \cos \theta$, $y = b \sin \theta$.

3. In polar coordinates, the differential area strip is a circular sector with area $\rho^2 \, d\theta/2$, where $\rho =$ radius vector and $\theta =$ central angle.   Find the area of the circle $\rho = a \sin \theta$.

4. Determine the moment of inertia n of the circle $x^2 + y^2 = r^2$ about the y axis.   (*Hint:* Multiply a differential area strip by x^2 and integrate.)

5. Compute the finite area between the parabola $y^2 = x + 1$ and the line $x + y = 1$.

6. Determine the polar moment of inertia J of the circle $x^2 + y^2 = r^2$. (*Hint:* Multiply a polar differential area $\rho \, d\rho \, d\theta$ by ρ^2, where $\rho =$ radius vector and $\theta =$ central angle, then integrate.)

7. Using Eq. (1-21), compute the length of the catenary $y = a \cosh x/a$ from $x = 0$ to $x = 1$.

8. In polar coordinates, the differential length of curve is $ds = (\rho^2 \, d\theta^2 + d\rho^2)^{1/2}$, where $\rho =$ radius vector and $\theta =$ central angle.   Compute the perimeter of the cardioid $\rho = a(1 + \cos \theta)$.

9. Find the surface area generated by rotating the hypocycloid $x^{2/3} + y^{2/3} = a^{2/3}$ about the x axis.   (*Hint:* Multiply a differential curve length by $2\pi r$, where r is its distance from the axis of rotation, then integrate.)

10. Find the volume of the ellipsoid generated by revolving $x^2/a^2 + y^2/b^2 = 1$ about the x axis.   (*Hint:* Take a differential volume with area πy^2 and thickness dx, then integrate.)

11. Find the volume of the torus obtained by rotating the area bounded by the circle $x^2 + y^2 = a^2$ about the line $x = b$, where $b > a$. (*Hint:* Multiply a differential area strip parallel to the axis of rotation by $2\pi r$, where r is the distance from the strip to the axis, then integrate.)

12. Find the volume of the paraboloid $x^2 + 4y^2 = z$ between the planes $z = 0$ and $z = 1$. (*Hint:* Integrate a differential volume strip $A\,dz$, where $A = f(z) = $ cross-sectional area.)

13. Compute the volume in the first octant bounded by the cylinder $x^2 + y^2 = a^2$, the coordinate planes, and the plane $z = x + y$. (*Hint:* Take a differential volume $z = x + y$ high, dy wide, and dx thick, and integrate.)

14. Water impounded by a dam exerts pressure on it over a trapezoidal area. When the water behind the dam is 15 feet deep, the area is 225 ft long at the top and 150 ft at the bottom. If the water weighs $w = 62.4$ lb per cu ft, what is the total water pressure on the dam?

15. A force F of 25 lb stretches a 10-in.-long spring 0.25 in. How much work is done in stretching the spring from 11 to 12 in. (Assume that force and deformation are proportional.)

16. Locate, with respect to the axes, the centroid of the area under $y = \sin x$ from $x = 0$ to $x = \pi$.

17. If 200 cu ft of air at 15 psi is compressed to 80 psi, what is the final volume? How much work is done? (Assume $pV = $ constant.)

18. Evaluate

$$y = \int_0^1 \frac{x^\alpha - 1}{\log_e x}\, dx \qquad \text{for } \alpha > -1$$

ANSWERS

1. By Eq. (2-10a)

$$A = \int_0^\pi y\, dx = \int_0^\pi \sin x\, dx = -\cos x\Big]_0^\pi = 2$$

See Arts. 2-2 and 2-3.

2. You may integrate over the area inside the ellipse or along the curve. If you integrate over the area, take advantage of symmetry by computing the area in the first quadrant and multiplying by 4. For the limits: when $x = 0$, $\theta = \pi/2$; when $x = a$, $\theta = 0$. From $x = a \cos \theta$, obtain $dx = -a \sin \theta\, d\theta$.

$$A = 4\int_0^a y\, dx = -4\int_{\pi/2}^0 ab \sin^2 \theta\, d\theta = 4ab \int_0^{\pi/2} \frac{1}{2}(1 - \cos 2\theta)\, d\theta$$

$$= 2ab\left[\theta - \frac{1}{2}\sin 2\theta\right]_0^{\pi/2} = \pi ab$$

Integration along the ellipse yields

$$A = \frac{1}{2} \int_C (-y\, dx + x\, dy) = \frac{1}{2} \int_0^{2\pi} [(-b \sin\theta)(-a \sin\theta\, d\theta)$$
$$+ (a \cos\theta)(b \cos\theta\, d\theta)]$$
$$= \frac{1}{2} \int_0^{2\pi} ab(\sin^2\theta + \cos^2\theta)\, d\theta = \frac{ab}{2} [\theta]_0^{2\pi} = \pi ab$$

See Arts. 2-2, 2-3, and 2-6.

3. Take advantage of symmetry by computing the area of the semicircle in the first quadrant and multiplying by 2.

$$A = \frac{2}{2} \int_0^{\pi/2} \rho^2\, d\theta = \int_0^{\pi/2} a^2 \sin^2\theta\, d\theta = \frac{a^2}{2} \int_0^{\pi/2} (1 - \cos 2\theta)\, d\theta$$
$$= \frac{a^2}{2} \left[\theta - \frac{1}{2} \sin 2\theta \right]_0^{\pi/2} = \frac{\pi a^2}{4}$$

See Arts. 2-2 and 2-3.

4. With a differential area $y\, dx$, the moment of inertia is given by

$$I = 4 \int_0^r x^2 y\, dx = 4 \int_0^r x^2 \sqrt{r^2 - x^2}\, dx$$

Let $x = r \sin\theta$, $dx = r \cos\theta\, d\theta$. When $x = 0$, $\theta = 0$; when $x = r$, $\theta = \pi/2$.

$$I = 4 \int_0^{\pi/2} r^2 \sin^2\theta \sqrt{r^2 - r^2 \sin^2\theta}\, r \cos\theta\, d\theta = 4r^4 \int_0^{\pi/2} \sin^2\theta \cos^2\theta\, d\theta$$
$$= r^4 \int_0^{\pi/2} \sin^2 2\theta\, d\theta = r^4 \int_0^{\pi/2} \frac{1}{2}(1 - \cos 4\theta)\, d\theta = r^4 \left[\frac{\theta}{2} - \frac{1}{8} \sin 4\theta \right]_0^{\pi/2} = \frac{\pi r^4}{4}$$

See Arts. 2-2 and 2-3 and also Probs. 2 and 3.

5. The points of intersection are $(3, 2)$ and $(0,1)$. Select a differential area $dx\, dy$. First, integrate with respect to x, holding y constant, to obtain the area of a horizontal strip, dy high, between the parabola and the line. Then, integrate with respect to y to add the strips between $(3,-2)$ and $(0,1)$.

$$A = \int_{-2}^1 \int_{y^2-1}^{1-y} dx\, dy = \int_{-2}^1 (1 - y - y^2 + 1)\, dy$$
$$= \left[2y - \frac{1}{2} y^2 - \frac{1}{3} y^3 \right]_{-2}^1 = 4\frac{1}{2}$$

See Arts. 2-2 and 2-5.

6. The polar moment of inertia equals

$$J = 4 \int_0^{\pi/2} \int_0^r \rho^3\, d\rho\, d\theta = 4 \int_0^{\pi/2} \left[\frac{1}{4} \rho^4 \right]_0^r = \int_0^{\pi/2} r^4\, d\theta = r^4\theta \Big]_0^{\pi/2} = \frac{\pi r^4}{2}$$

See Arts. 2-2 and 2-5.

7. Length of curve can be obtained by integrating Eq. (1-21).

$$y' = \sinh \frac{x}{a} \quad \text{and} \quad (1 + y'^2)^{1/2} = \left(1 + \sinh^2 \frac{x}{a}\right)^{1/2} = \cosh \frac{x}{a}$$

$$s = \int_0^1 \cosh \frac{x}{a} \, dx = a \sinh \frac{x}{a}\bigg]_0^1 = a \sinh \frac{1}{a}$$

See Arts. 2-2 and 2-3.

8. Take advantage of symmetry by computing the length of curve above the axis and then multiplying by 2. Differentiation yields $d\rho/d\theta = -a \sin \theta$.

$$s = 2 \int_0^\pi \left[\rho^2 + \left(\frac{d\rho}{d\theta}\right)^2\right]^{1/2} d\theta = 2 \int_0^\pi [a^2(1 + \cos \theta)^2 + a^2 \sin^2 \theta]^{1/2} \, d\theta$$

$$= 2a \int_0^\pi (1 + 2\cos\theta + \cos^2\theta + \sin^2\theta)^{1/2} \, d\theta = 2a\sqrt{2}\int_0^\pi (1+\cos\theta)^{1/2} \, d\theta$$

$$= 4a \int_0^\pi \cos\frac{\theta}{2} \, d\theta = 4a\left[2\sin\frac{\theta}{2}\right]_0^\pi = 8a$$

9. Take advantage of symmetry by computing the area from $x = 0$ to $x = a$ and multiplying by 2.

$$y = (a^{2/3} - x^{2/3})^{3/2} \quad y' = -\frac{y^{1/3}}{x^{1/3}} \quad (1 + y'^2)^{1/2} = \left(1 + \frac{y^{2/3}}{x^{2/3}}\right)^{1/2} = \frac{a^{1/3}}{x^{1/3}} = \frac{ds}{dx}$$

Then, the surface area is

$$S = 2\pi \int y \, ds = 4\pi \int_0^a (a^{2/3} - x^{2/3})^{3/2} \left(\frac{a^{1/3}}{x^{1/3}}\right) dx$$

To integrate, let $z^{2/3} = a^{2/3} - x^{2/3}$. Then, $\frac{2}{3}z^{-1/3} \, dz = -\frac{2}{3}x^{-1/3} \, dx$. When $x = 0$, $z = a$; when $x = a$, $z = 0$.

$$S = 4\pi \int_a^0 z \left(\frac{a^{1/3}}{x^{1/3}}\right)(-z^{-1/3}x^{1/3} \, dz) = 4\pi a^{1/3}\int_0^a z^{2/3} \, dz = 4\pi a^{1/3}\left[\frac{3}{5}z^{5/3}\right]_0^a = \frac{12\pi a^2}{5}$$

See Arts. 2-2 and 2-3.

10. Take advantage of symmetry by computing the volume on one side of the y axis and doubling it.

$$V = 2\pi \int_0^a y^2 \, dx = 2\pi \int_0^a \frac{b^2}{a^2}(a^2 - x^2) \, dx = \frac{2\pi b^2}{a^2}\left[a^2 x - \frac{1}{3}x^3\right]_0^a = \frac{4\pi ab^2}{3}$$

See Arts. 2-2 and 2-3.

11. Take a differential area of height $2y$ and width dr parallel to $x = b$. $r = b - x$ and $dr = -dx$.

$$V = 2\pi \int r(2y) \, dr = 4\pi \int_{-a}^a (b - x)\sqrt{a^2 - x^2}\,(-dx) = 2\pi^2 a^2 b$$

See Arts. 2-2 and 2-3.

12. For $z = z_i$, a constant, $A = \pi \sqrt{z_i} \left(\sqrt{z_i}/2\right) = \pi z_i/2$.

$$V = \int_0^1 \frac{\pi}{2} z \, dz = \frac{\pi}{2} \left[\frac{1}{2} z^2\right]_0^1 = \frac{\pi}{4}$$

See Arts. 2-2 and 2-3.

13. Integrate, with x held constant, from $y = 0$ to the boundary of the cylinder $y - (a^2 - x^2)^{1/2}$. This yields a differential volume strip parallel to the yz plane. Then, integrate from $x = 0$ to $x = a$.

$$V = \int_0^a \int_0^{\sqrt{a^2-x^2}} (x + y) \, dy \, dx = \int_0^a \left[xy + \frac{1}{2} y^2\right]_0^{\sqrt{a^2-x^2}} dx$$

$$= \int_0^a \left[x \sqrt{a^2 - x^2} + \frac{1}{2} (a^2 - x^2)\right] dx = \frac{2}{3} a^3$$

See Arts. 2-2, 2-3, and 2-5.

14. At a depth x below the water surface, the unit pressure is wx. Assume that wx acts on a horizontal, differential area on the dam with depth dx and length $225 - [(225 - 150)/15]x = 225 - 5x$. Then, the total pressure is

$$P = \int_0^{15} wx(225 - 5x) \, dx = w \left[\frac{225}{2} x^2 - \frac{5}{3} x^3\right]_0^{15} = 19,690w = 1,229,000 \text{ lb}$$

See Arts. 2-2 and 2-3.

15. Let $F = kx$, where k is a constant. $k = F/x = 25/0.25 = 100$. Hence, $F = 100x$. Initial stretch $= 11 - 10 = 1$. Final stretch $= 12 - 10 = 2$.

$$\text{Work} = \int F \, dx = \int_1^2 100x \, dx = 100 \left.\frac{x^2}{2}\right]_1^2 = 150 \text{ in.-lb}$$

See Arts. 2-2 and 2-3.

16. Select a differential area strip $y \, dx$. Take the moment of the strip about each axis and integrate. $(A\bar{x} = M_y$ and $A\bar{y} = M_x.)$

$$M_x = \int_0^\pi \frac{y}{2} y \, dx = \frac{1}{2} \int_0^\pi \sin^2 x \, dx = \frac{\pi}{4}$$

$$M_y = \int_0^\pi xy \, dx = \int_0^\pi x \sin x \, dx = \pi$$

The area under $y = \sin x$ is

$$A = \int_0^\pi y \, dx = \int_0^\pi \sin x \, dx = 2$$

Hence, $\bar{x} = M_y/A = \pi/2$ and $\bar{y} = M_x/A = \pi/8$. See Arts. 2-2 and 2-3.

17. $pV = C = 200 \times 15 \times 144 = 432,000$.   So the volume at 80 psi = 80 $\times$ 144 psf is

$$V_{80} = \frac{432,000}{80 \times 144} = 37.5 \text{ cu ft}$$

$$\text{Work} = \int p \, dV = \int_{200}^{37.5} \frac{432,000}{V} \, dV = 432,000[\log_e V]_{200}^{37.5} = -723,000 \text{ ft-lb}$$

18. Differentiation under the integral sign yields

$$\frac{dy}{d\alpha} = \int_0^1 x^\alpha \, dx = \frac{x^{\alpha+1}}{\alpha+1}\Big]_0^1 = \frac{1}{\alpha+1}$$

Integration yields $y = \log_e (\alpha + 1) + C$.   When $\alpha = 0, y = 0$; therefore, $C = 0$. So $y = \log_e (\alpha + 1)$.   (See Table 1-2 and Art. 2-7.)

THREE

Ordinary Differential Equations

Many engineering problems involving rates of change may be solved with differential equations. These are equations involving one or more derivatives of a function. Their solutions are relations between dependent and independent variables that are free of derivatives and satisfy identically the given equations.

The problems you solved in Chap. 2 are all examples of differential equations. Before integration, these problems were expressible in the forms $y' = f(x)$ or $dy = f(x)\,dx$, where y is the dependent variable, x the independent variable, and $y' = dy/dx$. Either form represents the simplest type of differential equation. It has a solution if $\int f(x)\,dx$ exists. But, as you may have discovered, finding the integral may not be easy.

More complex differential equations may or may not have a solution. If a solution exists, it may or may not be attainable in closed form. Sometimes, you may have to express the solution as an infinite series or as sets of numbers obtained by approximate methods (numerical integration). Fortunately, many of the equations encountered in engineering practice have relatively easily determined solutions.

But there is no general procedure for finding solutions. Instead, techniques have been developed for solving specific types of equations. So to solve differential equations, you must learn to recognize these types and to apply the appropriate method to each type.

Differential equations are divided into two main classes, ordinary and partial. If the derivatives that appear in an equation are total derivatives, the equation is called an ordinary differential equation. If partial derivatives occur, the equation is called a partial differential equation.

This chapter will present methods for solving types of ordinary differential equations important in engineering. Nearly all these methods will give solutions in closed form. Methods of operational calculus are given in the next chapter. Subsequent chapters show how to obtain solutions in infinite series and nonelementary forms and by numerical integration. Partial differential equations will be discussed in Chap. 7.

3-1 *Initial-value and Boundary-value Problems.* Article 2-1 pointed out that $\int f(x)\ dx = F(x) + C$, where C is an arbitrary constant. In specific problems, a given condition determines the value to be assigned to C. The presence of arbitrary constants in a solution is typical of differential equations.

When one or more conditions are imposed for one value of the independent variable for determining the arbitrary constants of a solution, the problem is called an initial-value problem.

Suppose, for example, that $y = c_1 \sin x + c_2 \cos x$, where c_1 and c_2 are constants, results from integration of a differential equation. Suppose also that given conditions require that $y = 0$ and $y' = 1$ when $x = 0$. The solution must contain specific values of c_1 and c_2. Since the conditions are imposed for only one value of the independent variable, $x = 0$, this is an initial-value problem.

The condition $y = 0$ when $x = 0$ requires that $c_2 = 0$. The condition $y'(0) = 1$ requires that $c_1 = 1$. Hence, the solution is $y = \sin x$.

When solutions of a differential equation are required to satisfy conditions imposed at more than one point, or for more than one value of the independent variable, the problem is called a boundary-value problem. If solutions exist, they need not be unique.

Suppose that $y = c_1 e^x + c_2 e^{-x}$ results from integration of a differential equation. Suppose also that given conditions require that $y = 0$ when $x = 0$ and $y = e^2 - 1$ when $x = 1$. Since the conditions are imposed for two values of the independent variable (or at two points), this is a boundary-value problem.

The condition $y(0) = 0$ requires that $c_2 = -c_1$. The condition $y(1) = e^2 - 1$ requires that

$$y = c_1 e + c_2 e^{-1} = c_1(e - e^{-1}) = c_1 \left(\frac{e^2 - 1}{e} \right) = e^2 - 1$$

Hence, $c_1 = e$, $c_2 = -e$, and the solution is

$$y = ee^x - ee^{-x} = e^{x+1} - e^{1-x}$$

3-2   General Characteristics of Ordinary Differential Equations.
As previously defined, an ordinary differential equation contains one
independent variable and derivatives with respect to it.   We define the
order of such an equation as the order of the highest derivative present.
Thus, $y'' + y'^2 = x^4$ is a second-order ordinary differential equation,
because the highest-order derivative of y in it is the second derivative, y''.

This is also called a first-degree equation, because the exponent of y''
is 1.   The **degree** of a differential equation is the exponent of the highest-
order derivative present after the equation has been rationalized and
cleared of fractions with regard to all derivatives present.   Not all
differential equations have a degree.   For example, $y'' = (1 + y')^{\frac{1}{2}}$ is a
second-degree equation, because after you rationalize it, you have
$y''^2 = 1 + y'$.   But $y'' = (1 + y)^{\frac{1}{2}}$ is a first-degree equation.   And
$\log y'' = 1 + y'$ has no degree.

A solution of an ordinary differential equation is any relation between
the variables that reduces the equation to an identity.   When an equation of order n has a solution, it contains n arbitrary constants; it is called
a **general solution.**   One with values assigned to the constants is called a
particular solution.

Some differential equations may have **singular solutions**—solutions
that cannot be obtained from the general solution by specifying values
for the arbitrary constants.   For example, if the general solution has an
envelope, its equation is a singular solution.   [An envelope of a family
of curves consists of one or more curves tangent at each point to a curve
of the family.   The equation of the envelope is obtained by eliminating
the parameter c from the equation of the family $\phi(x,y,c) = 0$ and
$\partial \phi / \partial c = 0$.]

There is no general method for finding the solutions of all types of
ordinary differential equations.   But methods of solutions have been
developed for certain types of equations.   These are described in the
following articles.

3-3   First-order Differential Equations. A typical first-order equation
may be written in the forms $y' = f(x,y)$ or $M(x,y) \, dx + N(x,y) \, dy = 0$.
This is one of the simplest types of differential equations, yet there is no
general solution.   In fact, a first-order equation may not have a solution.
Or if the equation has a solution, it may not be unique.   You can be
assured that a unique solution exists only if

1. $f(x,y)$ is real, finite, single-valued, and continuous at all points
(x,y) within a region R of the xy plane or the whole plane.

2. $\partial f / \partial y$ is real, finite, single-valued, and continuous in R.

Then, in R there exists one and only one solution $y = F(x)$, which passes through any given point of R.

To find the solution, you may have to investigate several possible methods of solution. See if

1. You can write the equation in the form $y' = f(x)$
2. You can separate the variables
3. The functions present form an exact differential
4. Multiplication by a factor makes the functions an exact differential
5. The equation is a Clairaut equation
6. Substitution of variables will yield a solution
7. Successive approximations will produce a solution

Method 1.   If you can write the first-order equation as $y' = f(x)$, then the solution can be found directly by integration by the methods discussed in Chap. 2.   The solution is $y = \int f(x)\, dx = F(x) + c$.

Method 2.   If the variables are separable, the equation can be written

$$P(x)\, dx + Q(y)\, dy = 0$$

Then the solution is

$$\int P(x)\, dx + \int Q(y)\, dy = c$$

where c is an arbitrary constant.

As an example of methods 1 and 2, let us examine the vertical motion of a body on a parachute.   Let us assume that it is dropped from a height below which air resistance at any time is proportional to the velocity at that time.   Let us also assume that the parachute opens at time $t = 0$, when the distance traveled $y = 0$ and the velocity $v = dy/dt = 0$.   (This is an initial-value problem.)

Let W be the combined weight of body and parachute.   Let the air resistance be kv, where k is a constant.   Then, from Newton's law, $F = Ma$,

$$W - kv = \frac{W}{g}\frac{dv}{dt}$$

where $g = 32.2$ ft per sec^2.   This is a first-order equation with variables separable.   It can be rewritten as

$$\frac{dv}{W - kv} = \frac{g}{W}\, dt$$

Integrating both sides of the equation, we obtain

$$-\frac{1}{k}\log_e (W - kv) = \frac{gt}{W} + c_1$$

where c_1 is a constant to be determined from the initial conditions.   For convenience, let us write c_1 as $(1/k) \log_e c_2$.   Let us transpose it to the left side of the equation and multiply both sides of the equation by $-k$. This yields

$$\log_e c_2 + \log_e (W - kv) = \log_e c_2(W - kv) = \frac{-kgt}{W}$$

$$c_2(W - kv) = e^{-kgt/W}$$

Since $v = 0$ when $t = 0$, $c_2 = 1/W$.   Thus, we obtain

$$v = \frac{W}{k} (1 - e^{-kgt/W})$$

The result indicates that as $t \to \infty$, $v \to W/k$, a constant velocity.

We now can apply method 1 to determine the distance traveled at time t.   Since $v = dy/dt$, we obtain by integration

$$y = \frac{W}{k} \left(t + \frac{W}{kg} e^{-kgt/W} \right) + c_3$$

where c_3 is a constant.   From $y = 0$ when $t = 0$, we find $c_3 = -W^2/k^2g$. Therefore,

$$y = \frac{W}{k} \left(t + \frac{W}{kg} e^{-kgt/W} - \frac{W}{kg} \right)$$

Method 3.   You may recall that $d(xy) = x\, dy + y\, dx$.   Thus, $x\, dy + y\, dx$ is an **exact differential**.   Similarly, $(2x + y)\, dx + (x - 3)\, dy$ is an exact differential, for it equals $d(x^2 + xy - 3y)$.   So if you have a differential equation of the first order in which the variables are not separable, write it in the form $M(x,y)\, dx + N(x,y)\, dy = 0$ and see if the left-hand side is an exact differential.   It will be if $\partial M/\partial y = \partial N/\partial x$.

The equation then has the solution $F(x,y) = c$, where c is a constant and

$$F = \int(M\, dx + N\, dy)$$

This holds because $dF = (\partial F/\partial x)\, dx + (\partial F/\partial y)\, dy = M\, dx + N\, dy$, so that $\partial M/\partial y = \partial^2 F/\partial x\, \partial y = \partial^2 F/\partial y\, \partial x = \partial N/\partial x$.

While $M\, dx + N\, dy$ forms an exact differential, the integral may not be obvious to you.   In that case, you will have to use the above relations between M, N, and F to evaluate F.

As an example of the method, let us solve $dy/dx = \frac{1}{2}(x/y - y/x)$. First, write it in the form $M\, dx + N\, dy = 0$.   You can do this by clearing the equation of fractions and transposing all terms to the left-hand side: $(x^2 - y^2)\, dx - 2xy\, dy = 0$.   Next, test to see if we have an exact differential equation:

$$\frac{\partial M}{\partial y} = \frac{\partial}{\partial y}(x^2 - y^2) = -2y \qquad \frac{\partial N}{\partial x} = \frac{\partial}{\partial x}(-2xy) = -2y$$

Since the partial derivatives are equal, the equation is exact. Let us try to determine the integral from the relation between M and F.

$$M = \frac{\partial F}{\partial x} = x^2 - y^2 \qquad F = \int (x^2 - y^2)\, dx = \frac{1}{3}x^3 - y^2x + f(y)$$

where $f(y)$ is an arbitrary function of y. To evaluate $f(y)$, we can use the relation between N and F.

$$N = -2xy = \frac{\partial F}{\partial y} = \frac{\partial}{\partial y}\left[\frac{1}{3}x^3 - xy^2 + f(y)\right] = -2xy + f'(y)$$

This indicates that $f'(y) = 0$ and therefore that $f(y) = c$, a constant. Hence, the solution of the given differential equation is $F = x^3/3 - xy^2 + c = 0$. You can verify this result by differentiating and solving for y'.

Method 4. When you have written the differential equation in the form $M\, dx + N\, dy$, you may find that you do not have an exact differential. In that case, you should try to convert the equation into an exact one by multiplying it by an appropriate function of the variables. Such a function is called an **integrating factor.**

When a first-order equation is linear, you usually can find an integrating factor easily. By definition, a **linear equation** is of the first degree in the dependent variable and its derivatives. A first-order linear equation can be written in the form

$$y' + Py = Q \tag{3-1}$$

where P and Q are functions of x. Then, e^z, where $z = \int P\, dx$, is an integrating factor. And the solution of the equation is

$$y = e^{-z}\int Qe^z\, dx + ce^{-z} \tag{3-2}$$

where c is a constant.

As an example, consider $y' - y = e^x$. Here, $P = -1$ and $z = \int -1\, dx = -x$. Hence, e^{-x} is an integrating factor. Multiplying both sides of the equation by e^{-x}, we obtain

$$e^{-x}y' - e^{-x}y = e^{-x}e^x = 1$$

The left-hand side of the equation is now an exact derivative. It equals $d(e^{-x}y)/dx$. Integration of both sides of the equation therefore yields the solution

$$e^{-x}y = x + c$$

For nonlinear equations, you have to find an integrating factor by inspection and knowledge of derivatives. For example, x^2, y^2, $x^2 + y^2$, and $x^2 - y^2$ can be used as integrating factors for $x\,dy - y\,dx$.

$$\frac{x\,dy - y\,dx}{x^2} = d\left(\frac{y}{x}\right) \tag{3-3}$$

$$\frac{x\,dy - y\,dx}{y^2} = d\left(-\frac{x}{y}\right) \tag{3-4}$$

$$\frac{x\,dy - y\,dx}{x^2 + y^2} = d\left(\tan^{-1}\frac{y}{x}\right) \tag{3-5}$$

$$\frac{x\,dy - y\,dx}{x^2 - y^2} = d\left(\frac{1}{2}\log_e\frac{x + y}{x - y}\right) \tag{3-6}$$

The factor to select is the one that makes other terms present integrable. Similarly, x^2 and y^2 are integrating factors for $2xy\,dy - y^2\,dx$ and $2xy\,dx - x^2\,dy$, respectively.

$$\frac{2xy\,dy - y^2\,dx}{x^2} = d\left(\frac{y^2}{x}\right) \tag{3-7}$$

$$\frac{2xy\,dx - x^2\,dy}{y^2} = d\left(\frac{x^2}{y}\right) \tag{3-8}$$

Method 5. You may sometimes recognize the differential equation you have to solve as one for which there is a known solution. The Clairaut equation is such a type. It has the form

$$y = xy' + f(y') \qquad \text{or} \qquad F(y - xy', y') = 0 \tag{3-9}$$

And the general solution is

$$F(y - cx, c) = 0$$

Notice that the solution can be obtained from the given equation by substituting a constant c for y'. But the solution may not be unique. There also may be a singular solution. This is found by differentiating the given equation with respect to y' and using the result to eliminate y' from the given equation.

For example, let us solve $xy'^2 - yy' + 3 = 0$. By rewriting it in the form $y - xy' - 3/y' = 0$, you may more readily recognize it as a Clairaut equation. The general solution then is $y - cx - 3/c = 0$, where c is a constant (or using the given equation, $c^2x - cy + 3 = 0$). For the singular solution, differentiation of the given equation gives $2xy' - y = 0$, from which $y' = y/2x$. Substitution of this value of y' in the given equation yields the singular solution $y^2 = 12x$, a parabola. It is the envelope of the family of lines $y = cx + 3/c$.   A line determined by a specific value of c intersects the parabola only at a point of tangency.

Method 6.   When the preceding methods do not seem to work, try substituting another variable for one of the given variables.   Suppose, for example, that the equation can be written in the form $y' = f(y/x)$. Try the substitution $y = vx$.   Or if the equation can be written as $y' = f(x/y)$, try $x = vy$.

Solve $x^2y \, dx = (x^3 - y^3) \, dy$.   Divide both sides of the equation by $x^2y \, dy$ to obtain

$$\frac{dx}{dy} = \frac{x}{y} - \frac{y^2}{x^2}$$

Let $x = vy$, then $dx/dy = y(dv/dy) + v$.   Substitution in the equation gives

$$y\frac{dv}{dy} + v = \frac{vy}{y} - \frac{y^2}{v^2y^2} = v - \frac{1}{v^2}$$

The variables now are separable.

$$-v^2 \, dv = \frac{dy}{y}$$

Integration yields

$$-\frac{1}{3}v^3 = \log_e y + \log_e c = \log_e cy$$

Since $v = x/y$, we have the general solution

$$\log_e cy = -\frac{x^3}{3y^3}$$

You can sometimes reduce the given equation to a linear equation by an appropriate substitution.   For instance, the substitution $y = z^{1/(1-n)}$ reduces the Bernoulli equation

$$y' + P(x)y = Q(x)y^n \tag{3-10}$$

to a linear equation.

For example, solve $y' + y = xy^3$.   Here, $n = 3$.   So let $y = z^{1/(1-3)} = z^{-1/2}$, $y' = -z^{-3/2}z'/2$.   Substitution in the equation yields

$$-\frac{1}{2}z^{-3/2}z' + z^{-1/2} = xz^{-3/2}$$

Multiplying the equation by $-2z^{3/2}$, we obtain

$$z' - 2z = -2x$$

This is a linear equation, with e^{-2x} as an integrating factor.

$$e^{-2x}z' - 2e^{-2x}z = \frac{d}{dx}(e^{-2x}z) = -2e^{-2x}x$$

Integration yields

$$e^{-2x}z = xe^{-2x} + \frac{1}{2}e^{-2x} + c$$

Since $z = y^{-2}$, we have the general solution

$$y^2\left(x + \frac{1}{2} + ce^{2x}\right) = 1$$

Another useful substitution technique is illustrated in the following example. Let us solve $y' = (x - y - 1)/(x + y + 3)$. Let us try $x = u + a, y = v + b$, where a and b are constants to be determined later. Substitution in the equation yields

$$y' = \frac{dv}{du} = \frac{u - v + (a - b - 1)}{u + v + (a + b + 3)}$$

Inspection indicates that we can simplify the right-hand side of the equation by setting $a - b - 1 = 0$ and $a + b + 3 = 0$. For this purpose, we must take $a = -1$ and $b = -2$. The resulting equation can be written in the form $dv/du = f(u/v)$. Hence, it can be solved by the substitution $v = zu$, with $dv/du = u\,dz/du + z$. On making the substitution, we get

$$u\frac{dz}{du} + z = \frac{u - zu}{u + zu} = \frac{1 - z}{1 + z}$$

The variables now are separable, and we obtain

$$\frac{1 + z}{1 - 2z - z^2}\,dz = \frac{du}{u}$$

Integration of both sides then produces

$$-\frac{1}{2}\log_e (1 - 2z - z^2) = \log_e u + \log_e c = \log_e cu$$

Hence, $1 - 2z - z^2 = (cu)^{-2}$. If we now make use of $z = v/u, u = x + 1$, and $v = y + 2$, we get the general solution

$$(x + 1)^2 - 2(x + 1)(y + 2) - (y + 2)^2 = k$$

where k is a constant.

Method 7. When other methods do not work, you can resort to Picard's method of solving $y' = f(x,y)$ by successive approximations. Assume that a condition imposed on the solution is that $y = b$ when $x = a$. Then the differential equation has the solution

$$y = b + \int_a^x f(x,y)\, dx \tag{3-11}$$

As a first approximation, take y in the integral equal to b (or any other reasonable value). Compute the new value of y and substitute that in the integral. Repeat as many times as desired.

Let us use Picard's method to solve $y' = 1 + y^2$, even though the variables are separable. And let us take the initial conditions as $y = 0$ when $x = 0$. For a first approximation, assume $y_1 = 0$.

$$y_2 = 0 + \int_0^x (1 + y^2)\, dx = \int_0^x dx = x$$

As a second approximation, substitute x for y in the integral:

$$y_3 = \int_0^x (1 + x^2)\, dx = x + \frac{1}{3} x^3$$

Now, substitute $x + x^3/3$ for y in the integral:

$$y_4 = \int_0^x \left[1 + \left(x + \frac{1}{3} x^3 \right)^2 \right] dx = \int_0^x \left(1 + x^2 + \frac{2}{3} x^4 + \frac{1}{9} x^6 \right) dx$$

$$= x + \frac{1}{3} x^3 + \frac{2}{15} x^5 + \frac{1}{63} x^7$$

Continue in the same manner to get

$$y_5 = x + \frac{1}{3} x^3 + \frac{2}{15} x^5 + \frac{17}{315} x^7 + \frac{38}{2,835} x^9 + \frac{134}{51,927} x^{11} + \cdots$$

(This solution will be a converging series only when $|x| < \pi/2$.)

Chapter 6 presents additional methods for numerical integration.

3-4 *Differential Operators.* You will be introduced here to a device useful in the solution of differential equations. We will delve into the technique in greater detail in later articles and in the next chapter.

We will use the symbol D from now on to indicate that the function following is to be differentiated with respect to the independent variable. Thus, $Dy = y' = dy/dx$. Consequently, $D(y - 3)^2 = 2(y - 3)Dy$.

More generally, we will use D^n to indicate that the function following it is to be differentiated n times. Thus, $D^2y = d^2y/dx^2$. Also,

$$(D^3 + 3D^2 - 4D + 1)y = \frac{d^3y}{dx^3} + 3\frac{d^2y}{dx^2} - 4\frac{dy}{dx} + y$$

Just as f may be used to represent a function of x, $f(x)$, so we may use a symbol L to represent a function of D, $L(D)$. Thus, Ly may represent $(D^3 + 3D^2 - 4D + 1)y$ in the expression above.

A polynomial in D with constant coefficients can be manipulated in addition, subtraction, and multiplication as if D were an algebraic quantity. For example,

$$(D^2 + D - 2) \sin x$$
$$\equiv (D - 1)(D + 2) \sin x \equiv (D - 1)(\cos x + 2 \sin x)$$
$$\equiv - \sin x + 2 \cos x - \cos x - 2 \sin x$$
$$\equiv \cos x - 3 \sin x$$

Differential operators are of particular value in the solution of linear differential equations with constant coefficients because of the simple results obtained when the operators operate on powers of e. For example, $D(e^{mx}y) \equiv e^{mx}(D + m)y$, where m is a constant. Similarly, $D^2(e^{mx}y) \equiv e^{mx}(D + m)^2y$. In general,

$$D^r(e^{mx}y) \equiv e^{mx}(D + m)^ry \tag{3-12}$$

where r is any integer. Consequently,

$$L(D)e^{mx}y \equiv e^{mx}L(D + m)y \tag{3-13}$$

We will examine applications of these identities in later articles.

3-5 Wronskian. In Art. 3-2, we observed that when an nth-order differential equation has a solution, it contains n arbitrary constants. Actually, these must be essential constants. For example, if a solution contained $c_1x + c_2x$, we would have only one essential constant not two, for $c_1x + c_2x = (c_1 + c_2)x = cx$. A criterion for determining whether constants in a solution are essential is that they be associated with terms that are linearly independent. You can determine whether functions are independent from the definition of independence or by evaluating their Wronskian.

Functions are linearly independent if there is no set of constants not all zero such that the sum of the products of the constants and the functions is zero.

A Wronskian W is a determinant in which the elements of each column are a function and its successive derivatives with respect to the independent variable.

The necessary and sufficient condition that a given set of functions of one variable, such as the solutions y_1, y_2, . . . , y_n of an nth-order linear differential equation, be linearly independent is that their Wronskian should not be zero.   If it is zero, the functions are linearly dependent.

$$W = \begin{vmatrix} y_1 & y_2 & \cdots & y_n \\ y_1' & y_2' & \cdots & y_n' \\ y_1'' & y_2'' & \cdots & y_n'' \\ \cdots & \cdots & \cdots & \cdots \\ y_1^{(n-1)} & y_2^{(n-1)} & \cdots & y_n^{(n-1)} \end{vmatrix} \neq 0 \qquad (3\text{-}14)$$

For example, let us show that e^x and e^{-x} are linearly independent for all values of x.   The Wronskian for these functions is

$$\begin{vmatrix} e^x & e^{-x} \\ e^x & -e^{-x} \end{vmatrix} = -1 - 1 = -2$$

Since the Wronskian is not zero for any value of x, the functions are independent for all values of x.

3-6   *Linear Differential Equations.* A more restrictive requirement is placed on linear differential equations than on first-degree equations. For a differential equation to be linear, the dependent variable and all its derivatives must be of the first degree.   Thus, a linear differential equation of the nth order can be written

$$a_0(x) \frac{d^n y}{dx^n} + a_1(x) \frac{d^{n-1} y}{dx^{n-1}} + \cdots + a_{n-1}(x) \frac{dy}{dx} + a_n(x)y = f(x)$$

$$(3\text{-}15)$$

Equation (3-2) gives the general solution of a linear first-order equation. No such formula is available when $n > 1$.   We will examine methods of solution of linear equations with constant coefficients in Art. 3-7.

If $f(x)$ in the general linear equation is zero, the resulting equation is called **homogeneous**.   Its solution is called the **complementary function**. The solution when $f(x) \neq 0$ is called a **particular integral** of the equation.

If a linear differential equation has y_c as a complementary function and y_p as a particular integral, its solution, containing n arbitrary constants, is $y = y_c + y_p$.

Consider, for example, the linear differential equation $(D^2 + 1)y = x^3 - x$.   Its homogeneous equation is $(D^2 + 1)y = 0$.   You can verify by substitution in the homogeneous equation that the complementary function is $y_c = c_1 \sin x + c_2 \cos x$.   You can also verify by substitution that $x^3 - 7x$ satisfies the given equation and therefore is a particular integral. Therefore, the solution of the given equation is

$$c_1 \sin x + c_2 \cos x + x^3 - 7x$$

If $y = y_1$ is any solution of the homogeneous equation, then $y = cy_1$, where c is a constant, is also a solution. If $y = y_2, y_3, \ldots, y_n$ are also solutions, and all are linearly independent, then the linear combination

$$y_c = c_1 y_1 + c_2 y_2 + \cdots + c_n y_n \tag{3-16}$$

is the complementary function, or general solution, of the homogeneous equation.

For example, when you find that $\sin x$ is a solution of $(D^2 + 1)y = 0$, then you know that $c_1 \sin x$ is also a solution. Similarly, since $\cos x$ is also a solution, so is $c_2 \cos x$ and $c_1 \sin x + c_2 \cos x$.

The solution of an nth-order linear differential equation is sometimes obvious. For example, $D^n y = f(x)$ may be solved by n successive integrations. By setting $D^{n-1}y$ equal to a new variable p, you reduce the equation to $Dp = dp/dx = f(x)$, for which the solution is $D^{n-1}y = p = \int f(x)\, dx$. Next, by setting $D^{n-2}y = q$, you further reduce the equation to $Dq = \int f(x)\, dx$, for which the solution is $D^{n-2}y = q = \int\int f(x)(dx)^2$. You can continue in this manner until the solution is obtained.

If the equation has constant coefficients, you can use the methods of Art. 3-7 or those of Chap. 4. If the coefficients are variable, look for a substitution that will convert the equation to one with constant coefficients.

For example, the Cauchy or Euler equation

$$(b_0 x^n D^n + b_1 x^{n-1} D^{n-1} + \cdots + b_{n-1} xD + b_n)y = f(x) \tag{3-17}$$

where the b's are constants, can be transformed into a linear equation with constant coefficients by the substitution $x = e^z$.

Substitutions sometimes may be used to reduce the order of a linear equation. Several possibilities of doing this are discussed in Art. 3-8 for differential equations in general. In particular, a substitution that may work for a second-order linear equation is $y = y_1 v$, where v is a new dependent variable and y_1 is a function of x that simplifies the resulting differential equation. For instance, if a specific integral can be found for the corresponding homogeneous differential equation

$$y'' + Py' + Qy = 0 \tag{3-18}$$

let y_1 be that integral.

Sometimes $y = e^z v$, where $z = -\frac{1}{2}\int P\, dx$, converts Eq. (3-18) to a linear equation with constant coefficients or to a Cauchy equation.

3-7   Linear Equations with Constant Coefficients. For constant coefficients, Eq. (3-15) becomes

$$Ly = (a_0D^n + a_1D^{n-1} + \cdots + a_{n-1}D + a_n)y = f(x) \qquad (3\text{-}19)$$

where $a_0, a_1, \ldots , a_n$ = constants
$\qquad\qquad L$ = linear operator representing $f(D)$
$\qquad\qquad D = d/dx$

The solutions have the characteristics of solutions of linear equations discussed in Art. 3-6.

Thus, to solve such nth-order equations, first find the complementary function y_c, which is the general solution of the homogeneous equation $Ly = 0$ and contains n arbitrary constants. Then, obtain the particular integral y_p that satisfies $Ly = f(x)$. The complete solution will be $y_c + y_p$.

Complementary Function

The clue to solution of $Ly = 0$ is Eq. (3-13), because it turns out that the solution has the form $y = c_ie^{m_ix}$ (summed for $i = 1$ to n). Consequently, we can obtain m_i from the **auxiliary polynomial equation** $Lm = 0$. If none of the roots $m_1, m_2, \ldots , m_n$ are repeated, then

$$y_c = c_1e^{m_1x} + c_2e^{m_2x} + \cdots c_ne^{m_nx} \qquad (3\text{-}20)$$

As an example, let us solve $(D^2 + 4D + 3)y = 0$. The auxiliary equation is $m^2 + 4m + 3 = (m + 3)(m + 1) = 0$. It has two roots, $m = -3, -1$. Hence, the complementary function is

$$y_c = c_1e^{-3x} + c_2e^{-x}$$

The solution must contain as many arbitrary constants as roots. Furthermore, all terms must be linearly independent. So if any root m_r occurs r times, the corresponding term in the complementary function is not just ce^{m_rx}. It becomes $(c_1 + c_2x + c_3x^2 + \cdots + c_rx^{r-1})e^{m_rx}$.

For example, let us solve $(D^4 - D^2)y = 0$. The auxiliary equation is $m^4 - m^2 = m^2(m^2 - 1) = 0$. It has roots $m = \pm1$ and the twice-repeated root $m = 0$. Therefore,

$$y_c = c_1e^x + c_2e^{-x} + e^0(c_3 + c_4x) = c_1e^x + c_2e^{-x} + c_3 + c_4x$$

Notice that the hyperbolic functions $\cosh x$ and $\sinh x$ can be substituted for the first two terms, since $\cosh x = (e^x + e^{-x})/2$ and $\sinh x = (e^x - e^{-x})/2$.

Quite often in engineering problems, Lm has complex roots. If Lm has one complex root $\alpha + \beta i$, it also has a conjugate root $\alpha - \beta i$, where α and β are constants and $i = \sqrt{-1}$. Thus, the complementary function has terms

$$y = c_1 e^{(\alpha+\beta i)x} + c_2 e^{(\alpha-\beta i)x} = c_1 e^{\alpha x} e^{\beta i x} + c_2 e^{\alpha x} e^{-\beta i x}$$
$$= e^{\alpha x}(c_1 e^{\beta i x} + c_2 e^{-\beta i x}) \tag{3-21}$$

Since the constants c_1 and c_2 may be imaginary numbers, the terms in the parenthesis may be replaced by the trigonometric functions $\sin \beta x$ and $\cos \beta x$. [$\sin \beta x = (e^{\beta i x} - e^{-\beta i x})/2i$ and $\cos \beta x = (e^{\beta i x} + e^{-\beta i x})/2$. See Arts. 8-1, 8-3, and 8-5.] Hence, we can rewrite Eq. (3-21) in any one of the following ways:

$$y = e^{\alpha x}(A \cos \beta x + B \sin \beta x) \tag{3-22a}$$
$$y = c e^{\alpha x} \sin (\beta x + \phi) \tag{3-22b}$$
$$y = c e^{\alpha x} \cos (\beta x + \theta) \tag{3-22c}$$

where A, B, c, ϕ, and θ are constants.

Just as for real roots of Lm, modifications must be made if a set of complex roots is repeated. For example, if $\alpha \pm \beta i$ is an r-fold root, the corresponding terms in the complementary function are

$$y = e^{\alpha x}(A_1 + A_2 x + \cdots + A_r x^{r-1}) \cos \beta x$$
$$+ (B_1 + B_2 x + \cdots + B_r x^{r-1}) \sin \beta x \tag{3-23}$$

Terms such as those in Eqs. (3-22) arise often in solution of vibration problems. Consider, for example, the problem of determining the natural period of vibration of a mass on a spring. (This frequently is used as an idealized representation of an actual construction in solving vibration problems.) The force exerted by the spring on the mass is proportional to the amount of stretch, or compression, of the spring. Furthermore, this force is always directed toward the equilibrium position, which we shall choose as our origin for locating the mass. Hence, we can set the spring force equal to $-ky$, where k is the force required to stretch the spring 1 in. and y is the distance of the mass from the origin. If the mass weighs W lb, it also will be subjected to an inertia force $(W/g)(d^2y/dt^2)$, where $g = 32.2$ ft per sec². For dynamic equilibrium

$$\frac{W}{g}\frac{d^2y}{dt^2} + ky = 0$$

This may be written more conveniently as

$$\frac{d^2y}{dt^2} + \frac{kg}{W}y = \frac{d^2y}{dt^2} + \omega^2 y = (D^2 + \omega^2)y = 0 \tag{3-24}$$

The auxiliary equation is $m^2 + \omega^2 = 0$, which has roots $m = \pm\omega i$, where

$$\omega = \sqrt{\frac{kg}{W}} \tag{3-25}$$

(ω is called the natural circular frequency of the system.)   Hence, the solution of Eq. (3-24) is

$$y = A \sin \omega t + B \cos \omega t \tag{3-26}$$

where A and B are constants to be determined from the initial conditions of the system.   The motion is called harmonic.

Natural period of vibration is the time required for the mass to go through one cycle of free vibration; that is, vibration after the disturbance causing the motion has ceased.   From Eq. (3-26), the natural period in seconds is

$$T = \frac{2\pi}{\omega} = 2\pi \sqrt{\frac{W}{gk}} \tag{3-27}$$

The natural frequency in cycles per second is $1/T$.

If at time $t = 0$ the mass has an initial displacement y_0 and velocity v_0, we find by substituting these values in Eq. (3-26) and its derivative with respect to t that $A = v_0/\omega$ and $B = y_0$.   Hence, at any time t, the mass is completely located by

$$y = \frac{v_0}{\omega} \sin \omega t + y_0 \cos \omega t \tag{3-28}$$

Particular Integrals

Any of several methods may be used in search of the particular integral y_p of $Ly = f(x)$.   If $f(x)$ consists of the sum of two or more terms, a different method may be used to find the particular integral corresponding to each term.   The complete particular integral is the sum of those integrals.

The method of undetermined coefficients is usually the simplest method when $f(x)$ is any one or a sum of the following terms:

$$cx^p \qquad ce^{qx} \qquad cx^pe^{qx} \qquad c \sin \beta x \qquad c \cos \beta x$$
$$cx^pe^{\alpha x} \sin \beta x \qquad cx^pe^{\alpha x} \cos \beta x$$

where p = positive integer or zero
c, q, α, β = constants
For the particular integral, select $f(x)$ and all its derivatives that are different from each other.   Arrange the terms in groups so that all terms derived from a single term of $f(x)$ appear in only one group.   If any group contains a term that is in the complementary function y_c, multiply all terms in the group by the lowest integral power of x that will make every term different from any term in y_c.   The particular integral then consists of the sum of the products of each term and a coefficient to be determined by the following procedure.

Substitute the particular integral in Ly and equate the resulting expression to $f(x)$. Equate the coefficients of terms on the left-hand side of the equation to the corresponding terms on the right and solve for the coefficients.

As an example, suppose that the mass on a spring in the preceding example was initially at rest in the equilibrium position when a force $F(t)$ that varies with time was applied to it. The equation of motion then becomes

$$(D^2 + \omega^2)y = F(t)\frac{g}{W} = F(t)\frac{\omega^2}{k} \tag{3-29}$$

Equation (3-26) remains the complementary function.

Suppose, as a simple case, that the applied force is constant. Then,

$$(D^2 + \omega^2)y = \frac{F\omega^2}{k} \tag{3-30}$$

Let us try aF as a particular integral. Substitution in Eq. (3-30) requires that $\omega^2 aF \equiv \omega^2 F/k$. Hence, $a = 1/k$, and

$$y = A \sin \omega t + B \cos \omega t + \frac{F}{k} \tag{3-31}$$

Since $y(0) = 0$, $B = -F/k$. From $y'(0) = 0$, $A = 0$. Therefore, for constant force, F,

$$y = \frac{F}{k}(1 - \cos \omega t) \tag{3-32}$$

Notice that F/k is the deflection of the spring with F as a static load. Since the maximum value of $y = 2F/k$, the constant dynamic load produces twice the deflection under static load.

Suppose now that we take $F(t) = F_0 \sin \phi t$, where F_0 is the maximum value of the force and $\phi/2\pi$ is the frequency of the force. Then,

$$(D^2 + \omega^2)y = \frac{F_0\omega^2}{k} \sin \phi t \tag{3-33}$$

Let us try $a \sin \phi t + b \cos \phi t$ as a particular integral. Substitution in Eq. (3-33) after differentiating twice requires that

$$-a\phi^2 \sin \phi t - b\phi^2 \cos \phi t + \omega^2(a \sin \phi t + b \cos \phi t) \equiv \frac{F_0\omega^2}{k} \sin \phi t$$

Since the coefficients of $\cos \phi t$ must be zero, $b = 0$. Equating the coefficients of $\sin \phi t$, we get $a\omega^2 - a\phi^2 = F_0\omega^2/k$, from which we find

$a = F_0\omega^2/k(\omega^2 - \phi^2)$.   Hence,

$$y = A \sin \omega t + B \cos \omega t + \frac{F_0\omega^2}{k(\omega^2 - \phi^2)} \sin \phi t \tag{3-34}$$

Since $y(0) = 0$, $B = 0$.   From $y'(0) = 0$, $A = -F_0\phi\omega/k(\omega^2 - \phi^2)$. Therefore, for $F(t) = F_0 \sin \phi t$,

$$\begin{aligned}y &= \frac{F_0\omega^2}{k(\omega^2 - \phi^2)} \left(\sin \phi t - \frac{\phi}{\omega} \sin \omega t \right) \\ &= \frac{F_0}{k \left(1 - \dfrac{\phi^2}{\omega^2} \right)} \left(\sin \phi t - \frac{\phi}{\omega} \sin \omega t \right)\end{aligned} \tag{3-35}$$

Some useful conclusions can be reached from analysis of Eq. (3-35). Note that F_0/k is the deflection of the spring due to a static load F_0. If ϕ is relatively small, so that the load changes slowly, the mass moves with substantially the same frequency as the load.   And the position of the mass at any moment is about the same as it would be if the instantaneous load $F_0 \sin \phi t$ were applied statically.   But if ϕ is relatively very large, the mass remains practically stationary, because its inertia prevents it from responding to the rapid fluctuations of load.   Now, if $\omega = \phi$, Eq. (3-35) indicates that $y = 0/0$.   But if you apply L'Hôpital's rule and differentiate numerator and denominator with respect to ϕ, you find that $y \to F_0(\sin \omega t - \omega t \cos \omega t)/2k$, which increases indefinitely with time. In practice, infinite deflections may not occur due to resonance, when ω and ϕ are nearly equal, because of energy losses due to friction, hysteresis, and other causes.   Nevertheless deformations usually will be very large, often intolerable.   Hence, you should see that construction is such as to preclude resonance from occurring.

In some engineering problems, such as those dealing with mechanical oscillations and electric circuits, the dependent variable in the differential equation governing the system represents the output of the system for a given input.   If the differential equation is linear with constant coefficients, it and the system it governs are considered **completely stable** if and only if all roots of the auxiliary equation have negative real parts.   Then, the effects of small changes in initial conditions tend to vanish with increase in the independent variable.   In such a case, all solutions of the homogeneous equation (no input or forcing function) are transients.

A function $y(t)$ is a **transient** if $\lim_{t \to \infty} y(t) = 0$.   If t is time, a transient becomes insignificant with passage of time.

If the system has an input, the portion of the solution of the governing differential equation that remains when transients become insignificant is called the **steady-state solution.**

The transient part of the solution is given by the complementary function. The steady-state part is determined by the particular integrals.

Consider, for example, a series electric circuit with a switch, inductance L, resistance R, capacitance C, and input voltage $E_0 \cos \phi t$. The voltage drop through the inductance for current I is $L \, dI/dt$, through the resistance IR, and through the capacitance Q/C, where $I = dQ/dt$. By Kirchhoff's law, $\Sigma E = 0$, the governing equation is

$$E_0 \cos \phi t - L \frac{d^2Q}{dt^2} - R \frac{dQ}{dt} - \frac{Q}{C} = 0$$

To obtain the steady-state solution, we find a particular integral and then differentiate to get the current I. The result is $I = [E_0 \cos (\phi t - \theta)]/Z$, where $\theta = \tan^{-1} (\phi L/R - 1/\phi CR)$ and $Z = \sqrt{R^2 + (\phi L - 1/\phi C)^2}$, the impedance. Thus, the steady-state current produced by a sinusoidal input has the same frequency as the input voltage but differs from it in phase.

Let us consider now some other methods of finding particular integrals when $f(x)$ contains terms X that do not permit use of the method of undetermined coefficients. One alternative is to try to obtain the particular integrals corresponding to X from

$$y_p = e^{m_n x} \int e^{(m_{n-1} - m_n)x} \int \cdots \int e^{(m_1 - m_2)x} \int e^{-m_1 x} X (dx)^n \qquad (3\text{-}36)$$

where $m_1, m_2, \ldots, m_n$ = roots of auxiliary equation

n = order of equation

Select the roots in an order that will simplify the repeated integration.

For example, let us solve $(D^2 - 2D + 1)y = e^x/x^2$. The auxiliary equation is $m^2 - 2m + 1 = (m - 1)^2 = 0$. The root $m = 1$ occurs twice. So the complementary function is

$$y_c = c_1 x e^x + c_2 e^x$$

From Eq. (3-36), the particular integral is

$$y_p = e^x \int e^{(1-1)x} \int e^{-x} e^x \frac{(dx)^2}{x^2} = e^x \int - e^0 \frac{dx}{x} = e^x \log_e x$$

Hence, the solution is

$$y = c_1 x e^x + c_2 e^x + e^x \log_e x$$

Lagrange's method of **variation of parameters** is another way of finding a particular integral for X. If $y_1, y_2, \ldots, y_n$ are terms of the complementary function, with arbitrary constants omitted, seek a particular integral in the form

$$y_p = v_1 y_1 + v_2 y_2 + \cdots + v_n y_n \qquad (3\text{-}37)$$

where $v_1, v_2, \ldots, v_n$ are unknown functions. Since the given equation is of the nth order, differentiate y_p successively $n - 1$ times. Set up $n - 1$ equations by equating to zero the sum of all terms in each derivative that involve the derivatives of $v_1, v_2, \ldots, v_n$. Obtain an nth equation by substituting the derivatives of y_p, less the terms equated to zero, in the differential equation. Then, solve the n equations for $v_1, v_2, \ldots, v_n$.

As an example, let us solve $(D^2 + 1)y = \sec x$. The complementary function is $y_c = c_1 \sin x + c_2 \cos x$. To find the particular integral, let us take

$$y_p = v_1 \sin x + v_2 \cos x$$

We can obtain our first equation from

$$Dy_p = v_1 \cos x - v_2 \sin x + v_1' \sin x + v_2' \cos\cdot x$$

by setting $v_1' \sin x + v_2' \cos x = 0$. Differentiating the remaining terms, we get

$$D^2 y_p = -v_1 \sin x - v_2 \cos x + v_1' \cos x - v_2' \sin x$$

When this is substituted in the given differential equation, it simplifies to

$$v_1' \cos x - v_2' \sin x = \sec x$$

Simultaneous solution of this equation and

$$v_1' \sin x + v_2' \cos x = 0$$

yields $v_1' = 1$, $v_1 = x$, $v_2' = -\tan x$, and $v_2 = \log_e \cos x$. Therefore, the particular integral is

$$y_p = x \sin x + \cos x \log_e \cos x$$

If you cannot solve a linear differential equation with constant coefficients by the methods in this article, try those in the following chapters.

3-8 *Reduction of Order.* With a general differential equation when a method of solution is not apparent, you can look for a way to lower the order of the equation, in the hope of obtaining a new equation that you can handle. Sometimes, for example, the order can be reduced by the substitution $y' = p$, where p is a new variable. We considered a simple case in Art. 3-6. Following are some other types of equations for which the method works.

If y is absent from an nth-order equation, it has the form

$$f(x,y',y'', \ldots ,y^{(n)}) = 0$$

Substitution of $y' = p$, $y'' = p'$, $\ldots$, $y^{(n)} = p^{(n-1)}$ transforms the equation to

$$f(x,p,p', \ldots ,p^{(n-1)}) = 0$$

reducing the order by 1. After solution for p, integrate to obtain y.

For example, solve $x^2y'' = 1 - x$. Let $y' = p$, $y'' = p'$. Then,

$$x^2p' = 1 - x$$

The variables are separable. Integration yields

$$p = -\frac{1}{x} - \log_e x + c_1 = y'$$

Integrating again, we obtain the solution

$$y = -\log_e x - x \log_e x + c_3x + c_2$$

If x is absent from the given differential equation, let y' be the new dependent variable p. Take y as the independent variable.

$$y'' = \frac{dp}{dx} = \frac{dp}{dy}\frac{dy}{dx} = p\frac{dp}{dy} = pp'$$
$$y^{(3)} = p^2p'' + pp'^2$$
$$y^{(4)} = p^3p^{(3)} + 4p^2p'p'' + pp'^3$$

Substitute these in the given equation, solve for p, and integrate to find y.

Let us solve $yy'' + y'^2 + 1 = 0$ by this method. Let $y' = p$ and $y'' = pp'$, where $p' = dp/dy$. Then,

$$ypp' + p^2 + 1 = 0$$

The variables are separable, and integration leads to

$$p = \sqrt{\frac{c_1}{y^2} - 1} = \frac{dy}{dx}$$

Integrating again, we obtain the solution

$$c_1 - y^2 = (x + c_2)^2$$

3-9 Simultaneous Linear Differential Equations. The differential operators introduced in Art. 3-4 facilitate solution of a system of linear differential equations with constant coefficients. To solve a system of n linear equations involving one independent variable t and n independent variables x, y, z, . . . , treat the operator $D = d/dt$ as an algebraic quantity.

To illustrate, let us solve

$$(D + 1)x + (D + 3)y = t \qquad (3\text{-}38)$$
$$x + (D + 1)y = e^t \qquad (3\text{-}39)$$

Multiply Eq. (3-39) by $D + 1$ and subtract the result from Eq. (3-38).

$$(D + 3)y - (D + 1)^2 y = t - (D + 1)e^t = t - 2e^t$$
$$(D^2 + D - 2)y = 2e^t - t$$

This is a linear equation with constant coefficients in the variables y and t. It has the solution

$$y = \left(c_1 + \frac{2}{3}t\right)e^t + c_2 e^{-2t} + \frac{t}{2} + \frac{1}{4}$$

Now, multiply Eq. (3-38) by $D + 1$ and Eq. (3-39) by $-(D + 3)$ and add. This yields

$$(D + 1)^2 x - (D + 3)x = (D + 1)t - (D + 3)e^t = 1 + t - 4e^t$$

We now have a linear equation with constant coefficients in the variables x and t. When we obtain a solution, we must be certain that the arbitrary constants are essential. You can determine relations between the constants by substitution in Eq. (3-38) or (3-39). Thus, the solution is

$$x = \left(\frac{1}{3} - 2c_1 - \frac{4}{3}t\right)e^t + c_2 e^{-2t} - \frac{t}{2} - \frac{3}{4}$$

3-10 *Bibliography*

R. P. Agnew, "Differential Equations," McGraw-Hill Book Company, New York.

W. E. Boyce and R. D. Di Prima, "Elementary Differential Equations and Boundary Value Problems," John Wiley & Sons, Inc., New York.

L. Brand, "Differential and Difference Equations," John Wiley & Sons, Inc., New York.

M. Golomb and M. E. Shanks, "Elements of Ordinary Differential Equations," McGraw-Hill Book Company, New York.

L. M. Kells, "Elementary Differential Equations," McGraw-Hill Book Company, New York.

PROBLEMS

1. Given the length of a curve as $ds = \sqrt{1 + y'^2}\, dx = x\, dx$.
   (a) What is the order of the differential equation?

(b) How many essential arbitrary constants should the solution have?

(c) What is the degree of the equation?

(d) If $y(\sqrt{2}) = 0$, what is the solution?

2. Given the radius of curvature of a curve as $(1 + y'^2)^{3/2}/y'' = R$, a constant.

(a) What is the order of the differential equation?

(b) How many essential arbitrary constants should the solution have?

(c) What is the degree of the equation?

(d) If $y(0) = \pm R$, what is the solution?

3. In a precipitation chamber, observations indicate that the resistance of a liquid to particles falling through it is proportional to the square of the downward velocity. Assume particles entering have zero vertical velocity.

(a) How far does a particle fall in time t?

(b) What is the maximum depth h to be permitted for the liquid, if the length of the chamber is L, the average horizontal velocity is V, and all particles are to be precipitated to the bottom?

4. The orthogonal trajectories of a family of curves defined by $f(x,y) = c$, such as the lines of force about a pole of a magnet, form another family of curves, such as the lines of equal potential about the same pole. Each of the orthogonal curves is normal to a curve of $f(x,y) = c$. Find the orthogonal trajectories of $y^2 - x^2 = cx$. (*Hint:* If the slope of f is y', the slope of a trajectory is determined by $-1/y'$, but with the parameter c absent.)

5. Given an electric circuit consisting of a switch, a resistance R, an inductance L, and a source of electromotive force $E(t)$ in series, where t is time. If I is the current after the switch is closed, the voltage drop through the resistance is IR and through the inductance, $L\, dI/dt$.

(a) Determine the current if the switch is closed when $t = 0$, if E is constant.

(b) Determine the current if $E(t) = E_0 \sin \phi t$, where E_0 and ϕ are constant.

6. Find the shape of a reflector that emanates light from a point source parallel to a fixed line.

7. Solve $xy' - y = \sqrt{y'^2 + 1}$.

8. Solve $x + y \cos y/x = xy' \cos y/x$.

9. Solve by Picard's method to a fourth approximation $y' = x^2 - y$, where $y(0) = 0$.

10. A mass weighing W lb is suspended on a weightless string of length L. If the system oscillates through a small angle 2θ (θ can replace $\sin \theta$) in a plane, what is the period of the vibration?

11. A mass weighing W lb, suspended on a weightless spring with spring constant k is subjected, when vibrating, to a damping force proportional to the velocity.

(a) Assume that the mass initially has a velocity v_0 and is at a distance y_0 from its equilibrium position. Find the location y of the mass at any time t when no other external forces are acting on the system.

(b) Find the response of the damped oscillating system when a constant external force F acts on it. Assume the mass initially is at rest in the equilibrium position and that the undamped natural frequency and damped natural frequency are nearly equal, so either can be substituted for the other.

12. Given an electric circuit consisting of a switch, resistance R, inductance L, capacitance C, and electromotive force $E(t)$ in series, where t is time. If I is the current after the switch is closed, the voltage drop through the resistance is IR, through the inductance $L\,dI/dt$, and through the capacitance $Q/C = (1/C)\int I\,dt$, where Q is the charge on the capacitance. Find the charge and current at any time after the switch is closed, if E is constant and the capacitor initially was uncharged.

13. Given a circular-cross-section wire suspended in air at constant temperature. Suppose that one end is maintained at a temperature θ_0 relative to the air; and suppose also that the wire is sufficiently long that the other end may be assumed to be at air temperature (relative temperature $= 0$), what is the distribution of temperature along the wire? (Heat flow by conduction is proportional to wire area A and change in temperature. Heat dissipated by convection and radiation may be taken proportional to surface area and temperature relative to ambient air.)

14. Bending moment $M(x)$ and curvature of a beam are related by $EI = MR$, where R is the radius of curvature, E the modulus of elasticity of beam material, and I the moment of inertia of beam cross section.

    (a) For a constant moment M, due to couples at the beam ends, determine the shape of the beam if y', the slope of the elastic (deflection) curve, is assumed negligible, as is conventionally done.

    (b) Determine the shape if y' is not neglected.

15. Given a vertical column subjected to a concentric vertical load and free to rotate at top and bottom, while translation is prevented. What is the smallest load P under which the column will be in equilibrium with a small deflection y at midspan? (Neglect y' in the equation for curvature.)

16. (a) What shape does a flexible cable take when hung from two points a distance L apart and at the same level if subjected to a uniform vertical load w lb per ft of horizontal length? (Bending moment at any point is zero.)

    (b) What slope does the cable take if the load w is uniform along the cable?

17. A mass weighing W lb is suspended on a weightless spring from a mass of the same weight also suspended on a spring. Both springs have the same spring constant k. At time $t = 0$, both masses have no velocity; the lower mass is at its equilibrium position $y_2 = 0$, but the upper mass is at a distance $y_1 = -a$ from its equilibrium position. Locate the masses at any time t.

ANSWERS

    1. (a) First order. See Art. 3-2.

       (b) One. See Art. 3-2.

       (c) Second degree. See Art. 3-2.

       (d) Integration of $y' = (x^2 - 1)^{1/2}$ yields

$$y = \frac{1}{2}\,[x\,\sqrt{x^2 - 1} - \log_e\,(x + \sqrt{x^2 - 1})] + c$$

Use of initial conditions gives $c = -\tfrac{1}{2}[\sqrt{2} - \log_e\,(1 + \sqrt{2})]$.

2. (a) Second order.   See Art. 3-2.

(b) Two.   See Art. 3-2.

(c) Second degree.   See Art. 3-2.

(d) Let $p = y'$ and $p' = y''$.   The equation become first order, $Rp' = (1 + p^2)^{3/2}$ with variables separable.   Integration gives $Rp(1 + p^2)^{-1/2} = x + c_1$. Solution for $p = dy/dx$, followed by integration, gives

$$(x + c_1)^2 + (y + c_2)^2 = R^2$$

The given conditions require that $c_1 = c_2 = 0$.   See Arts. 3-3 and 3-8.

3. (a) From $F = Ma = (W/g)(dv/dt)$, the equation of motion is $dv/dt = k^2v^2 - g$, where $v = $ downward velocity, k is a constant, and $g = 32.2$ ft per sec^2. The variables are separable.   Integration gives

$$v = - \frac{\sqrt{g}}{k} \tanh kt \sqrt{g} + c_1$$

Since $v = 0$ when $t = 0$, $c_1 = 0$.   A second integration yields

$$x = - \frac{1}{k^2} \log_e \cosh kt \sqrt{g} + c_2$$

If $x = x_0$ at $t = 0$, $c_2 = x_0$.   Hence, in time t a particle falls

$$x_0 - x = \frac{1}{k^2} \log_e \cosh kt \sqrt{g}$$

See Art. 3-3.

(b) For the topmost particle to reach the bottom, the length of fall must be $h = x_0 - x$.   Liquid entering the tank leaves in time $t = L/V$.   So the particle must fall a distance h in this time.   Hence,

$$h = \frac{1}{k^2} \log_e \cosh \frac{kL}{V} \sqrt{g}$$

4. Differentiation of $y^2 - x^2 = cx$ gives $2yy' - 2x = c$.   Elimination of c between this and the given equation results in $y' = (x^2 + y^2)/2xy$.   Hence, the differential equation of the trajectories is $y' = -2xy/(x^2 + y^2)$ or $2xy\ dx + (x^2 + y^2)\ dy = 0$.   The equation is an exact differential because $\partial 2xy/\partial y = 2x$ and $\partial(x^2 + y^2)/\partial x = 2x$.   Hence the solution is $x^2y + y^3/3 = k$.   See method 2, Art. 3-3.

5. (a) By Kirchhoff's law, $\Sigma E = 0$, the equation for the circuit is

$$E - L\frac{dI}{dt} - IR = 0 \qquad \text{or} \qquad (D + \omega)I = \frac{E}{L}$$

where $\omega = R/L$.   An integrating factor is $e^{\omega t}$.   Integration gives

$$e^{\omega t}I = \frac{E}{L} \int e^{\omega t}\ dt = \frac{E}{\omega L} e^{\omega t} + c$$

When $t = 0$, $I = 0$; so $c = -E/\omega L$. Therefore,

$$I = \frac{E}{\omega L} (1 - e^{-\omega t}) = \frac{E}{R} (1 - e^{-Rt/L})$$

(b) The equation for the circuit now becomes

$$(D + \omega)I = \frac{E_0}{L} \sin \phi t$$

With $e^{\omega t}$ as an integrating factor, we get on integrating

$$e^{\omega t}I = \frac{E_0}{L} \int e^{\omega t} \sin \phi t \, dt = \frac{E_0}{L} \left(\frac{\omega \sin \phi t - \phi \cos \phi t}{\omega^2 + \phi^2} e^{\omega t} \right) + c$$

When $t = 0$, $I = 0$; so $c = E_0\phi/L(\omega^2 + \phi^2)$. Let $\cos \theta = \omega/\sqrt{\omega^2 + \phi^2}$ and $\sin \theta = \phi/\sqrt{\omega^2 + \phi^2}$. Then,

$$I = \frac{E_0}{L \sqrt{\omega^2 + \phi^2}} [\sin (\phi t - \theta) + e^{-\omega t} \sin \theta]$$

See method 4, Art. 3-3.

6. Since the light is to consist of parallel rays, we deduce that the reflector must be a surface of revolution. Therefore, we need only find the shape in a plane to obtain the solution. Let us take the origin of coordinates at the point source and the x axis parallel to the fixed line. The reflected rays, then, must be parallel to the x axis. If y is an ordinate to a point on the reflector and y' is the slope of the tangent at that point, then the angle of reflection of a light ray there must be $\pi/2 - \tan^{-1} y'$. From the geometry of point source and reflector, the angle of incidence at the point equals $\pi/2 - \tan^{-1} y/x + \tan^{-1} y'$. Since angle of incidence equals angle of reflection, $\tan^{-1} y' = \tan^{-1} y/x - \tan^{-1} y'$. Thus, $2 \tan^{-1} y' = \tan^{-1} y/x$. Taking the tangent of both sides of the equation, we get

$$2y'/(1 - y'^2) = y/x \qquad \text{or} \qquad yy'^2 + 2xy' - y = 0$$

Solution for y' yields $y' = -x/y \pm \sqrt{x^2/y^2 + 1}$, from which

$$\pm \sqrt{x^2 + y^2} \, dx + x \, dx + y \, dy = 0$$

$\pm 1/\sqrt{x^2 + y^2}$ is an integrating factor. Integration yields $x \pm \sqrt{x^2 + y^2} = c$. Transposition of x to the right-hand side and squaring leads to $y^2 = 2c(x + c/2)$. This represents a family of parabolas with foci at the origin, or light source. See method 4, Art. 3-3.

7. This is a Clairaut equation. The general solution is

$$cx - y = \sqrt{c^2 + 1}$$

For the singular solution, differentiate the given equation with respect to y', to obtain $x = y'/\sqrt{y'^2 + 1}$. Then, $y' = x/\sqrt{1 - x^2}$. Substitution in the given differential equation yields the singular solution $y = -\sqrt{1 - x^2}$. See method 5, Art. 3-3.

8. Divide through by x to obtain

$$1 + \frac{y}{x} \cos \frac{y}{x} = y' \cos \frac{y}{x}$$

Let $y = vx$ and $y' = xv' + v$. Then,

$$1 + \frac{vx}{x} \cos \frac{vx}{x} = (xv' + v) \cos \frac{vx}{x}$$

which simplifies to

$$1 + v \cos v = (xv' + v) \cos v \qquad \text{or} \qquad x \frac{dv}{dx} \cos v = 1$$

Dividing both sides by x/dx and integrating, we get $\sin v = \log_e cx$. Substitution of $v = y/x$ gives the solution $\sin y/x = \log_e cx$. See method 6, Art. 3-3.

9. The solution is $y = 0 + \int_0^x (x^2 - y)\, dx$. Start with $y_1 = 0$.

$$y_2 = \int_0^x x^2\, dx = \frac{1}{3} x^3$$

$$y_3 = \int_0^x \left(x^2 - \frac{1}{3} x^3 \right) = \frac{1}{3} x^3 - \frac{1}{12} x^4$$

$$y_4 = \int_0^x \left(x^2 - \frac{1}{3} x^3 + \frac{1}{12} x^4 \right)$$

$$= \frac{1}{3} x^3 - \frac{1}{12} x^4 + \frac{1}{60} x^5 = 2 \left(\frac{x^3}{3!} - \frac{x^4}{4!} + \frac{x^5}{5!} - \cdots \right)$$

10. From $F = Ma$, the equation of motion is $-W \sin \theta = (W/g)(d^2s/dt^2)$, where $ds = L\, d\theta =$ differential arc length along the path of the mass. Since θ can replace $\sin \theta$, the equation can be written

$$\frac{d^2\theta}{dt^2} + \frac{g}{L} \theta = \left(D^2 + \frac{g}{L} \right) \theta = 0$$

The auxiliary equation is $m^2 + g/L = 0$, from which $m = \pm i \sqrt{g/L} = \pm \omega i$. The general solution is $\theta = A \sin \omega t + B \cos \omega t$. Hence the period is $2\pi/\omega = 2\pi \sqrt{L/g}$.

11. (a) The equation of motion is

$$\frac{W}{g} \frac{d^2y}{dt^2} + c \frac{dy}{dt} + ky = 0 \qquad \text{or} \qquad \left(D^2 + \frac{gc}{W} D + \frac{gk}{W} \right) y = 0$$

where c is the damping constant of proportionality. Let $\beta = gc/2W$ and $\omega = \sqrt{gk/W}$. The equation then becomes

$$(D^2 + 2\beta D + \omega^2)y = 0$$

The auxiliary equation is $m^2 + 2\beta m + \omega^2 = 0$. It has roots $m_1 = -\beta + \sqrt{\beta^2 - \omega^2}$ and $m_2 = -\beta - \sqrt{\beta^2 - \omega^2}$. If $\beta > \omega$, the solution is

$$y = c_1 e^{m_1 t} + c_2 e^{m_2 t} \qquad \beta > \omega$$

From the initial conditions,

$$c_1 = \frac{v_0 - m_2 y_0}{m_1 - m_2} = \frac{v_0 - m_2 y_0}{2\sqrt{\beta^2 - \omega^2}} \qquad c_2 = -\frac{v_0 - m_1 y_0}{2\sqrt{\beta^2 - \omega^2}}$$

There is no oscillation. (This type of system is called overdamped.) If $\beta < \omega$, let $\omega_d = \sqrt{\omega^2 - \beta^2}$, so that $m_1 = -\beta + \omega_d i$ and $m_2 = -\beta - \omega_d i$. Then, the general solution is

$$y = e^{-\beta t}(A \sin \omega_d t + B \cos \omega_d t) \qquad \beta < \omega$$

From the initial conditions, $A = (v_0 + \beta y_0)/\omega_d$ and $B = y_0$. This solution represents a vibration with decreasing amplitude. ω_d is the natural frequency of the damped system. [Compare with Eq. (3-26).] If $\beta = \omega$, the general solution is $y = (c_3 + c_4 t)e^{-\omega t}$. Taking into account initial conditions, we get

$$y = e^{-\omega t}[v_0 t + (1 + \omega t)y_0] \qquad \beta = \omega$$

This solution represents critical damping, at which oscillation disappears.

(b) The equation of motion is $(D^2 + 2\beta D + \omega^2)y = F\omega^2/k$. The complementary function is

$$y_c = e^{-\beta t}(A \sin \omega t + \beta \cos \omega t)$$

with $\omega_d = \omega$. For the particular integral, try $y_p = aF$. Substitution in the differential equation gives $\omega^2 aF = \omega^2 F/k$, from which $a = 1/k$. We now have

$$y = e^{-\beta t}(A \sin \omega t + B \cos \omega t) + \frac{F}{k}$$

From the initial conditions, $A = (-F/k)(\beta/\omega)$ and $B = -F/k$. So the particular solution is

$$y = \frac{F}{k}\left[1 - e^{-\beta t}\left(\cos \omega t + \frac{\beta}{\omega}\sin \omega t\right)\right]$$

Compare with Eq. (3-32). See Art. 3-7.

12. Since $Q = \int I\,dt$, $I = dQ/dt$. By Kirchhoff's law, $\Sigma E = 0$, the equation for the circuit is

$$E - \frac{Q}{C} - R\frac{dQ}{dt} - L\frac{d^2Q}{dt^2} = 0 \quad \text{or} \quad \left(D^2 + \frac{R}{L}D + \frac{1}{CL}\right)Q = \frac{E}{L}$$

(Compare with answers to Probs. 5a and 11.) Let $\beta = R/2L$ and $\omega^2 = 1/CL$. The auxiliary equation then is $m^2 + 2\beta + \omega^2 = 0$. The roots are $m_1 = -\beta + \sqrt{\beta^2 - \omega^2}$ and $m_2 = -\beta - \sqrt{\beta^2 - \omega^2}$. If $\beta > \omega$, the complementary function is

$$Q_c = c_1 e^{m_1 t} + c_2 e^{m_2 t} \qquad \beta > \omega$$

For the particular integral try $Q_p = aE/L$. Substitution in the differential equation gives $\omega^2 aE/L = E/L$, from which $a = 1/\omega^2$. So the general solution is

$$Q = c_1 e^{m_1 t} + c_2 e^{m_2 t} + \frac{E}{\omega^2 L} \qquad \beta > \omega$$

From initial conditions, $c_1 = m_2 E/\omega^2 L(m_1 - m_2)$ and $c_2 = -m_1 E/\omega^2 L(m_1 - m_2)$. The current $I = dQ/dt$ is then

$$I = \frac{m_1 m_2 E}{\omega^2 L(m_1 - m_2)} e^{m_1 t} - \frac{m_1 m_2 E}{\omega^2 L(m_1 - m_2)} e^{m_2 t}$$

$$= \frac{E}{2L \sqrt{\beta^2 - \omega^2}} (e^{m_1 t} - e^{m_2 t}) \qquad \beta > \omega$$

If $\beta < \omega$, let $\omega' = \sqrt{\omega^2 - \beta^2}$, so that $m_1 = -\beta + \omega' i$ and $m_2 = -\beta - \omega' i$. Then the general solution is

$$Q = e^{-\beta t}(A \sin \omega' t + B \cos \omega' t) + \frac{E}{\omega^2 L} \qquad \beta < \omega$$

From the initial conditions, $A = (-\beta/\omega')(E/\omega^2 L)$ and $B = -E/\omega^2 L$. The current then is

$$I = \frac{E}{\omega' L} e^{-\beta t} \sin \omega' t \qquad \beta < \omega$$

If $\beta = \omega$, the general solution is $Q = (c_3 + c_4 t)e^{-\omega t} + E/\omega^2 L$. Taking into account initial conditions, we get

$$Q = \frac{E}{\omega^2 L} [1 - e^{-\omega t}(1 + \omega t)] \qquad \beta = \omega$$

The current then is

$$I = \frac{E}{\omega^2 L} [\omega e^{-\omega t}(1 + \omega t) - \omega e^{-\omega t}] - \frac{Ee^{-\omega t} t}{L} \qquad \beta = \omega$$

See Art. 3-7.

13. Consider a length dx of wire at a distance x from the heated end. Heat in, Q_i, equals heat out, Q_o. $Q_i \Big]_x = -kA \, d\theta/dx$, where k is a constant. $Q_o = c\theta C \, dx + Q_i \Big]_{x+dx}$, where c is a constant and C the circumference of the wire. Using a Taylor's series for expansion at $x + dx$ $\{f(x + dx) = f(x) + f'(x) \, dx + (1/2!)[f''(x)] \, dx^2 + \cdots\}$, we have for Q_i at $x + dx$

$$Q_i = -kA \left(\frac{d\theta}{dx} + \frac{d^2\theta}{dx^2} \, dx + \cdots \right)$$

Let us neglect powers of dx greater than one. Hence, for heat balance,

$$-kA\frac{d\theta}{dx} = c\theta C\,dx - kA\frac{d\theta}{dx} - kA\frac{d^2\theta}{dx^2}$$

Let $\mu^2 = cC/kA$. The differential equation then becomes

$$(D^2 - \mu^2)\theta = 0$$

The auxiliary equation is $m^2 - \mu^2 = 0$. The roots are $\pm\mu$. Hence, the general solution is

$$\theta = c_1 e^{\mu x} + c_2 e^{-\mu x}$$

From the initial conditions, $c_1 = 0$ and $c_2 = \theta_0$. Hence, the particular solution is $\theta = \theta_0 e^{-\mu x}$. See Art. 3-7.

14. (a) The shape of the elastic curve of the beam may be determined from $EI = M/y''$, where y is the beam deflection. Hence,

$$\frac{d^2 y}{dx^2} = \frac{M}{EI}$$

Integration yields $y' = Mx/EI + c_1$ and $y = Mx^2/2EI + c_1 x + c_2$. Take the origin at midspan of the undeflected beam. Then, since $y = 0$ when $x = \pm L/2$, where L is the span, $c_1 = 0$ and $c_2 = -ML^2/8EI$. Therefore, the particular solution is a parabola

$$y = \frac{M}{2EI}x^2 - \frac{ML^2}{8EI} = -\frac{M}{2EI}\left(\frac{L^2}{4} - x^2\right)$$

(b) The shape of elastic curve may be found from $EI = M(1 + y'^2)^{3/2}/y''$. Hence, $EIy'' = M(1 + y'^2)^{3/2}$. Since this equation does not contain y, let $y' = p$ and $y'' = p'$. We then have a first-order equation $EIp' = M(1 + p^2)^{3/2}$. The variables are separable:

$$\frac{dp}{(1 + p^2)^{3/2}} = \frac{M}{EI}dx$$

Integration yields $p/\sqrt{1 + p^2} = Mx/EI + c_1$. Take the origin at midspan of the undeflected beam. From symmetry, $p = y' = 0$ when $x = 0$. Hence, $c_1 = 0$. We now have

$$y' = p = \frac{Mx/EI}{\sqrt{1 - M^2 x^2/E^2 I^2}}$$

Integration and use of $y = 0$ at $x = \pm L/2$, where L is the span, leads to the solution, the circle with radius EI/M,

$$x^2 + (y + c_2)^2 = \frac{E^2 I^2}{M^2}$$

where $c_2 = -\sqrt{E^2 I^2/M^2 - L^2/4}$. See Art. 3-8.

15. From Prob. 14, $EI = MR = -Py/y''$, or $(D^2 + P/EI)y = 0$. The auxiliary equation is $m^2 + P/EI = 0$, from which $m = \pm i\sqrt{P/EI}$. Hence, the general solution is

$$y = A \sin\sqrt{\frac{P}{EI}}\, x + B \cos\sqrt{\frac{P}{EI}}\, x$$

When $x = 0$, $y = 0$; so $B = 0$. When $x = L$, the column length, $y = 0$. Since A cannot also be zero, $\sin\sqrt{P/EI}\, L = 0$. (This is an eigenvalue or characteristic-value problem. See Art. 6-11.) Consequently, $\sqrt{P/EI}\, L = k\pi$, where k is an integer, and $P = k^2\pi^2 EI/L^2$. The smallest load occurs for $k = 1$. Then, $P = \pi^2 EI/L^2$. See Art. 3-7.

16. (a) Take the origin of coordinates at the midpoint of the cable. Since the stress at any point is directed along the cable, to keep the bending moment zero, the slope of the cable is $y' = V/H$, where V is the shear at any point and H is the horizontal component of the cable tension, a constant. $V = wL/2 - w(L/2 - x) = wx$. Integration yields $y = wx^2/2H + c$. Since $y(0) = 0$, $c = 0$, and $y = wx^2/2H$, a parabola.

(b) $V = wS/2 - w(S/2 - s) = ws$, where S = length of cable and s = distance along cable from midspan $= \int\sqrt{1 + y'^2}\, dx$. Hence,

$$y' = \frac{V}{H} = \frac{w}{H}\int\sqrt{1 + y'^2}\, dx$$

Differentiating both sides of the equation, we get

$$y'' = \frac{w}{H}\sqrt{1 + y'^2}$$

Since y is not present in this equation, let $p = y'$ and $p' = y''$. Then, $p' = (w/H)\sqrt{1 + p^2}$. The variables are separable

$$\frac{dp}{\sqrt{1 + p^2}} = \frac{w}{H}\, dx$$

Integration gives $\sinh^{-1} p = wx/H + c_1$. From symmetry, when $x = 0$, $y' = p = 0$; so $c_1 = 0$. Then, $\sinh wx/H = p = y'$. Integrating again, we get $y = (H/w)\cosh wx/H + c_2$. Let us take the origin a distance $a = H/w$ below the cable low point. Then, when $x = 0$, $y = a = a + c_2$; so $c_2 = 0$. Therefore, the shape of the cable is $y = a \cosh x/a$, a catenary. See Art. 3-8.

17. From $F = Ma$, the equation for the upper mass is

$$\frac{W}{g}\frac{d^2y_1}{dt^2} = k(y_2 - y_1) - ky_1 = k(y_2 - 2y_1)$$

The equation for the lower mass is

$$\frac{W}{g}\frac{d^2y_2}{dt^2} = -k(y_2 - y_1)$$

Let $\omega^2 = gk/W$.   We then have the simultaneous equations

$$(D^2 + 2\omega^2)y_1 - \omega^2 y_2 = 0$$
$$(D^2 + \omega^2)y_2 - \omega^2 y_1 = 0$$

Operate on the first equation with $D^2 + \omega^2$, multiply the second by ω^2, and then add to eliminate y_2.   We get

$$(D^4 + 3\omega^2 D^2 + \omega^4)y_1 = 0$$

Auxiliary equation is $m^4 + 3\omega^2 m^2 + \omega^4 = 0$, whence $m^2 = (-3 \pm \sqrt{5})\omega^2/2$. Thus, the roots are approximately $\pm 0.62\omega i$ and $\pm 1.62\omega i$.   Hence, the general solution is

$$y_1 = c_1 \sin 0.62\omega t + c_2 \cos 0.62\omega t + c_3 \sin 1.62\omega t + c_4 \cos 1.62\omega t$$

Similarly, by eliminating y_1 from the simultaneous equations, we get

$$y_2 = c_5 \sin 0.62\omega t + c_6 \cos 0.62\omega t + c_7 \sin 1.62\omega t + c_8 \cos 1.62\omega t$$

Substitution of y_1 and y_2 in the original equations yields

$$c_5 = 1.62c_1 \qquad c_6 = 1.62c_2 \qquad c_7 = -0.62c_3 \qquad c_8 = -0.62c_4$$

From the initial conditions, $c_1 = c_3 = 0$, $c_2 = -0.28a$, and $c_4 = -0.72a$.   Therefore, the solution is

$$y_1 = -a(0.28 \cos 0.62\omega t + 0.72 \cos 1.62\omega t)$$
$$y_2 = -a(0.45 \cos 0.62\omega t - 0.45 \cos 1.62\omega t)$$

FOUR

Operational Calculus

Under certain conditions, we can simplify problems by switching from one mathematical model to another. Operational calculus applies this concept to simplify the solution of differential and integral equations. By transformations, shifting of mathematical models, operational calculus converts the problem of solving such equations into a problem of algebraic equations. When a solution is obtained for the algebraic equations, another transformation converts it to the solution of the original equations.

We used operational calculus in Chap. 3 to find the complementary functions of linear differential equations with constant coefficients. Article 3-4 introduced the linear operator D. Article 3-7 then showed how an auxiliary algebraic equation could be set up by substituting an algebraic variable m for D and also provided rules for transforming the roots of the auxiliary equation into the complementary function of the original differential equation.

To introduce you to the concepts of operational calculus, this chapter deals mainly with one linear operator, the Laplace transform, and its application to the solution of linear ordinary differential equations with constant coefficients. It can, however, be used to solve partial differential equations and integral equations.   We will discuss other types of transformations only briefly.

The Laplace transformation provides an alternative method of solving linear differential equations from those given in Art. 3-7.   It has several advantages.   It converts the problem to that of solving algebraic equations.   For initial-value problems, it automatically takes initial conditions into account.   It is effective in solving differential equations with discontinuous input or forcing functions, on the right-hand side of the equations.   And the work can be expedited by the use of tables of transforms, in the same manner that tables of integrals are used for integration.

4-1   *General Principles of the Laplace Transformation.* The Laplace transformation is defined by

$$\mathcal{L}f(t) = F(s) = \int_0^\infty f(t)e^{-st}\,dt = \lim_{\substack{a \to 0 \\ b \to \infty}} \int_a^b f(t)e^{-st}\,dt \tag{4-1}$$

It associates a unique image function $F(s)$ of the complex variable $s = \alpha + \beta i$ with a single-valued function $f(t)$ of a real variable t (when the integral in the definition exists).   Usually in engineering problems the correspondence is essentially one-to-one.   Hence, the Laplace transformation lets us substitute simpler operations on $F(s)$ for operations on $f(t)$.

$F(s)$ is called the Laplace transform of $f(t)$.   [The lower limit of the defining integral may also be taken as $-\infty$, in which case $F(s)$ is called the two-sided Laplace transform.]   The operator $\mathcal{L}$ is used to signify the transformation of f into F.   The operator $\mathcal{L}^{-1}$ signifies the inverse Laplace transformation.   You will usually find it expedient to find an inverse from a knowledge of the properties of the Laplace transform (Table 4-1) and with the aid of transform tables (Table 4-2 or more extensive tables, such as those listed in Art. 4-6).   But if $f(t)$ is continuous, you may be able to compute the inverse for $t > 0$ from

$$f(t) = \mathcal{L}^{-1}F(s) = \frac{1}{2\pi i} \int_{\alpha-i\infty}^{\alpha+i\infty} F(s)e^{st}\,ds \tag{4-2}$$

providing the integral exists (see Art. 4-2).

The Laplace transform is a linear operator.   As indicated in Table 4-1,

$$\mathcal{L}c_1 f(t) + \mathcal{L}c_2 g(t) = \mathcal{L}[c_1 f(t) + c_2 g(t)] \tag{4-3a}$$

where c_1 and c_2 are constants.   Thus, addition of multiples of original functions corresponds to addition of the same multiples of the image functions.   Also,

$$\mathcal{L}cf(t) = c\mathcal{L}f(t) \tag{4-3b}$$

TABLE 4-1   Important Characteristics of Laplace Transforms

$$\mathcal{L}f = F(s) = \int_0^\infty f(t)e^{-st}\,dt$$

$$\mathcal{L}(f + g) = \mathcal{L}(f) + \mathcal{L}(g) \qquad f = f(t),\ g = g(t)$$

$$\mathcal{L}(cf) = c\mathcal{L}f \qquad c = \text{constant}$$

$$\mathcal{L}[f(t - c)] = e^{-cs}F(s) \qquad t > c,\ c = \text{positive constant}$$

$$\mathcal{L}[e^{ct}f(t)] = F(s - c) \qquad c = \text{constant}$$

$$\mathcal{L}f(ct) = \frac{1}{c}F\left(\frac{s}{c}\right) \qquad c = \text{constant}$$

$$\mathcal{L}f' = sF(s) - f(0) \qquad f' = \frac{df(t)}{dt}$$

$$\mathcal{L}f'' = s^2F(s) - sf(0) - f'(0) \qquad f'' = \frac{d^2f(t)}{dt^2}$$

$$\mathcal{L}f^{(n)} = s^nF(s) - s^{n-1}f(0) - s^{n-2}f'(0) - \cdots - f^{(n-1)}(0)$$

$$\mathcal{L}(tf) = -F'(s) \qquad F'(s) = \frac{dF(s)}{ds}$$

$$\mathcal{L}(t^nf) = (-1)^nF^{(n)}(s)$$

$$\mathcal{L}\left(\frac{1}{t}f\right) = \int_s^\infty F(s)\,ds$$

$$\mathcal{L}\int_a^t f\,dt = \frac{1}{s}F(s) + \frac{1}{s}\int_a^0 F(s)\,ds$$

$$\mathcal{L}f\mathcal{L}g = \mathcal{L}f * g = \mathcal{L}g * f \qquad f * g = \int_0^t f(u)g(t - u)\,du \text{ (Convolution theorem)}$$

When certain conditions are met (Art. 4-2), the transform of the derivative of $f(t)$ with respect to t is given by

$$\mathcal{L}f'(t) = s\mathcal{L}f(t) - f(0) \tag{4-4}$$

Similarly, the transform of the second derivative is

$$\mathcal{L}f''(t) = s^2\mathcal{L}f(t) - sf(0) - f'(0) \tag{4-5}$$

And, in general, the nth derivative is given by

$$\mathcal{L}f^{(n)}(t) = s^n\mathcal{L}f(t) - s^{n-1}f(0) - s^{n-2}f'(0) - \cdots - f^{(n-1)}(0) \tag{4-6}$$

Furthermore,

$$\mathcal{L}\int_0^t f(t)\,dt = \frac{1}{s}F(s) \tag{4-7}$$

The preceding equations provide the essential relations for solving linear differential equations with constant coefficients.   For example, let us solve $y'' + y = e^{-t}$ when $y(0) = y'(0) = 0$.   With $\mathcal{L}y = \mathcal{L}f(t) = F(s)$, let us take the transform of the equation term by term.   From Eq. (4-5),

TABLE 4-2   Some Common Laplace Transforms

$$\mathcal{L}c = \frac{c}{s} \qquad c = \text{constant}$$

$$\mathcal{L}t = \frac{1}{s^2}$$

$$\mathcal{L}t^n = \frac{n!}{s^{n+1}} \qquad n = \text{integer} > 0 \text{ (see also Art. 5-6)}$$

$$\mathcal{L}t^{-\frac{1}{2}} = \sqrt{\frac{\pi}{s}}$$

$$\mathcal{L}t^{\frac{1}{2}} = \frac{1}{2}\sqrt{\pi}\, s^{-\frac{3}{2}}$$

$$\mathcal{L}e^{ct} = \frac{1}{s-c} \qquad c = \text{constant}$$

$$\mathcal{L}t^n e^{ct} = \frac{n!}{(s-c)^{n+1}} \qquad n = \text{integer} > 0$$

$$\mathcal{L}\sin ct = \frac{c}{s^2 + c^2}$$

$$\mathcal{L}t \sin ct = \frac{2cs}{(s^2 + c^2)^2}$$

$$\mathcal{L}e^{at} \sin ct = \frac{c}{(s-a)^2 + c^2} \qquad a = \text{constant}$$

$$\mathcal{L}\cos ct = \frac{s}{s^2 + c^2}$$

$$\mathcal{L}t \cos ct = \frac{s^2 - c^2}{(s^2 + c^2)^2}$$

$$\mathcal{L}e^{at} \cos ct = \frac{s-a}{(s-a)^2 + c^2}$$

$$\mathcal{L}\sinh ct = \frac{c}{s^2 - c^2}$$

$$\mathcal{L}\cosh ct = \frac{s}{s^2 - c^2}$$

$$\mathcal{L}f(t) = \frac{1}{1 - e^{-Ts}} \int_0^T f(t)e^{-st}\, dt \qquad f(t) \text{ periodic, period } T$$

$$\mathcal{L}f(t) \sin ct = \frac{1}{2i}[F(s - ci) - F(s + ci)] \qquad F(s) = \mathcal{L}f(t)$$

$$\mathcal{L}f(t) \cos ct = \frac{1}{2}[F(s - ci) + F(s + ci)] \qquad F(s) = \mathcal{L}f(t)$$

$\mathcal{L}y'' = s^2\mathcal{L}y$. From Table 4-2, $\mathcal{L}e^{-t} = 1/(s+1)$. Thus, the differential equation transforms into the algebraic equation

$$s^2\mathcal{L}y + \mathcal{L}y = (s^2 + 1)\mathcal{L}y = \frac{1}{s+1}$$

From it we obtain

$$\mathcal{L}y = \frac{1}{(s^2 + 1)(s + 1)}$$

The solution of the differential equation is then the function $f(t)$ whose transform is $\mathcal{L}y$. To find it, we first simplify $F(s)$ by expressing it in partial fractions.

$$\mathcal{L}y = \frac{1}{2(s+1)} - \frac{s-1}{2(s^2+1)} = \frac{1}{2(s+1)} - \frac{s}{2(s^2+1)} + \frac{1}{2(s^2+1)}$$

We next look up the inverse transforms term by term in Table 4-2 to get

$$y = \frac{1}{2}e^{-t} - \frac{1}{2}\cos t + \frac{1}{2}\sin t$$

As another example, consider an electric circuit consisting of a switch, an inductance L of 1 henry, a resistance R of 100 ohms, a capacitance C of 0.0001 farad, and a constant electromotive source providing 100 volts— all in series. What is the current I after the switch is closed? From Kirchhoff's law, $\Sigma E = 0$, we can write

$$100 - L\frac{dI}{dt} - IR - \frac{Q}{C} = 100 - \frac{dI}{dt} - 100I - 10,000 \int_0^t I\,dt = 0$$

Taking transforms of the terms, we obtain, from Eqs. (4-4) and (4-7) and Table 4-2,

$$\frac{100}{s} - s\mathcal{L}I - 100\mathcal{L}I - \frac{10,000}{s}\mathcal{L}I = 0$$

From this algebraic equation, we find that

$$\mathcal{L}I = \frac{100}{s^2 + 100s + 10,000} = \frac{2\sqrt{3}}{3}\frac{50\sqrt{3}}{(s+50)^2 + (50\sqrt{3})^2}$$

Using Table 4-2 to get the inverse, we obtain the solution to the original equation

$$I = \frac{2\sqrt{3}}{3}e^{-50t}\sin 50\sqrt{3}\,t$$

4-2 *Existence and Uniqueness of Laplace Transforms and Inverses.* Success of operational calculus with Laplace transforms in solving an engineering problem depends on whether the transforms and inverses involved exist and are unique.   Following are some rules to guide you in determining existence and uniqueness.   But generally, you will find it advisable to check solutions by substitution in the original equations.

One requirement that may be placed on $f(t)$ in Eq. (4-1) is that it be of **exponential order.**   A function $f(t)$ is of exponential order if constants α, K, and T exist such that

$$|f(t)| < Ke^{\alpha t} \qquad \text{for all } t > T \tag{4-8}$$

Many values of α may satisfy this inequality.   For if $f(t)$ meets the requirement for α_1, it also does so for all values of α greater than α_1.   The greatest lower bound α_0 of the set α is called the **abscissa of convergence** of $f(t)$.   Thus, $f(t) = t$ is of exponential order, with 0 as the abscissa of convergence; and so is $f(t) = e^{-2t}$, with -2 as the abscissa of convergence. But e^{t^2} is not of exponential order, because it increases with increase in t much more rapidly than $e^{\alpha t}$ for large values of t.

Another requirement that may be imposed on $f(t)$ is that it be continuous or at least piecewise continuous.   A function is piecewise continuous in an interval if the interval contains a finite number of maxima, minima, and discontinuities and if limits exist at both sides of each discontinuity.

Suppose s in Eq. (4-1) is a complex number $\alpha + \beta i$.   Then, the existence of the Laplace transform is assured for all $\alpha > \alpha_0$ if $f(t)$ is of exponential order, with abscissa of convergence α_0, and is piecewise continuous in every finite interval $0 \le t < T$.   These conditions, however, are sufficient but not necessary; the Laplace transform of a function may exist even though they are not satisfied.   This is one reason why when you believe you have found a solution, you should see if it satisfies the given equations.

Certain conditions also have to be met for Eq. (4-4) for the transform of the derivative of $f(t)$ with respect to t to hold.   Both $f(t)$ and $f'(t)$ must be of exponential order and piecewise continuous.   If $s = \alpha + \beta i$, α must be greater than the abscissa of convergence of $f(t)$.   And $f(t)$ must approach $f(0)$ as a limit as t approaches zero from the positive side. Similar conditions must be satisfied for Eqs. (4-5) and (4-6) to hold.   And Eq. (4-7) applies when $f(t)$ is of exponential order and piecewise continuous.

In solving differential equations by the method described in Art. 4-1, you solve an algebraic equation for $F(s)$ and then obtain the inverse of that solution to get the solution to the given equations.   You can be assured that the inverse exists if the following conditions are satisfied:

$F(s)$ must be differentiable for all α equal to or greater than α_0. $s^n F(s)$ must be bounded as $s \to 0$ for some n greater than 1.

Under these conditions, Eq. (4-2) holds. Furthermore, $\mathcal{L}^{-1}F(s) = 0$, for $t \leq 0$, is continuous for all t, and of order $e^{\alpha_0 t}$ as $t \to \infty$. Again, the conditions are sufficient but not necessary.

A necessary condition for the existence of an inverse is that $\lim_{s \to \infty} F(s) = 0$.

The Laplace transform $F(s)$ is unique for each function $f(t)$ that has such a transform. But the inverse may not be unique. Different discontinuous functions may have the same transform. Such functions, however, differ at most by a null function N, which for all $t > 0$ satisfies $\int_0^t N(u) \, du = 0$. They will be equal where both are continuous (Lerch's theorem). Thus, a specific $F(s)$ cannot have more than one inverse transform continuous for all $t > 0$. Consequently, a unique $f(t)$ corresponds to a transform for almost all $t > 0$.

4-3 *Methods of Finding Laplace Transforms and Inverses.* The procedure given in Art. 4-1 for solving linear differential equations with constant coefficients is simple in concept. You substitute the Laplace transform for each term in the equation, solve the resulting algebraic equation for the transform of the independent variable, and find the inverse of the solution. You will usually find the first and second steps easy. Equations (4-4) to (4-7) give the transforms of the derivatives. You can find the transforms of terms containing the independent variable from tables such as Tables 4-1 and 4-2, known properties of transforms, or by integration of Eq. (4-1). However, the third step, finding the inverse transforms of the solution of the algebraic equation, may be troublesome. You may have to try several methods for complicated transforms.

As a first step, you should simplify the transform as much as possible. If transform tables are available, try to put the transform in forms that duplicate those in the tables.

In simplifying a transform, take advantage of Eq. (4-3). If a transform consists of a sum of terms, its inverse is the sum of the inverses of each term. A common device for employing this relation is to break complicated fractions into partial fractions. This was done in the first example of solution of a differential equation by Laplace transforms in Art. 4-1. The following example will again demonstrate the technique. In addition, it will indicate how a term may be rearranged to match one in a table. Also, the example will show how to handle a boundary-value problem, one for which the initial conditions do not apply to only one value of the independent variable.

Let us solve $(D^3 - 2D^2 + 5D)x = 0$, where $D = d/dt$, given $x(0) = 0$, $x'(0) = 1$, and $x(\pi/2) = 0$. We first take transforms of each term, using Eqs. (4-4) to (4-6). According to Eq. (4-6), the transform of D^3x contains a term $x''(0)$, which is not covered by the initial conditions.   Let us call this term d.   We can use the third initial condition $x(\pi/2) = 0$ to determine d later.   The transform of the differential equation is then

$$s^3\mathcal{L}x - s - d - 2s^2\mathcal{L}x + 2 + 5s\mathcal{L}x = 0$$

This yields

$$\mathcal{L}x = \frac{s + d - 2}{s(s^2 - 2s + 5)}$$

To simplify the transform, we break it up into partial fractions:

$$\frac{s + d - 2}{s(s^2 - 2s + 5)} \equiv \frac{A}{s} + \frac{Bs + C}{s^2 - 2s + 5}$$

Multiplying both sides of the identity by $s(s^2 - 2s + 5)$, we get

$$s + d - 2 \equiv A(s^2 - 2s + 5) + Bs^2 + Cs$$

Hence, $A + B = 0$; $-2A + C = 1$; and $5A = d - 2$.   So

$$A = \frac{d - 2}{5} \qquad B = -\frac{d - 2}{5} \qquad C = \frac{2d + 1}{5}$$

We now have

$$\mathcal{L}x = \frac{d - 2}{5s} - \frac{(d - 2)s}{5(s^2 - 2s + 5)} + \frac{2d + 1}{5(s^2 - 2s + 5)}$$

We can match the first term on the right with the first entry in Table 4-2. To match the other terms on the right with entries in the table, we must put the denominator in the form $(s - a)^2 + c^2$.   Thus, the factor in the denominator becomes $(s - 1)^2 + 4$.   Also, to match the second term with the entry in Table 4-2 for $e^{at} \cos ct$, we need $s - 1$ in the numerator of the second term.   So we must add $d - 2$ to the numerator of the second term and subtract $d - 2$ from the numerator of the third term.   This leads to

$$\mathcal{L}x = \frac{d - 2}{5}\left[\frac{1}{s} - \frac{s - 1}{(s - 1)^2 + 4}\right] + \frac{d + 3}{10}\frac{2}{(s - 1)^2 + 4}$$

Picking the inverses from Table 4-2, we get

$$x = \frac{d - 2}{5} + e^t\left(\frac{d + 3}{10}\sin 2t - \frac{d - 2}{5}\cos 2t\right)$$

Since the initial conditions require $x = 0$ when $t = \pi/2$, $d = 2$. Therefore, the solution is

$$x = \frac{1}{2} e^t \sin 2t$$

Knowledge of the important properties of Laplace transforms listed in Table 4-1 can be very useful in obtaining transforms and inverses. We previously have made use of $\mathfrak{L}(f + g) = \mathfrak{L}f + \mathfrak{L}g$ and $\mathfrak{L}cf = c\mathfrak{L}f$, where c is a constant. You also should bear in mind the translation or shifting relations:

$$\mathfrak{L}[f(t - c)] = e^{-cs}\mathfrak{L}f(t) \qquad t > c, \ c = \text{positive constant} \qquad (4\text{-}9)$$
$$\mathfrak{L}[e^{ct}f(t)] = F(s - c) \qquad c = \text{constant} \qquad (4\text{-}10)$$

Equation (4-9) states that decreasing the independent variable by a constant c is equivalent to multiplication of the transform by e^{-cs}. Also, if a transform contains e^{-cs} as a factor, we can obtain the inverse for $t > c$ by first getting the inverse of the transform without e^{-cs} and then substituting $t - c$ for t.

Equation (4-10) states that if the original function contains e^{ct} as a factor, we can obtain the transform by first writing the transform of the original function without e^{ct} and then substituting $s - c$ for s. Also, if a transform is a function of $s - c$, we can get the inverse by first finding the inverse of the transform with $c = 0$ and then multiplying by e^{ct}. Examples: Compare $\mathfrak{L}t^n$ and $\mathfrak{L}t^n e^{ct}$; and $\mathfrak{L} \sin ct$ and $\mathfrak{L}e^{at} \sin ct$ in Table 4-2.

Another set of relations to remember are

$$\mathfrak{L}f(ct) = \frac{1}{c} F\left(\frac{s}{c}\right) \qquad c = \text{constant} \qquad (4\text{-}11)$$

$$\mathfrak{L}\left[\frac{1}{c} f\left(\frac{t}{c}\right)\right] = F(cs) \qquad (4\text{-}12)$$

They expedite the determination of transforms and inverses when t or s are multiplied or divided by a constant. Table 4-2 contains several examples of these relations.

When the original function contains a power of t as a factor, you can obtain the transform by differentiating the transform of the original function without that factor. For

$$\mathfrak{L}t^n f(t) = (-1)^n F^{(n)}(s) \qquad F^{(n)}(s) = \frac{d^n F(s)}{dt^n} \qquad (4\text{-}13)$$

For example, let us find the transform of $t^2 e^{ct}$. $\mathcal{L}e^{ct} = 1/(s - c)$.   Differentiating twice and multiplying by $(-1)^2$, we obtain

$$\mathcal{L}t^2 e^{ct} = \frac{2}{(s - c)^3}$$

Similarly, when the object function contains a power of $1/t$ as a factor, you can get the transform by multiple integration of the transform of the object function without that factor.   For

$$\mathcal{L}\frac{1}{t}f(t) = \int_s^\infty F(s)\ ds \tag{4-14}$$

For example, let us find $\mathcal{L}(1 - e^{-ct})/t$. $\mathcal{L}(1 - e^{-ct}) = 1/s - 1/(s + c)$. Integration yields

$$\mathcal{L}\frac{1 - e^{-ct}}{t} = \left[\log_e \frac{s}{s + c}\right]_s^\infty = \log_e\left(1 + \frac{c}{s}\right)$$

To obtain $\mathcal{L}(1 - e^{-ct})/t^2$, we would integrate again but would learn that the transform does not exist.

Still another useful relation to remember is the convolution theorem

$$\mathcal{L}f\mathcal{L}g = \mathcal{L}f * g = \mathcal{L}g * f \tag{4-15}$$

$$f * g = \int_0^t f(u)g(t - u)\ du \qquad g * f = \int_0^t g(u)f(t - u)\ du \tag{4-16}$$

This often will enable you to obtain inverses when the transforms consist of the product of two factors.   For example, let us obtain the inverse transform of $1/s(s^2 + 1)$.   From Table 4-2, we see that $\mathcal{L}^{-1}1/s = 1$ and $\mathcal{L}^{-1}1/(s^2 + 1) = \sin t$.   Therefore, by Eq. (4-16),

$$\mathcal{L}^{-1}\frac{1}{s}(s^2 + 1) = \int_0^t (1)\sin(t - u)\ du = [\cos(t - u)]_0^t = 1 - \cos t$$

When other methods fail, you can try expanding $F(s)$ in an infinite series and finding the inverse of each term.

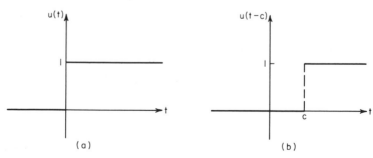

Fig. 4-1

4-4 *Step and Pulse Functions.* In many engineering problems, $t = 0$ marks the beginning of a change in a system, such as application of a force or start of an input. For $t < 0$, the force or input is zero. For such problems, the Heaviside unit step function is useful. This function (Fig. 4-1a) is defined by

$$u(t) = \begin{cases} 0 & t < 0 \\ 1 & t > 0 \end{cases} \tag{4-17}$$

Its transform is simply $1/s$.

Suppose, for example, that a mass weighing W lb, suspended on a spring with spring constant k, is at rest at $t = 0$ when a force F is applied. From $F = Ma$, the motion of the mass for $t > 0$ is governed by

$$\left[\left(\frac{W}{g}\right) D^2 + k\right] y = Fu(t)$$

where $D = d/dt$ and y is the position of the mass with respect to the rest position. If we multiply both sides of the equation by g/W and set $\omega^2 = kg/W$, the equation becomes $(D^2 + \omega^2)y = Fgu(t)/W$. [Multiplication of a function by $u(t)$ causes it to vanish for $t < 0$.]

Suppose now that F is a constant force. If we take the Laplace transforms of both sides of the equation, we get $s^2 \mathcal{L}y + \omega^2 \mathcal{L}y = Fg/Ws$, from which

$$\mathcal{L}y = \frac{Fg}{W} \frac{1}{s(s^2 + \omega^2)} = \frac{F}{k} \frac{\omega^2}{s(s^2 + \omega^2)}$$

With partial fractions or the convolution theorem and Table 4-2, we find the inverse, $y = (F/k)(1 - \cos \omega t)$.

In some cases, a change in a system occurs at $t = c$. In that case, you will find useful the unit step function $u(t - c)$ shown in Fig. 4-1b and defined by

$$u(t - c) = \begin{cases} 0 & t < c \\ 1 & t > c \end{cases} \tag{4-18}$$

Its transform is given by e^{-cs}/s.

Suppose in the preceding example that the mass at $t = 0$ was at y_0 and had a velocity v_0. Then, the force F was applied at $t = c$. The equation governing the motion is $(D^2 + \omega^2)y = Fgu(t - c)/W$. If we take Laplace transforms of both sides of the equation, we get $s^2 \mathcal{L}y - sy_0 - v_0 + \omega^2 \mathcal{L}y = Fge^{-cs}/Ws$, from which

$$\mathcal{L}y = \frac{Fge^{-cs}}{Ws} \frac{1}{s^2 + \omega^2} + y_0 \frac{s}{s^2 + \omega^2} + \frac{v_0}{\omega} \frac{\omega}{s^2 + \omega^2}$$

The inverse of the first term on the right is obtained from the preceding example without e^{-cs}, then $t - c$ is substituted for t. You can obtain the other terms from Table 4-2. The result is

$$y = \frac{F}{k}[1 - \cos \omega(t - c)] + y_0 \cos \omega t + \frac{v_0}{\omega} \sin \omega t \qquad t \geq c$$

If we write $Fu(t - c)$ for F, we eliminate the need for the separate requirement $t > c$. Multiplication of a function by $u(t - c)$ causes it to vanish for $t < c$.

In this solution, only the first term includes the influence of the input or force function. Therefore, if a different force is applied, only this term changes.

If force F is not constant, we need the transform of $f(t - c)u(t - c)$ or of $f(t)u(t - c)$ to solve the differential equation. These are given by

$$\mathcal{L}[f(t - c)u(t - c)] = e^{-cs}\mathcal{L}f(t) \qquad c \geq 0 \tag{4-19a}$$
$$\mathcal{L}[f(t)u(t - c)] = e^{-cs}\mathcal{L}f(t + c) \tag{4-19b}$$

When an inverse is required, if $\mathcal{L}^{-1}\phi(s) = f(t)$, then

$$\mathcal{L}^{-1}e^{-cs}\phi(s) = f(t - c)u(t - c) \tag{4-20}$$

If the force acting on the mass on the spring at $t = c$ is $F_0 \sin \phi(t - c)$, the differential equation becomes

$$(D^2 + \omega^2)y = u(t - c)\left(\frac{F_{0}g}{W}\right)\sin \phi(t - c)$$

From Eq. (4-19a), the transform of the right-hand side is

$$\left(\frac{F_{0}g}{W}\right)e^{-cs}\mathcal{L}\sin \phi t$$

And the transform of the first term of the solution becomes

$$\mathcal{L}y = \frac{F_{0}g}{W}e^{-cs}\frac{\phi}{s^2 + \phi^2}\frac{1}{s^2 + \omega^2}$$

We can obtain the inverse without e^{-cs} by using the convolution theorem; we can then apply Eq. (4-20). The first term of the solution results in

$$y = u(t - c)\frac{F_{0}g}{W}\left[\frac{\sin \phi(t - c)}{\omega^2 - \phi^2} - \frac{\phi}{\omega}\frac{\sin \omega(t - c)}{\omega^2 - \phi^2}\right]$$

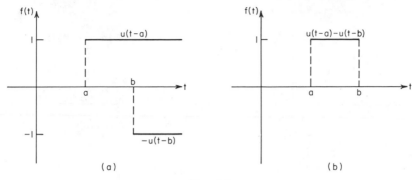

Fig. 4-2

Discontinuous functions of various types can often be represented with the use of $u(t - c)$. For example, the rectangular pulse, or filter function, of unit magnitude and duration $b - a$ (Fig. 4-2b) is given by $u(t - a) - u(t - b)$.

If the mass on a spring is acted on at $t = a$ by such a pulse with magnitude F, the differential equation becomes

$$(D^2 + \omega^2)y = \frac{Fg}{W}[u(t - a) - u(t - b)]$$

From Eq. (4-19), the transform of the right-hand side of this equation is $(Fg/W)(e^{-ca}/s - e^{-cb}/s)$. And the transform of the first term of the solution becomes

$$\mathcal{L}y = \frac{Fg}{W}\left(\frac{e^{-ca}}{s}\frac{1}{s^2 + \omega^2} - \frac{e^{-cb}}{s}\frac{1}{s^2 + \omega^2}\right)$$

Without the exponential terms the transforms have the same form as that in the solution for a constant force. Hence, we can write

$$y = \frac{F}{k}\{u(t - a)[1 - \cos \omega(t - a)] - u(t - b)[1 - \cos \omega(t - b)]\}$$

Thus, from $t = a$ to $t = b$, the motion is described by

$$\frac{F}{k}[1 - \cos \omega(t - a)]$$

From $t = b$ on, the motion is described by

$$\frac{F}{k}[\cos \omega(t - b) - \cos \omega(t - a)]$$

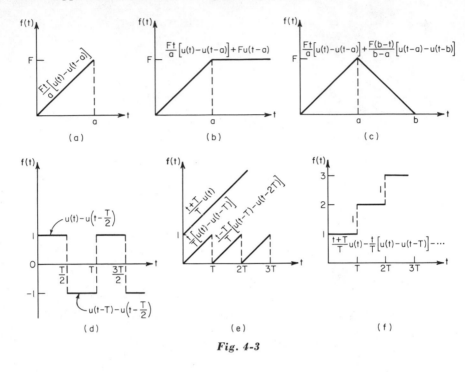

Fig. 4-3

Figure 4-3 shows how several discontinuous functions may be represented by $u(t - c)$. For the triangular pulse in Fig. 4-3a, the function is $(Ft/a)[u(t) - u(t - a)]$. It is added to $Fu(t - a)$ to produce the function for Fig. 4-3b. And it is added to

$$\frac{F(b - t)}{(b - a)}[u(t - a) - u(t - b)]$$

to produce the function for Fig. 4-3c. The rectangular alternating wave in Fig. 4-3d is developed from the function for Fig. 4-2. The function for the sawtooth wave in Fig. 4-3e is produced from the function for Fig. 4-3a. And the function for the staircase function in Fig. 4-3f is obtained by subtracting the function for the sawtooth wave from the function for the sloping line $[(t + T)/T]u(t)$ in Fig. 4-3e.

As indicated in Table 4-2, the transforms for the periodic functions can be determined from

$$\mathcal{L}f(t) = \frac{1}{1 - e^{-Ts}} \int_0^T f(t)e^{-st}\, dt \qquad (4\text{-}21)$$

where T is the period. For the alternating wave in Fig. 4-3d, for example,

$$\mathcal{L}f(t) = \frac{1}{1 - e^{-Ts}} \left[\int_0^{T/2} (1)e^{-st}\, dt + \int_{T/2}^T (-1)e^{-st}\, dt \right]$$

$$= \frac{1}{1 - e^{-Ts}} \left(\frac{-e^{-sT/2} + 1 + e^{-sT} - e^{-sT/2}}{s} \right)$$

$$= \frac{(1 - e^{-Ts/2})^2}{s(1 - e^{-Ts/2})(1 + e^{-Ts/2})} = \frac{1 - e^{-Ts/2}}{s(1 + e^{-Ts/2})} = \frac{1}{s} \tanh \frac{Ts}{4}$$

Let us return now to the rectangular pulse in Fig. 4-2. Suppose that we let b approach a and allow the magnitude of the pulse to increase so that the area between it and the t axis is always equal to 1. In the limit, we obtain the Dirac distribution $\delta(t - a)$, sometimes called the unit impulse function even though it is not a function. It is defined by

$$\delta(t - a) = 0 \qquad t \neq a$$
$$\int_{-\infty}^{\infty} \delta(t - a)\, dt = 1 \tag{4-22}$$

It has the interesting property that the integral of the product of the Dirac δ and a function $f(t)$ equals the value of the function for $t = a$; that is,

$$\int_{-\infty}^{\infty} f(t)\delta(t - a)\, dt = f(a) \tag{4-23}$$

Notice that in the above integrals, the Dirac delta is treated as if it were a function. For practical purposes it may be handled that way in solving linear differential equations where the input or force function is an impulse.

The Laplace transform of $\delta(t - a) = e^{-as}$, for $a \geq 0$.

Let us see what happens when an impulse $F\delta(t)$ is applied at $t = 0$ to the mass on a spring when the mass is at rest. The differential equation becomes $(D^2 + \omega^2)y = Fg\delta(t)/W$. The transform of $\delta(t)$ is 1. Hence, the transform of the right-hand side of the equation is simply Fg/W. So

$$\mathcal{L}y = \frac{Fg}{W} \frac{1}{s^2 + \omega^2}$$

and the solution is $y = (Fg/W\omega) \sin \omega t$. Thus, the motion is harmonic with frequency $\omega/2\pi$.

This result can be generalized. Let $y_\delta =$ response to a unit impulse of a linear system initially at rest. Also, let $f(t)$ be the actual input or force function applied at $t = 0$. Then, if y_δ is known, the response to

$f(t)$ can be determined from

$$y(t) = \int_0^t y_\delta(t - r)f(r)\, dr \qquad \text{(Duhamel formula)} \qquad (4\text{-}24)$$

For example, from the previously derived result for an impulse applied to a mass on a spring, we have the response to a unit impulse $y_\delta = (g/W\omega) \sin \omega t$. Hence, the response to a constant force F applied at $t = 0$ is, from Eq. (4-24),

$$y = \int_0^t \frac{g}{W\omega} \sin \omega(t - r)F\, dr = \left[\frac{Fg}{W\omega} \frac{1}{\omega} \cos \omega(t - r) \right]_0^t$$

$$= \frac{F}{k} (1 - \cos \omega t)$$

Similarly, if the response y_u of a linear system to the unit step function $u(t)$ is known, perhaps determined experimentally, the response to an actual input or force function $f(t)$ can be obtained from the Duhamel formula as

$$y(t) = \int_0^t y_u(t - r)f'(r)\, dr + y_u(t)f(0) \qquad (4\text{-}25)$$

For example, from the previously determined result for a constant force applied to a mass on a spring, we have the response to a unit step function $y_u = (1/k)(1 - \cos \omega t)$. Hence, the response to a force $F_0 \sin \phi t$ applied at time $t = 0$ is, from Eq. (4-25),

$$y = \int_0^t \frac{1}{k} [1 - \cos \omega(t - r)]F_0\phi \cos \phi r\, dr$$

Integration leads to the result obtained before for this loading.

4-5 *Fourier Transforms.* Operational calculus with types of transforms other than the Laplace also may be used to solve differential equations. Fourier transforms, for example, can be a powerful tool. Since the techniques of using these transforms do not differ greatly from those for Laplace transforms, the intention of this article is only to acquaint you with the forms Fourier transforms may take.

As before, we will associate an original function with an image function. Then, in the solution of differential equations, we substitute simpler operations on the image function for complex operations on the original function.

Fourier transformations may be based on sines, cosines, or exponentials. In any case, the transform is a linear operator.

The function $\mathfrak{F}(n)$ given by

$$\mathfrak{F}(n) = \int_0^\pi f(x) \sin nx \, dx \qquad (4\text{-}26)$$

where n is an integer greater than zero, is called the finite Fourier sine transform of $f(x)$.   The inverse is

$$f(x) = \frac{2}{\pi} \sum_{n=1}^\infty \mathfrak{F}(n) \sin nx \qquad 0 < x < \pi \qquad (4\text{-}27)$$

The Fourier sine transformation of $f(x)$ on the half axis is defined by

$$\mathfrak{F}(r) = \int_0^\infty f(x) \sin rx \, dx \qquad r \geq 0 \qquad (4\text{-}28)$$

and its inverse is given by

$$f(x) = \frac{2}{\pi} \int_0^\infty \mathfrak{F}(r) \sin rx \, dr \qquad x > 0 \qquad (4\text{-}29)$$

Similarly, the finite Fourier cosine transformation of $f(x)$ is defined by

$$\mathfrak{F}(c) = \int_0^\pi f(x) \cos cx \, dx \qquad (4\text{-}30)$$

where c is an integer greater than zero.   The inverse is

$$f(x) = \frac{1}{\pi} \mathfrak{F}(0) + \frac{2}{\pi} \sum_{c=1}^\infty \mathfrak{F}(c) \cos cx \qquad 0 < x < \pi \qquad (4\text{-}31)$$

The Fourier cosine transformation of $f(x)$ on the half axis is defined by

$$\mathfrak{F}(c) = \int_0^\infty f(x) \cos cx \, dx \qquad c \geq 0 \qquad (4\text{-}32)$$

and its inverse is given by

$$f(x) = \frac{2}{\pi} \int_0^\infty \mathfrak{F}(r) \cos rx \, dr \qquad x > 0 \qquad (4\text{-}33)$$

The Fourier exponential transformation of $f(x)$ is defined by

$$\mathfrak{F}(t) = \int_{-\infty}^\infty f(x) e^{itx} \, dx \qquad (4\text{-}34)$$

The inverse is given by

$$f(x) = \frac{1}{2\pi} \int_{-\infty}^\infty \mathfrak{F}(t) e^{-itx} \, dt \qquad (4\text{-}35)$$

Tables of transforms and operational properties simplify application of the transforms.

4-6 *Bibliography*

R. Bracewell, "Fourier Transform and Its Applications," McGraw-Hill Book Company, New York.

R. V. Churchill, "Operational Mathematics," McGraw-Hill Book Company, New York.

A. Erdelyi, "Tables of Integral Transforms," vol. I, McGraw-Hill Book Company, New York.

D. S. Jones, "Generalized Functions," McGraw-Hill Book Company, New York.

F. E. Nixon, "Handbook of Laplace Transformations," Prentice-Hall, Inc., Englewood Cliffs, N.J.

A. Papoulis, "Fourier Integral and Its Applications," McGraw-Hill Book Company, New York.

C. J. Savant, "Fundamentals of the Laplace Transformation," McGraw-Hill Book Company, New York.

R. D. Strum and J. R. Ward, "Laplace Transform Solution of Differential Equations," Prentice-Hall, Inc., Englewood Cliffs, N.J.

PROBLEMS

1. Use Eq. (4-1) to obtain the Laplace transform of
   (a) 1.
   (b) t.
   (c) e^{ct}.
   (d) sin ct.
   (e) $e^{ict} = \cos ct + i \sin ct$.

2. Use Eq. (4-7) and the transforms obtained in Prob. 1 to obtain the Laplace transform of
   (a) t^2.
   (b) cos ct.

3. Use Eq. (4-4) and the transforms obtained in Prob. 1 to obtain the Laplace transform of
   (a) t^2.
   (b) te^{ct}.

4. Use Eq. (4-10) to find the Laplace transform of $e^{3t} \cos 4t$, given cos ct in Table 4-2.

5. Use Eq. (4-13) to find the Laplace transform of
   (a) $t \sin t$, given sin ct in Table 4-2.
   (b) $t^2 \sin t$, given sin ct in Table 4-2.

6. Using Laplace transforms, solve the following differential equations:
   (a)

$$(D^2 + 1)x = 4 \cos t \qquad Dx = \frac{dx}{dt}, \; x(0) = x'(0) = 0$$

(b)
$$(D^2 + 1)y = 0 \qquad Dy = \frac{dy}{dt}, \; y(0) = 4, \; y'(0) = -3$$

(c)
$$(D^2 + 2D + 2)y = 0 \qquad Dy = \frac{dy}{dx}, \; y(0) = 1, \; y'(0) = 0$$

(d)
$$(D^2 + 3D + 2)y = 2u(x) \qquad Dy = \frac{dy}{dx}, \; y(0) = 1, \; y'(0) = -3$$

(e)
$$(D^2 + 2D + 10)y = 150e^{-2t}(1 - 2t) \qquad Dy = \frac{dy}{dt}, \; y(0) = y'(0) = 0$$

(f)
$$(D^2 + 4D + 13)x = 18e^{-2t} \sin 3t \qquad Dx = \frac{dx}{dt}, \; x(0) = x'(0) = 0$$

(g)
$$(D^2 + 4)y = u(t) \qquad Dy = \frac{dy}{dt}, \; y(0) = y'(0) = 0$$

(h)
$$(D^2 - 1)x = \cosh t \qquad Dx = \frac{dx}{dt}, \; x(0) = x'(0) = 0$$

7. Using Laplace transforms solve

$$(D + 6)y = Dx$$
$$(D - 3)x = -2Dy$$

with $x(0) = 2$ and $y(0) = 3$, where $Dy = dy/dt$ and $Dx = dx/dt$.

8. An electric circuit consists of an inductance $L = 1$, resistance $R = 3$, and capacitance $C = 0.5$ with a current $I = 1$ at $t = 0$, where $t =$ time. During the interval $1 < t < 2$, a unit voltage is impressed on the circuit. What is the current before, during, and after the action?

9. A mass, weighing W lb, suspended on a spring with spring constant k, is at rest at $t = 0$. Give the equation of the motion of the mass if at $t = 0$ it is subjected to
   (a) The triangular pulse in Fig. 4-3a.
   (b) The triangular pulse in Fig. 4-3c, with $a = b/2$.

10. Determine the transform of
   (a) The sawtooth function in Fig. 4-3e.
   (b) The staircase function in Fig. 4-3f.

11. Solve $F(t) - t = \int_0^t F(r) \sin (t - r) \, dr$ for $F(t)$.

12. (a) Given the Fourier sine transform of $f(x)$ defined by Eq. (4-26), find the transform of $f''(x)$.
   (b) Given the Fourier cosine transform of $f(x)$ defined by Eq. (4-30), find the transform of $f''(x)$.

ANSWERS

1. (a)

$$\mathcal{L}1 = \int_0^\infty (1)e^{-st}\, dt = \left[-\frac{1}{s} e^{-st} \right]_0^\infty = \frac{1}{s} \qquad \text{for } s > 0$$

(b)

$$\mathcal{L}t = \int_0^\infty t e^{-st}\, dt = \left[\frac{e^{-st}}{s^2}(-st - 1) \right]_0^\infty = \frac{1}{s^2} \qquad \text{for } s > 0$$

(c)

$$\mathcal{L}e^{ct} = \int_0^\infty e^{ct}e^{-st}\, dt = \int_0^\infty e^{-(s-c)t}\, dt$$

$$= \left[-\frac{e^{-(s-c)t}}{s - c} \right]_0^\infty = \frac{1}{s - c} \qquad \text{for } s > c$$

(d)

$$\mathcal{L}\sin ct = \int_0^\infty e^{-st} \sin ct\, dt = \left[\frac{e^{-st}(-s \sin ct - c \cos ct)}{s^2 + c^2} \right]_0^\infty$$

$$= \frac{c}{s^2 + c^2}$$

(e) From Prob. 1c,

$$\mathcal{L}e^{ict} = \frac{1}{s - ic} = \frac{s + ic}{s^2 + c^2} = \frac{s}{s^2 + c^2} + i\frac{c}{s^2 + c^2} = \mathcal{L}\cos ct + i\mathcal{L}\sin ct$$

2. (a) Since $t^2 = 2\int_0^t t\, dt$, $\mathcal{L}t^2 = 2\mathcal{L}\int_0^t t\, dt = (2/s)\mathcal{L}t = 2/s^3$.

(b) $\cos ct = 1 - c\int_0^t \sin ct\, dt$. Hence,

$$\mathcal{L}\cos ct = \mathcal{L}1 - c\mathcal{L}\int_0^t \sin ct\, dt = \frac{1}{s} - \frac{c}{s}\frac{c}{s^2 + c^2} = \frac{s}{s^2 + c^2}$$

3. (a) $\mathcal{L}\, dt^2/dt = s\mathcal{L}t^2 = \mathcal{L}2t = 2/s^2$. Hence, $\mathcal{L}t^2 = 2/s^3$.

(b) $\mathcal{L}d(te^{ct})/dt = s\mathcal{L}te^{ct} = \mathcal{L}(e^{ct} + cte^{ct}) = 1/(s - c) + c\mathcal{L}te^{ct}$, from which $\mathcal{L}te^{ct} = 1/(s - c)^2$.

4. $\mathcal{L}\cos 4t = s/(s^2 + 16)$. By Eq. (4-10),

$$\mathcal{L}e^{3t} \cos 4t = \frac{s - 3}{(s - 3)^2 + 16} = \frac{s - 3}{s^2 - 6s + 25}$$

5. (a) $\mathcal{L}\sin t = (s^2 + 1)^{-1}$. $\mathcal{L}t \sin t = -(d/ds)(s^2 + 1)^{-1} = 2s/(s^2 + 1)^2$.

(b) $\mathcal{L}t^2 \sin t = (-1)^2(d^2/ds^2)(s^2 + 1)^{-1} = (6s^2 - 2)/(s^2 + 1)^3$.

6. (a) Transforming, we get $s^2\mathcal{L}x + \mathcal{L}x = 4s/(s^2 + 1)$, from which $\mathcal{L}x = 4s/(s^2 + 1)^2$. From Table 4-2, $x = 2t \sin t$. Or by Eq. (4-14),

$$\int_s^\infty 4s(s^2 + 1)^{-2}\, ds = 2[-(s^2 + 1)^{-1}]_s^\infty = \frac{2}{s^2 + 1} = \frac{\mathcal{L}x(t)}{t}$$

Hence, $x/t = 2 \sin t$ and $x = 2t \sin t$.

(b) By Eq. (4-5), $s^2 \mathcal{L}y - 4s + 3 + \mathcal{L}y = 0$, from which $\mathcal{L}y = (4s - 3)/(s^2 + 1) = 4s/(s^2 + 1) - 3/(s^2 + 1)$. Using Table 4-2, we find $y = 4 \cos t - 3 \sin t$.

(c) By Eqs. (4-4) and (4-5), $s^2 \mathcal{L}y - s + 2s \mathcal{L}y - 2 + 2 \mathcal{L}y = 0$, from which

$$\mathcal{L}y = \frac{s + 2}{s^2 + 2s + 2} = \frac{s + 2}{(s + 1)^2 + 1} = \frac{s + 1}{(s + 1)^2 + 1} + \frac{1}{(s + 1)^2 + 1}$$

From Table 4-2, $y = e^{-x} \cos x + e^{-x} \sin x$.

(d) By Eqs. (4-4) and (4-5), $s^2 \mathcal{L}y - s + 3 + 3s \mathcal{L}y - 3 + 2 \mathcal{L}y = 2/s$. (See also Art. 4-4.) This yields

$$\mathcal{L}y = \frac{s^2 + 2}{s(s^2 + 3s + 2)} = \frac{s^2 + 2}{s(s + 1)(s + 2)} = \frac{1}{s} - \frac{3}{s + 1} + \frac{3}{s + 2}$$

Therefore, $y = 1 - 3e^{-x} + 3e^{-2x}$.

(e) By Eqs. (4-4) and (4-5), $s^2 \mathcal{L}y + 2s \mathcal{L}y + 10 \mathcal{L}y = 150/(s + 2) - 300/(s + 2)^2 = 150s/(s + 2)^2$. This yields

$$\mathcal{L}y = \frac{150s}{(s + 2)^2(s^2 + 2s + 10)}$$

$$= \frac{9}{s + 2} - \frac{30}{(s + 2)^2} - 9 \frac{s + 1}{(s + 1)^2 + 9} + \frac{39}{(s + 1)^2 + 9}$$

Consequently, $y = 9e^{-2t} - 30te^{-2t} - 9e^{-t} \cos 3t + 13e^{-t} \sin 3t$.

(f) By Eqs. (4-4) and (4-5), $s^2 \mathcal{L}x + 4s \mathcal{L}x + 13 \mathcal{L}x = (18 \times 3)/[(s + 2)^2 + 9] = 54/(s^2 + 4s + 13)$. From this, we get $\mathcal{L}x = 54/(s^2 + 4s + 13)^2$. By Eq. (4-10), $x = \mathcal{L}^{-1} 54/[(s + 2)^2 + 9]^2 = 54e^{-2t} \mathcal{L}^{-1}(s + 9)^{-2}$. But $1/(s + 9) = \mathcal{L}(\sin 3t)/3$, and $(s + 9)^{-2} = [\mathcal{L}(\sin 3t)/3][\mathcal{L}(\sin 3t)/3]$. From Eqs. (4-15) and (4-16) then,

$$x - 54e^{-2t} \int_0^t \left(\frac{1}{3} \sin 3u\right) \left[\frac{1}{3} \sin 3(t - u)\right] du$$

$$= e^{-2t}(\sin 3t - 3t \cos 3t)$$

(g) By Eqs. (4-4) and (4-5), $s^2 \mathcal{L}y + 4 \mathcal{L}y = 1/s$, from which $\mathcal{L}y = (1/s)[1/(s^2 + 4)]$. By Eq. (4-7),

$$y = \int_0^t \frac{1}{2} \sin 2t \, dt = \frac{1 - \cos 2t}{4}$$

(h) By Eqs. (4-4) and (4-5), $s^2 \mathcal{L}x - \mathcal{L}x = s/(s^2 - 1)$, from which $\mathcal{L}x = s/(s^2 - 1)^2$. By Eq. (4-14),

$$x = t \mathcal{L}^{-1} \int_s^\infty s \frac{ds}{(s^2 - 1)^2} = \frac{t}{2} \sinh t$$

7. By Eq. (4-4), the equations transform to

$$s \mathcal{L}y - 3 + 6 \mathcal{L}y = (s + 6) \mathcal{L}y - 3 = s \mathcal{L}x - 2$$
$$s \mathcal{L}x - 2 - 3 \mathcal{L}x = (s - 3) \mathcal{L}x - 2 = -2s \mathcal{L}y + 6$$

Simultaneous solution yields

$$\mathcal{L}x = \frac{2s + 16}{s^2 + s - 6} = \frac{4}{s - 2} - \frac{2}{s + 3}$$

$$\mathcal{L}y = \frac{3s - 1}{s^2 + s - 6} = \frac{1}{s - 2} + \frac{2}{s + 3}$$

Hence, $x = 4e^{2t} - 2e^{-3t}$ and $y = e^{2t} + 2e^{-3t}$.

8. The impressed voltage is described by $u(t - 1) - u(t - 2)$. (See Art. 4-4.) By Kirchhoff's law,

$$u(t - 1) - u(t - 2) - \frac{dI}{dt} - 3I - 2\int_0^t I \, dt = 0$$

By Eqs. (4-4), (4-7), and (4-18), $e^{-s}/s - e^{-2s}/s - s\mathcal{L}I + 1 - 3\mathcal{L}I - 2/s\mathcal{L}I = 0$, from which

$$\mathcal{L}I = \frac{e^{-s} - e^{-2s} + s}{s^2 + 3s + 2}$$

$$= e^{-s}\left(\frac{1}{s + 1} - \frac{1}{s + 2}\right) - e^{-2s}\left(\frac{1}{s + 1} - \frac{1}{s + 2}\right) + \frac{2}{s + 2} - \frac{1}{s + 1}$$

To obtain the inverses of the first two terms, use Eq. (4-20). $I = (e^{-(t-1)} - e^{-2(t-1)})u(t - 1) - (e^{-(t-2)} - e^{-2(t-2)})u(t - 2) + 2e^{-2t} - e^{-t}$. Or

$$I = \begin{cases} 2e^{-2t} - e^{-t} & 0 < t < 1 \\ 2e^{-2t} - e^{-t} + e^{-(t-1)} - e^{-2(t-1)} & 1 < t < 2 \\ 2e^{-2t} - e^{-t} + e^{-(t-1)} - e^{-2(t-1)} - e^{-(t-2)} + e^{-2(t-2)} & t > 2 \end{cases}$$

9. (a) The equation of motion is $(D^2 + \omega^2)y = (Fgt/aW)[u(t) - u(t - a)]$, where $\omega^2 = kg/W$. By Eqs. (4-5) and (4-19b), $s^2\mathcal{L}y + \omega^2\mathcal{L}y = (Fg/aW)[1/s^2 - e^{-as}(1/s^2 + a/s)]$. This yields

$$\mathcal{L}y = \frac{Fg}{aW}\left\{\frac{1}{s^2(s^2 + \omega^2)} - e^{-as}\left[\frac{1}{s^2(s^2 + \omega^2)} + \frac{a}{s(s^2 + \omega^2)}\right]\right\}$$

$$= \frac{Fg}{aW}\left\{\frac{1}{\omega^2 s^2} - \frac{1}{\omega^2(s^2 + \omega^2)}\right.$$

$$\left. - e^{-as}\left[\frac{1}{\omega^2 s^2} - \frac{1}{\omega^2(s^2 + \omega^2)} + \frac{a}{\omega^2 s} - \frac{as}{\omega^2(s^2 + \omega^2)}\right]\right\}$$

Using Eq. (4-20), we obtain

$$y = \frac{F}{ak}\left\{\left(t - \frac{1}{\omega}\sin \omega t\right)u(t) - \left[t - a - \frac{1}{\omega}\sin \omega(t - a) + a\right.\right.$$

$$\left.\left. - \cos \omega(t - a)\right]u(t - a)\right\}$$

$$= \frac{F}{ak}\left(t - \frac{1}{\omega}\sin \omega t\right) \qquad\qquad 0 < t < a$$

$$= \frac{F}{ak\omega}\left[\sin \omega(t - a) - \sin \omega t + a\cos \omega(t - a)\right] \qquad t > a$$

(b) The equation of motion is

$$(D^2 + \omega^2)y = \frac{2Fgt}{Wb}\left[u(t) - u\frac{t-b}{2}\right]$$

$$+ \frac{2Fg(b-t)}{Wb}\left[u\left(\frac{t-b}{2}\right) - u(t-b)\right]$$

By Eqs. (4-5) and (4-19b),

$$s^2\mathcal{L}y + \omega^2\mathcal{L}y = \frac{2Fg}{Wb}\left[\frac{1}{s^2} - 2e^{-bs/2}\left(\frac{1}{s^2} + \frac{b}{2s}\right) + e^{-bs}\left(\frac{1}{s^2} + \frac{b}{s}\right)\right.$$

$$\left. + (e^{-bs/2} - e^{-bs})\frac{b}{s}\right] = \frac{2Fg}{Wb}(1 - 2e^{-bs/2} + e^{-bs})\frac{1}{s^2}$$

From this, we get

$$\mathcal{L}y = \frac{2Fg}{Wb}(1 - 2e^{-bs/2} + e^{-bs})\left[\frac{1}{\omega^2 s^2} - \frac{1}{\omega^2(s^2 + \omega^2)}\right]$$

Hence, using Eq. (4-20), we obtain

$$y = \frac{2F}{bk}\left\{u(t)\left(t - \frac{1}{\omega}\sin\omega t\right) - u\left(t - \frac{b}{2}\right)\left[2t - b - \frac{2}{\omega}\sin\omega\left(t - \frac{b}{2}\right)\right]\right.$$

$$\left. + u(t-b)\left[t - b - \frac{1}{\omega}\sin\omega(t-b)\right]\right\}$$

$$= \frac{2F}{bk}\left(t - \frac{1}{\omega}\sin\omega t\right) \qquad\qquad 0 < t < \frac{b}{2}$$

$$= \frac{2F}{bk}\left[b - t + \frac{2}{\omega}\sin\omega\left(t - \frac{b}{2}\right) - \frac{1}{\omega}\sin\omega t\right] \qquad \frac{b}{2} < t < b$$

$$= \frac{2F}{bk\omega}\left[-\sin\omega(t-b) + 2\sin\omega\left(t - \frac{b}{2}\right) - \sin\omega t\right] \qquad t > b$$

10. (a) Let $F_1 =$ the sawtooth function in Fig. 4-3e. By Eq. (4-21),

$$\mathcal{L}F_1 = \frac{1}{1 - e^{-sT}}\int_0^T \frac{t}{T}e^{-st}\,dt = \frac{1}{T(1 - e^{-sT})}\left[\frac{e^{-st}}{s^2}(-st - 1)\right]_0^T$$

$$= \frac{1 - e^{-sT}(sT + 1)}{s^2 T(1 - e^{-sT})} = \frac{1 + sT}{s^2 T} - \frac{1}{s(1 - e^{-sT})}$$

(b) Let $F_1 =$ the sawtooth function in Fig. 4-3e and $F_2 =$ the staircase function in Fig. 4-3f. Then,

$$\mathcal{L}F_2 = \mathcal{L}\frac{T + t}{T}u(t) - \mathcal{L}F_1 = \frac{1}{s} + \frac{1}{s^2 T} - \frac{1 + sT}{s^2 T} + \frac{1}{s(1 - e^{-sT})}$$

$$= \frac{1}{s(1 - e^{-sT})}$$

11. Using Eq. (4-16), we can write the equation as $F(t) - t = F(t) * \sin t$. Taking transforms, we get $\mathcal{L}F(t) - 1/s^2 = \mathcal{L} \sin t \, \mathcal{L}F(t) = \mathcal{L}F(t)/(s^2 + 1)$. This yields $\mathcal{L}F(t) = (s^2 + 1)/s^4 = 1/s^2 + 1/s^4$.   Hence, $F(t) = t + t^3/6$.

12. (a) Integration by parts [Eq. (2-5)] yields

$$\mathcal{F}f''(x) = \int_0^\pi f''(x) \sin nx \, dx = [f'(x) \sin nx]_0^\pi - n \int_0^\pi f'(x) \cos nx \, dx$$

$$= 0 - [nf(x) \cos nx]_0^\pi - n^2 \int_0^\pi f(x) \sin nx \, dx$$

$$= -n^2 \mathcal{F}f(x) + n[f(0) - (-1)^n f(\pi)]$$

(b) Integration by parts [Eq. (2-5)] yields

$$\mathcal{F}f''(x) = \int_0^\pi f''(x) \cos nx \, dx = [f'(x) \cos nx]_0^\pi + n \int_0^\pi f'(x) \sin nx \, dx$$

$$= (-1)^n f'(\pi) - f'(0) + n[f(x) \sin nx]_0^\pi - n^2 \int_0^\pi f(x) \cos nx \, dx$$

$$= -n^2 \mathcal{F}f(x) - f'(0) + (-1)^n f'(\pi)$$

FIVE

Nonelementary Solutions of Differential Equations

In brief, to solve a differential equation, you must first recognize the type and then apply an appropriate method of solution. To aid you in recognizing types of equations, Chap. 3 provided rules for classifying them. If, for example, you have to solve a first-order equation, you try the methods given in Art. 3-3. If the equation is of higher order, you try to reduce its order by the methods of Art. 3-8. Next, you decide whether the equation is linear and particularly whether the coefficients are constants. Linear differential equations with constant coefficients may be solved by the methods of Art. 3-7 or Chap. 4. By changing the variable, you can sometimes convert a given equation into one that these methods can handle.

In any event, you may or may not be able to get a solution in closed form or even in the form of elementary functions. **Elementary functions** are finite combinations, obtained by addition, subtraction, multiplication, and division, of powers, roots, exponentials, logarithms, trigonometric and inverse trigonometric functions, and hyperbolic and inverse hyperbolic functions of a variable and constants. x^2, e^x, $x^{1/3}$, $\log \cos x$, and $(1 + e^{-x}) \log \sin x$ are examples. Some differential equations involving only elementary functions and their derivatives have solutions that cannot be expressed as finite combinations of elementary functions.

In this chapter, we will examine methods of solving differential equations with infinite series and discuss solutions containing nonelementary functions. On occasion, some of the operations may not be mathematically justifiable. Therefore, when you obtain a possible solution, verify that it is a solution by substitution in the original differential equation. If you have a solution, determine whether it is the only solution. If it is not the only solution, look for others.

5-1 Solution by Taylor and Maclaurin Series. If $f(x)$ possesses derivatives of all orders in an interval including $x = 0$, it can be represented in that interval by the Maclaurin power series

$$f(x) = f(0) + f'(0)x + \frac{f''(0)}{2!} x^2 + \frac{f^{(3)}(0)}{3!} x^3 + \cdots$$

$$+ \frac{f^{(n)}(0)}{n!} x^n + \cdots \quad (5\text{-}1)$$

In this series, $f(0)$, $f'(0)$, $f''(0)$, $\ldots$, $f^{(n)}(0)$ represent, respectively, the value of the function, first derivative, second derivative, and nth derivative at $x = 0$.

For example, let us expand e^x by a Maclaurin series.

$$f(x) = e^x \qquad f(0) = 1$$
$$f^{(n)}(x) = e^x \qquad f^{(n)}(0) = 1$$

Hence, $e^x = 1 + x + x^2/2! + x^3/3! + \cdots + x^n/n! + \cdots$.

Similarly, if $f(x)$ possesses derivatives of all orders in an interval including $x = a$, it can be represented in that interval by the Taylor power series

$$f(x) = f(a) + f'(a)(x - a) + \frac{f''(a)}{2!} (x - a)^2 + \cdots$$

$$+ \frac{f^{(n)}(a)}{n!} (x - a)^n + \cdots \quad (5\text{-}2)$$

The Maclaurin series, Eq. (5-1), is the special case of the Taylor series when $a = 0$.

Some differential equations, particularly homogeneous linear equations with variable coefficients, can be solved by assuming a solution in the form of a Maclaurin series

$$y = \sum_{n=0}^{\infty} a_n x^n \qquad (5\text{-}3)$$

or a Taylor series

$$y = \sum_{n=0}^{\infty} a_n (x - x_0)^n \tag{5-4}$$

You can determine the coefficients a_n by substituting Eq. (5-3) or (5-4) in the differential equation and equating coefficients of like powers of x.

For example, let us solve $y'' - xy = 0$. We will assume that y is given by Eq. (5-3). Then, $y' = \Sigma n a_n x^{n-1}$ and $y'' = \Sigma n(n-1) a_n x^{n-2}$. Substitution in the given equation yields

$$\Sigma n(n-1) a_n x^{n-2} - x \Sigma a_n x^n = 0$$

When we equate coefficients of like powers of x to zero, we get

$$2a_2 = 0 \qquad a_0 = 3 \cdot 2 a_3 \qquad a_1 = 4 \cdot 3 a_4 \qquad 5 \cdot 4 a_5 = a_2 = 0$$

Hence,

$$a_2 = a_5 = a_8 = \cdots = 0$$

$$a_3 = \frac{a_0}{2 \cdot 3} \qquad a_6 = \frac{a_3}{5 \cdot 6} = \frac{a_0}{2 \cdot 3 \cdot 5 \cdot 6} \qquad \cdots$$

$$a_4 = \frac{a_1}{3 \cdot 4} \qquad a_7 = \frac{a_4}{6 \cdot 7} = \frac{a_1}{3 \cdot 4 \cdot 6 \cdot 7} \qquad \cdots$$

Thus, we have the solution

$$y = a_0 \left(1 + \frac{x^3}{2 \cdot 3} + \frac{x^6}{2 \cdot 3 \cdot 5 \cdot 6} + \cdots \right)$$
$$+ a_1 \left(x + \frac{x^4}{3 \cdot 4} + \frac{x^7}{3 \cdot 4 \cdot 6 \cdot 7} + \cdots \right)$$

5-2   Sine-, Cosine-, and Exponential-integral Functions. Suppose that we are given what looks like a simple equation: $xy' = \sin x$. Since the variables are separable, we get the solution $y = \int \sin x \, dx/x$. But investigation will soon indicate that the integral cannot be expressed as a combination of elementary functions. We can, however, evaluate the integral by expanding $\sin x$ in a Maclaurin series and integrating term by term. This nonelementary function is called the sine-integral function of x and is usually written $Si(x)$.

The Maclaurin series for $\sin x$ is $x - x^3/3! + x^5/5! - x^7/7! + \cdots$. Hence,

$$Si(x) = \int \frac{\sin x \, dx}{x} = \int dx - \int \frac{x^2}{3!} dx + \int \frac{x^4}{5!} dx - \int \frac{x^6}{7!} dx + \cdots$$
$$= x - \frac{x^3}{3 \cdot 3!} + \frac{x^5}{5 \cdot 5!} - \frac{x^7}{7 \cdot 7!} + \cdots \tag{5-5}$$

Similarly, the cosine-integral function is defined by

$$Ci(x) = \int \frac{\cos x \, dx}{x} = \log |x| - \frac{x^2}{2 \cdot 2!} + \frac{x^4}{4 \cdot 4!} - \frac{x^6}{6 \cdot 6!} + \cdots \quad (5\text{-}6)$$

And the exponential-integral function is defined by

$$Ei(x) = \int \frac{e^{-x} \, dx}{x} = \log |x| - x + \frac{x^2}{2 \cdot 2!} - \frac{x^3}{3 \cdot 3!} + \frac{x^4}{4 \cdot 4!} - \cdots$$
$$(5\text{-}7)$$

Books listed in the Bibliography in Art. 5-8 contain tables of values for these nonelementary functions.

5-3   *Frobenius' Method of Solution.* If a Maclaurin or Taylor series does not yield the general solution of a differential equation, try

$$y = x^c \Sigma a_n x^n \qquad n \geq 0 \tag{5-8}$$

The constant c may be determined by substitution of y in the differential equation.

For example, let us solve $4xy'' + 2y' - y = 0$. We will assume that the solution is $y = \Sigma a_n x^{n+c}$. Then, $y' = \Sigma(n + c)a_n x^{n+c-1}$, and $y'' = \Sigma(n + c)(n + c - 1)a_n x^{n+c-2}$. Substitution in the differential equation yields

$$4x\Sigma(n + c)(n + c - 1)a_n x^{n+c-2} + 2\Sigma(n + c)a_n x^{n+c-1} - \Sigma a_n x^{n+c} = 0$$

This reduces to

$$2\Sigma(n + c)(2n + 2c - 1)a_n x^{n+c-1} - \Sigma a_n x^{n+c} = 0$$

Noting that $n \geq 0$, we determine c by setting the coefficients of x^{c-1} equal to zero. Thus,

$$(0 + c)(0 + 2c - 1) = 0$$
$$c = 0, \frac{1}{2}$$

With $c = 0$, we equate coefficients of like power of x to zero. The result is

$$y = a_0 \left(1 + \frac{x}{2!} + \frac{x^2}{4!} + \frac{x^3}{6!} + \cdots \right) = a_0 \cosh \sqrt{x}$$

In this case, the solution is a Maclaurin expansion of an elementary function; but you may get solutions you do not recognize or that are nonelementary functions.

Next, with $c = \frac{1}{2}$, we equate coefficients of like powers of x to zero. The result is

$$y = b_0 \left(x^{\frac{1}{2}} + \frac{x^{\frac{3}{2}}}{3!} + \frac{x^{\frac{5}{2}}}{5!} + \cdots \right) = b_0 \sinh \sqrt{x}$$

Therefore, the complete solution is $y = a \cosh \sqrt{x} + b \sinh \sqrt{x}$.

5-4 *Fourier Series Solutions.* Fourier series expand a function in combinations of trigonometric functions. They thus offer an alternative method to those given in the preceding articles for solving differential equations.

Let $f(x)$ be single-valued and continuous, except possibly for a finite number of finite discontinuities in an interval of length 2π. Also, let us restrict $f(x)$ to a finite number of maxima and minima in this interval. Then $f(x)$ may be represented by the convergent Fourier series:

$$f(x) = \frac{1}{2} a_0 + a_1 \cos x + a_2 \cos 2x + \cdots + a_n \cos nx + \cdots$$
$$+ b_1 \sin x + b_2 \sin 2x + \cdots + b_n \sin nx + \cdots$$
$$= \frac{1}{2} a_0 + \sum_{n=1}^{\infty} (a_n \cos nx + b_n \sin nx) \tag{5-9}$$

where

$$a_n = \frac{1}{\pi} \int_c^{c+2\pi} f(x) \cos nx \, dx \tag{5-10}$$

$$b_n = \frac{1}{\pi} \int_c^{c+2\pi} f(x) \sin nx \, dx \tag{5-11}$$

c = value of x at start of interval

For example, let us express as a Fourier series $f(x)$ defined by

$$f(x) = \begin{cases} 1 & 0 < x < \pi \\ 2 & \pi < x < 2\pi \end{cases}$$

Notice that $f(x)$ is not defined at the midpoint of the interval. The Fourier series will yield $f(\pi) = \frac{3}{2}$. The series will also indicate that $f(0) = f(2\pi) = \frac{3}{2}$ because of the periodicity of the series.

The coefficients of the trigonometric functions in the series are

$$a_0 = \frac{1}{\pi} \int_0^{2\pi} f(x) \, dx = \frac{1}{\pi} \int_0^{\pi} 1 \, dx + \frac{1}{\pi} \int_{\pi}^{2\pi} 2 \, dx = 1 + 2 = 3$$

$$a_n = \frac{1}{\pi} \int_0^{\pi} 1 \cos nx \, dx + \frac{1}{\pi} \int_{\pi}^{2\pi} 2 \cos nx \, dx = 0$$

$$b_n = \frac{1}{\pi} \int_0^{\pi} 1 \sin nx \, dx + \frac{1}{\pi} \int_{\pi}^{2\pi} 2 \sin nx \, dx = \frac{1}{n\pi} (\cos n\pi - 1)$$

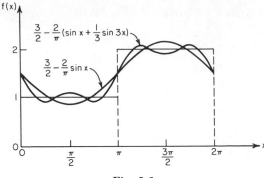

Fig. 5-1

Hence, $f(x) = \frac{3}{2} - (2/\pi)[\sin x + (\sin 3x)/3 + (\sin 5x)/5 + \cdots]$.
Curves representing the first two and three terms of the series are plotted
in Fig. 5-1 to indicate how the series approximates the given function.
[Since $f(x)$ converges to 1 for $0 < x < \pi$, then for $x = \pi/2$,

$$f(x) = 1 = \frac{3}{2} - \frac{2}{\pi}\left(1 - \frac{1}{3} + \frac{1}{5} - \frac{1}{7} + \cdots\right)$$

Hence, $1 - \frac{1}{3} + \frac{1}{5} - \frac{1}{7} + \cdots = \pi/4$.]

When an even function $[f(-x) \equiv f(x)]$ is expanded in a Fourier series
in the interval $-\pi < x < \pi$, the series will not contain sine terms. In
that case, the coefficients of the cosine terms are given by

$$a_n = \frac{2}{\pi}\int_0^\pi f(x)\ \cos nx\ dx \qquad\qquad (5\text{-}12)$$

When an odd function $[f(-x) \equiv -f(x)]$ is expanded in this interval, the
series will not contain cosine terms. The coefficients of the sine terms
are then given by

$$b_n = \frac{2}{\pi}\int_0^\pi f(x)\ \sin nx\ dx \qquad\qquad (5\text{-}13)$$

The Fourier series for an interval $-L < x < L$ is given by

$$f(x) = \frac{1}{2}a_0 + a_1\cos\frac{\pi x}{L} + a_2\cos\frac{2\pi x}{L} + \cdots + a_n\cos\frac{n\pi x}{L} + \cdots$$

$$+ b_1\sin\frac{\pi x}{L} + b_2\sin\frac{2\pi x}{L} + \cdots + b_n\sin\frac{n\pi x}{L} \quad (5\text{-}14)$$

where

$$a_n = \frac{1}{L} \int_{-L}^{L} f(x) \cos \frac{n\pi x}{L} \, dx \tag{5-15}$$

$$b_n = \frac{1}{L} \int_{-L}^{L} f(x) \sin \frac{n\pi x}{L} \, dx \tag{5-16}$$

For the half interval $0 < x < L$, the half-range sine series is

$$f(x) = b_1 \sin \frac{\pi x}{L} + b_2 \sin \frac{2\pi x}{L} + \cdots + b_n \sin \frac{n\pi x}{L} + \cdots \tag{5-17}$$

where

$$b_n = \frac{2}{L} \int_0^L f(x) \sin \frac{n\pi x}{L} \, dx \tag{5-18}$$

The half-range cosine series is

$$f(x) = \frac{1}{2} a_0 + a_1 \cos \frac{\pi x}{L} + a_2 \cos \frac{2\pi x}{L} + \cdots + a_n \cos \frac{n\pi x}{L} + \cdots \tag{5-19}$$

where

$$a_n = \frac{2}{L} \int_0^L f(x) \cos \frac{n\pi x}{L} \, dx \tag{5-20}$$

To demonstrate a technique of using Fourier series in solving differential equations, let us determine the equation of the elastic curve (deflection curve) of beam AB in Fig. 5-2. The beam has span L and carries a concentrated load P at a distance kL from the origin of coordinates at A. We will consider deflections y downward as positive. Then, the equation

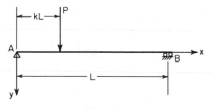

Fig. 5-2

of the elastic curve may be obtained from $y'' = -M/EI$, where M is the bending moment at a distance xL from A, E is the modulus of elasticity of the beam material, and I is the moment of inertia of the beam cross section.

Let us assume that the deflection is given by the half-range sine series

$$y = y_1 \sin \pi x + y_2 \sin 2\pi x + y_3 \sin 3\pi x + \cdots$$

We will choose the coefficients $y_1, y_2, y_3, \ldots$ so that y will satisfy the given differential equation.

The bending moment is given by

$$M = \begin{cases} (1 - k)PLx & 0 < x < k \\ kP(1 - x) & k < x < 1 \end{cases}$$

In this form, however, we will be unable to evaluate the coefficients in the Fourier series for y. Hence, we must also express M in a half-range sine series. This offers the incidental advantage that we will have a single representation of M for the whole span instead of the two functions above.

$$M = b_1 \sin \pi x + b_2 \sin 2\pi x + \cdots + b_n \sin n\pi x + \cdots$$

We can determine b_n from Eq. (5-18).

$$b_n = \frac{2}{L} \int_0^k (1 - k)PLx \sin n\pi x(L\,dx)$$

$$+ \frac{2}{L} \int_k^1 kPL(1 - x) \sin n\pi x(L\,dx)$$

$$= \frac{2PL}{n^2\pi^2} \sin n\pi k$$

Hence,

$$M = \frac{2PL}{\pi^2} \left(\sin \pi k \sin \pi x + \frac{1}{4} \sin 2\pi k \sin 2\pi x \right.$$

$$\left. + \frac{1}{9} \sin 3\pi k \sin 3\pi x + \cdots \right)$$

Notice that $M = 0$ when $x = 0$ or 1, as required.

We can test the accuracy of this representation by assuming the load is at midspan and computing the midspan bending moment. Taking only the first term of the series, we get $M = 2PL/\pi^2 = PL/4.92$. With three terms, we get $M = PL/4.42$. Actually, $M = PL/4$.

Now, if we substitute the Fourier series for M in the differential equation, we obtain after differentiating y twice with respect to xL:

$$y'' = -\frac{\pi^2}{L^2} \sum_{n=1}^{\infty} n^2 y_n \sin n\pi x$$

$$= -\frac{M}{EI} = -\frac{2PL}{\pi^2 EI} \sum_{n=1}^{\infty} \frac{1}{n^2} \sin n\pi k \sin n\pi x$$

Since for each value of n the coefficients of $\sin n\pi x$ must be identical, we obtain

$$y_n = \frac{2PL^3}{\pi^4 EI}\left(\frac{1}{n^4}\sin n\pi k\right)$$

Therefore,

$$y = \frac{2PL^3}{\pi^4 EI}\left(\sin \pi k \sin \pi x + \frac{1}{16}\sin 2\pi k \sin 2\pi x \right.$$

$$\left. + \frac{1}{81}\sin 3\pi k \sin 3\pi x + \cdots\right)$$

As before, we can check this result by computing the midspan deflection of the beam for a load at midspan. If we use only the first term of the series, we get $y = PL^3/48.5EI$, compared with the more exact solution $y = PL^3/48EI$.

5-5 Elliptic Integrals. When we apply the formula $s = \int(dx^2 + dy^2)^{1/2}$ to determine the circumference, or part of the circumference, of a circle, we readily obtain a solution in closed form, $s = 2\pi r$ or $s = r\theta$, where r is the radius and θ the central angle of the arc. But when we try to compute the circumference of an ellipse, we find that it cannot be expressed in finite combinations of elementary functions. The resulting nonelementary function and related functions are called elliptic integrals.

We will examine the problem of the ellipse shortly. First, let us consider the motion of a pendulum with a mass M concentrated at the end of a rigid rod of length L and swinging through an angle 2ψ (Fig. 5-3). When the mass is at Q, the top of its swing, it has no velocity, thus no kinetic energy. But its potential energy is at a maximum, Mgh, where $g = 32.2$ ft per sec² and h is the maximum rise of the mass. When at time t the mass reaches $P(x,y)$, its potential energy drops $Mg(h - y)$ and the kinetic energy rises to $Mv^2/2$, where v is its velocity. By the law of conservation of energy, the increase in kinetic energy equals the decrease in potential energy. Consequently,

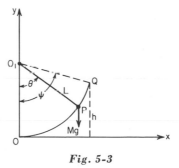

Fig. 5-3

$$\frac{Mv^2}{2} = Mg(h - y)$$

Let us substitute $L(d\theta/dt)$ for v and solve for dt.   We get

$$dt = \frac{L\,d\theta}{\sqrt{2gh}\,\sqrt{1-y/h}}$$

Now, we make some more significant substitutions.   Let

$$y = L - L\cos\theta = h\sin^2\varphi \tag{5-21}$$

Then, $dy = L\sin\theta\,d\theta = 2h\sin\varphi\cos\varphi\,d\varphi$.   From Fig. 5-3, you can observe that

$$L\sin\theta = \sqrt{L^2 - (L-y)^2} = \sqrt{2Ly - y^2}$$

Again letting $y = h\sin^2\varphi$, simplifying, and solving for $d\theta$, we get

$$d\theta = \sqrt{\frac{2h}{L}}\,\frac{\cos\varphi\,d\varphi}{\sqrt{1 - k^2\sin^2\varphi}}$$

where

$$k^2 = \frac{h}{2L} \tag{5-22}$$

Substituting this value of $d\theta$ in the expression for dt and using Eq. (5-21), we find that

$$t = \sqrt{\frac{L}{g}}\int_0^\varphi \frac{d\varphi}{\sqrt{1 - k^2\sin^2\varphi}} \tag{5-23}$$

The integral in Eq. (5-23) cannot be evaluated in finite combinations of elementary functions.   It is called an **elliptic integral of the first kind** and is designated by $F(k,\varphi)$.

Alternative forms for this function include

$$F(k,\varphi) = \int_0^\varphi \frac{d\beta}{\sqrt{1 - \sin^2\alpha\sin^2\beta}} \tag{5-24a}$$

$$F(k,x) = \int_0^x \frac{dz}{\sqrt{(1 - z^2)(1 - k^2z^2)}} \tag{5-24b}$$

where $\sin\alpha = k$ = modulus
$\quad\quad\ \ \varphi$ = amplitude
$\quad\quad\ \ x = \sin\varphi$

For the pendulum, when $y = h$, $\varphi = \pi/2$.   When $\varphi = \pi/2$, Eq. (5-24a) becomes

$$K = \int_0^{\pi/2} \frac{d\beta}{\sqrt{1 - \sin^2 \alpha \sin^2 \beta}}$$

$$= \frac{\pi}{2}\left[1 + (\tfrac{1}{2})^2 k^2 + \left(\frac{1 \cdot 3}{2 \cdot 4}\right)^2 k^4 + \left(\frac{1 \cdot 3 \cdot 5}{2 \cdot 4 \cdot 6}\right)^2 k^6 + \cdots\right] \quad (5\text{-}25)$$

It is called the **complete elliptic integral of the first kind.**

The elliptic integrals may be evaluated by use of infinite series.   Values of $F(k,\varphi)$ are plotted in Fig. 5-4a.   Books listed in the Bibliography (Art. 5-8) contain tables for these integrals.   Values need be tabulated only for $\varphi \le \pi/2$, because

$$F(k,\varphi) = 2K - F(k, \pi - \varphi) \quad\quad\quad\quad (5\text{-}26a)$$
$$F(k, m\pi + \varphi) = 2mK + F(k,\varphi) \quad\quad\quad\quad (5\text{-}26b)$$

where m is an integer.

Note that the period of the pendulum is given by $T = 4K \sqrt{L/g}$.   For a swing of $\psi = 60°$, for example, $k = \sqrt{h/2L} = \sqrt{(1 - \cos \psi)/2} = 0.5 = \sin \alpha$ or $\alpha = 30°$.   From Fig. 5-4a, for $\alpha = 30°$ and $\varphi = \pi/2$, $K = 1.7$.   Then, $T = 4 \times 1.7 \sqrt{L/g} = 6.8 \sqrt{L/g}$.   (In a commonly used approximate solution, where $\sin \theta$ is replaced by θ, $T = 2\pi \sqrt{L/g} = 6.3 \sqrt{L/g}$.)

Let us consider now the problem of determining the circumference of an ellipse.   Take the equation of the ellipse in parametric form: $x = a \sin \varphi$ and $y = b \cos \varphi$, where a is the semimajor axis; b, the semiminor axis; and φ, a central angle, measured clockwise from the positive semiminor axis.   Then, $dx = a \cos \varphi\, d\varphi$ and $dy = -b \sin \varphi\, d\varphi$.   Hence, the arc length, measured from the positive semiminor axis, is given by

$$s = \int \sqrt{dx^2 + dy^2} = \int_0^\varphi \sqrt{a^2 \cos^2 \varphi + b^2 \sin^2 \varphi}\, d\varphi$$

Substitute $1 - \sin^2 \varphi$ for $\cos^2 \varphi$ and let

$$k^2 = \frac{a^2 - b^2}{a^2} \quad\quad\quad\quad (5\text{-}27)$$

The result is

$$s = a \int_0^\varphi \sqrt{1 - k^2 \sin^2 \varphi}\, d\varphi \quad\quad\quad\quad (5\text{-}28)$$

The integral in Eq. (5-28) cannot be evaluated in finite combinations of elementary functions.   It is called an **elliptic integral of the second kind** and is designated by $E(k,\varphi)$.

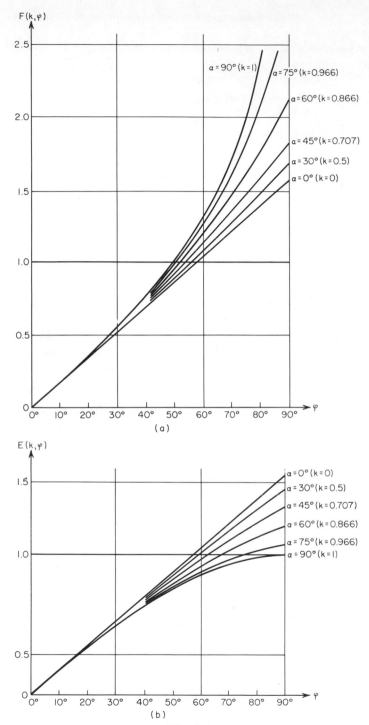

Fig. 5-4

Alternative forms for this function include

$$E(k,\varphi) = \int_0^\varphi \sqrt{1 - \sin^2 \alpha \sin^2 \beta} \; d\beta \qquad (5\text{-}29a)$$

$$E(k,x) = \int_0^x \sqrt{\frac{1 - k^2 z^2}{1 - z^2}} \; dz \qquad (5\text{-}29b)$$

Again, $k = \sin \alpha$ is the modulus, φ is the amplitude, and $x = \sin \varphi$.
When $\varphi = \pi/2$, Eq. (5-29a) becomes

$$E = \int_0^{\pi/2} \sqrt{1 - \sin^2 \alpha \sin^2 \beta} \; d\beta$$

$$= \frac{\pi}{2} \left[1 - \left(\frac{1}{2}\right)^2 k^2 - \left(\frac{1 \cdot 3}{2 \cdot 4}\right)^2 k^4 - \left(\frac{1 \cdot 3 \cdot 5}{2 \cdot 4 \cdot 6}\right)^2 \frac{k^6}{5} - \cdots \right] \qquad (5\text{-}30)$$

It is called the **complete elliptic integral of the second kind.**

Values of $E(k,\varphi)$ are plotted in Fig. 5-4b. Books listed in the Bibliography (Art. 5-8) contain tables for these integrals. Values need be tabulated only for $\varphi \leq \pi/2$, because

$$E(k,\varphi) = 2E - E(k, \pi - \varphi) \qquad (5\text{-}31a)$$

$$E(k, m\pi + \varphi) = 2mE + E(k,\varphi) \qquad (5\text{-}31b)$$

where m is an integer.

Note that the circumference of the ellipse is given by $C = 4Ea$. Suppose, for example, that an ellipse is given by $16x^2 + 25y^2 = 400$, or $x^2/25 + y^2/16 = 1$. Then, $a^2 = 25$, $b^2 = 16$, and the eccentricity of the ellipse $= k = \sqrt{25 - 16}/5 = 0.6 = \sin \alpha$, from which $\alpha = 37°$. From Fig. 5-4b, for $\alpha = 37°$ and $\varphi = \pi/2$, $E = 1.42$. Therefore, since $a = \sqrt{25} = 5$, $C = 4 \times 1.42 \times 5 = 28.4$.

5-6 Gamma Function and Gauss' Pi Function. In Chap. 4, we defined the Laplace transform of $f(t)$ as

$$\mathcal{L}f(t) = F(s) = \int_0^\infty f(t)e^{-st} \, dt$$

In Table 4-2, we listed the transform of t^n as $n!/s^{n+1}$ but restricted n to integers greater than zero. Though the table does not give a formula for transforms for other values of n, such transforms nevertheless exist. But they cannot be expressed in finite combinations of elementary functions. For the purpose, following Euler, we define a nonelementary function, called the gamma function, by

$$\Gamma(n) = \int_0^\infty t^{n-1}e^{-t} \, dt \qquad n > 0 \qquad (5\text{-}32)$$

Integration by parts reveals that

$$\Gamma(n + 1) = n\Gamma(n) \qquad (5\text{-}33)$$

Also, direct integration gives

$$\Gamma(1) = 1$$

Therefore, when n is a positive integer

$$\Gamma(2) = 1 \cdot \Gamma(1) = 1 \qquad \Gamma(3) = 2 \cdot \Gamma(2) = 2 \cdot 1$$
$$\Gamma(4) = 3 \cdot \Gamma(3) = 3 \cdot 2 \cdot 1$$
$$\Gamma(n + 1) = n(n - 1)(n - 2) \cdots 1 = n! \qquad (5\text{-}34)$$

Hence, the gamma function is a generalization of the factorial concept.

Values of the gamma function are plotted in Fig. 5-5. Notice that it is not necessary to restrict n only to positive numbers. Values of the gamma function exist for negative numbers other than integers. $\Gamma(n)$ becomes infinitely large when n is zero or a negative integer.

Tables are available giving values of the gamma function for the interval $1 < n < 2$ (see references in Art. 5-8). For other values of n, $\Gamma(n)$ can be computed by use of the recurrence formula, Eq. (5-33).

Another useful formula is

$$\frac{\Gamma(m)\Gamma(n)}{2\Gamma(m + n)} = \int_0^{\pi/2} \cos^{2m-1} \theta \sin^{2n-1} \theta \, d\theta \qquad m > 0, n > 0 \quad (5\text{-}35a)$$

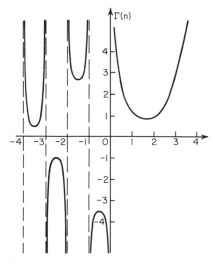

Fig. 5-5

For example, if we set $n = \frac{1}{2}$ and $m = \frac{1}{2}$, we find $\Gamma(\frac{1}{2}) = \sqrt{\pi}$. If $n = \frac{1}{2}$ and $m = (n + 1)/2$, we get

$$\int_0^{\pi/2} \cos^n \theta \, d\theta = \int_0^{\pi/2} \sin^n \theta \, d\theta = \frac{\Gamma(n+1)/2}{\Gamma(n/2+1)} \frac{\sqrt{\pi}}{2} \qquad n > -1$$

$$= \frac{(n-1)(n-3) \cdots 1}{n(n-2) \cdots 2} \frac{\pi}{2}$$

$$n = \text{positive even integer} \qquad (5\text{-}35b)$$

$$= \frac{(n-1)(n-3) \cdots 2}{n(n-2) \cdots 1}$$

$$n = \text{positive odd integer} > 1$$

Still another important formula results when we set $m = 1 - n$:

$$\Gamma(n)\Gamma(1-n) = \frac{\pi}{\sin n\pi} \qquad n \text{ not an integer} \qquad (5\text{-}35c)$$

As an example of the use of these formulas, let us determine $\Gamma(-\frac{1}{2})$. By Eq. (5-33), $\Gamma(\frac{1}{2}) = (-\frac{1}{2})\Gamma(-\frac{1}{2})$. So $\Gamma(-\frac{1}{2}) = -2\Gamma(\frac{1}{2})$. Now, in Eq. (5-35c), let $n = \frac{1}{2}$. Then, we find $\Gamma(\frac{1}{2})\Gamma(1 - \frac{1}{2}) = \pi/\sin \pi/2$, from which $\Gamma(\frac{1}{2}) = \sqrt{\pi}$. Therefore, $\Gamma(-\frac{1}{2}) = -2\sqrt{\pi}$.

Now we can give the transform of t^n when n is not an integer:

$$\mathcal{L}t^n = \frac{\Gamma(n+1)}{s^{n+1}} \qquad n > -1 \qquad (5\text{-}36)$$

Gauss' pi function is also a generalization of the factorial concept. It is related to the gamma function by

$$\Pi(n) = \Gamma(n+1) \qquad (5\text{-}37)$$

When n is a positive integer,

$$\Pi(n) = n! \qquad (5\text{-}38)$$

5-7 Bessel Functions. An important class of nonelementary functions often occurs in the solutions of partial differential equations for phenomena involving circular symmetry. German mathematicians call them die Zylinderfunktionen. We will call them Bessel functions, because they can be derived from Bessel's equation of order n

$$x^2 \frac{d^2y}{dx^2} + x \frac{dy}{dx} + (x^2 - n^2)y = 0 \qquad (5\text{-}39)$$

Bessel functions, however, also arise in the solution of other types of problems. And they may be used to expand a function in an infinite series just as trigonometric functions are used in Fourier series.

When n is not an integer, Bessel's equation has the general solution

$$y = c_1 J_n(x) + c_2 J_{-n}(x) \tag{5-40}$$

where c_1 and c_2 are constants. $J_n(x)$ and $J_{-n}(x)$ are called **Bessel functions of the first kind** of order n.   They are defined by the infinite series

$$J_n(x) = \sum_{m=0}^{\infty} \frac{(-1)^m}{\Pi(m)\Pi(m+n)} \left(\frac{x}{2}\right)^{2m+n} \tag{5-41}$$

$$J_{-n}(x) = \sum_{m=0}^{\infty} \frac{(-1)^m}{\Pi(m)\Pi(m-n)} \left(\frac{x}{2}\right)^{2m-n} \tag{5-42}$$

where $\Pi(m)$ is Gauss' pi function (Art. 5-6).

For some values of n (not an integer), the series is the expansion of an elementary function.   For example,

$$J_{\frac{1}{2}}(x) = \sqrt{\frac{2}{\pi x}} \sin x \tag{5-43}$$

$$J_{-\frac{1}{2}}(x) = \sqrt{\frac{2}{\pi x}} \cos x \tag{5-44}$$

Books listed in the Bibliography (Art. 5-8) contain tables of values of Bessel functions.

If n is an integer, the general solution of Bessel's equation is

$$y = c_1 J_n(x) + c_2 Y_n(x) \tag{5-45}$$

where

$$Y_n(x) = \lim_{r \to n} \frac{J_r(x) \cos r\pi - J_{-r}(x)}{\sin r\pi} \tag{5-46}$$

$Y_n(x)$ is called a **Bessel function of the second kind.**

The Bessel functions of the first kind, $J_1(x)$ and $J_0(x)$, resemble $\sin x$ and $\cos x$, respectively, with damped amplitudes (Fig. 5-6a).   Actually, for large values of x, $J_n(x)$ is given approximately by

$$J_n(x) \approx \sqrt{\frac{2}{\pi x}} \cos \left(x - \frac{\pi}{4} - \frac{n\pi}{2}\right) \tag{5-47}$$

Bessel functions of the second kind become infinitely large for $x = 0$. For large values of x, they are given approximately by

$$Y_n(x) \approx \sqrt{\frac{2}{\pi x}} \sin \left(x - \frac{\pi}{4} - \frac{n\pi}{2}\right) \tag{5-48}$$

$Y_0(x)$ and $Y_1(x)$ are plotted in Fig. 5-6b.

Just as recurrence formulas enable gamma functions to be generated from each other, the following recurrence formulas apply to Bessel

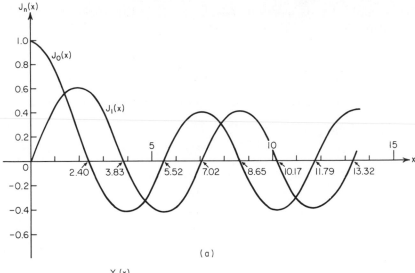

(a)

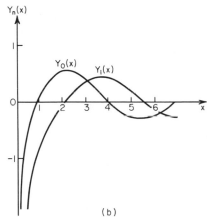

(b)

Fig. 5-6

functions:

$$J_n = \frac{x}{2n} (J_{n-1} + J_{n+1}) \tag{5-49}$$

$$J_0' = -J_1 \qquad J_0' = \frac{d}{dx} J_0(x) \tag{5-50}$$

$$J_n' = \frac{J_{n-1} - J_{n+1}}{2} \qquad J_n' = \frac{d}{dx} J_n(x) \tag{5-51}$$

$$xJ_n' = nJ_n - xJ_{n+1} = -nJ_n + xJ_{n-1} \tag{5-52}$$

$$\frac{d}{dx} (x^{-n}J_n) = -x^{-n}J_{n+1} \tag{5-53}$$

$$\frac{d}{dx} (x^n J_n) = x^n J_{n-1} \tag{5-54}$$

For example, let us evaluate $J_{3/2}(x)$. Solving Eq. (5-49) for J_{n+1} and using Eqs. (5-43) and (5-44), we get

$$J_{3/2}(x) = \frac{2}{x} - \frac{1}{2} J_{1/2}(x) - J_{-1/2}(x) = \sqrt{\frac{2}{\pi x}} \left(\frac{\sin x}{x} - \cos x \right)$$

A more general form of Bessel's equation incorporates a parameter k:

$$\frac{d^2y}{dx^2} + \frac{1}{x}\frac{dy}{dx} + \left(k^2 - \frac{n^2}{x^2} \right) y = 0 \qquad (5\text{-}55)$$

When n is not an integer, the general solution is

$$y = c_1 J_n(kx) + c_2 J_{-n}(kx) \qquad (5\text{-}56)$$

When n is an integer, the general solution is

$$y = c_1 J_n(kx) + c_2 Y_n(kx) \qquad (5\text{-}57)$$

Bessel functions may appear in the solutions to a wide variety of linear differential equations with variable coefficients. One important set has the form

$$z'' + (1 - 2m)x^{-1}z' + [(m^2 - n^2)x^{-2} + k^2]z = 0 \qquad (5\text{-}58)$$

as you may verify by setting $y = x^{-m}z$ in Eq. (5-55). Thus, Eq. (5-58) has the general solution $z = x^m y$, where y is given by Eq. (5-56) or (5-57). If $m = \frac{1}{2}$, the equation takes the form

$$z'' + [k^2 + (\frac{1}{4} - n^2)x^{-2}]z = 0 \qquad (5\text{-}59)$$

with the general solution $z = \sqrt{x}\, y$. Notice that for large values of x, the coefficient of z is approximately equal to k^2, a constant. Hence, the solution of Eq. (5-59) is approximately $y = c_1 \sin kx + c_2 \cos kx$.

Similarly,

$$w'' + c^2 x w = 0 \qquad (5\text{-}60)$$

has the general solution

$$w = x^{1/2}\left[c_1 J_{1/3}\left(\frac{2c}{3} x^{3/2} \right) x^{3/2} + c_2 J_{-1/3}\left(\frac{2c}{3} x^{3/2} \right) \right] \qquad (5\text{-}61)$$

There are also Bessel functions of the third kind and modified Bessel functions. The latter may be derived by setting $k = \sqrt{-1} = i$ in Eqs. (5-56) and (5-57). In addition, new functions, such as ber x, bei x, ker x, and kei x, are derived from their real and imaginary parts. The books listed in Art. 5-8 will provide you with detailed information on these as well as on Bessel functions of the first and second kinds.

5-8 *Bibliography*

M. Abramovitz and I. A. Stegun, "Handbook of Mathematical Functions with Formulas, Graphs, and Mathematical Tables," Superintendent of Documents, Government Printing Office, Washington, D.C.

R. P. Agnew, "Differential Equations," McGraw-Hill Book Company, New York.

M. L. Boas, "Mathematical Methods in the Physical Sciences," John Wiley & Sons, Inc., New York.

R. V. Churchill, "Fourier Series and Boundary Value Problems," McGraw-Hill Book Company, New York.

H. B. Dwight, "Tables of Integrals and Other Mathematical Data," The Macmillan Company, New York.

E. F. Jahnke, F. Emde, and F. Losch, "Tables of Higher Functions," McGraw-Hill Book Company, New York.

Y. L. Luke, "Integrals of Bessel Functions," McGraw-Hill Book Company, New York.

B. O. Peirce and R. M. Foster, "A Short Table of Integrals," Ginn and Company, Boston.

C. R. Wylie, Jr., "Advanced Engineering Mathematics," McGraw-Hill Book Company, New York.

PROBLEMS

1. (a) Expand $\cosh x$ in a Maclaurin series.

   (b) Expand $\log_e x$ in a Taylor series about $x = 1$.

2. Find a solution of $xy'' + y' + xy = 0$ by using a Maclaurin series.

3. Solve $y' = x + y + 1$ by using a Maclaurin series, given $y = 0$ when $x = 0$. (*Hint:* Use the given equation in forming the series.)

4. Solve $2x(1 - x)y'' + (1 + x)y' - y = 0$ by Frobenius' method.

5. Solve $xy'' - y' + x^3y = 0$ by Frobenius' method.

6. Expand in a Fourier series

   (a) $f(t) = \begin{cases} 0 & -\pi < t < 0 \\ 1 & 0 < t < \pi \end{cases}$

   (b) $f(x) = \begin{cases} 0.2x & 0 < x \le 5 \\ 0.2(10 - x) & 5 \le x < 10 \end{cases}$

   Also, show from the solution that $\pi^2 = 8(1 + 1/3^2 + 1/5^2 + 1/7^2 + \cdots)$.

   (c) $f(x) = \begin{cases} 0 & -\pi < x < 0 \\ \sin x & 0 < x < \pi \end{cases}$

7. Suppose that the beam in Fig. 5-2 is subjected to a compressive longitudinal force Q as well as the lateral force P. Find the equation of the elastic curve and approximate maximum deflection of the beam.

8. (a) The root-mean-square value $\bar{y}$ of a function is defined by

$$\bar{y}^2 = \frac{1}{b - a} \int_a^b y^2 \, dx$$

over an interval $a < x < b$. Show that, expanded in a Fourier series over an interval $c < x < c + 2\pi$,

$$\bar{y}^2 = \frac{a_0^2}{4} + \frac{1}{2} \sum_{n=1}^{\infty} (a_n^2 + b_n^2) \qquad n > 0$$

where a_n and b_n are given by Eqs. (5-10) and (5-11), respectively.

(b) In an electric circuit with an alternating current, the effective value of the current I (or voltage E) is the root-mean-square value of I (or E). What is the effective value of the current in terms of I_n if the current is given by $I = \sum_{n=1}^{\infty} I_n \sin (n\omega t + \theta_n)$?

9. (a) Show that if $k > 1$, the elliptic integral of the first kind $F(k,\varphi) = (1/k)F(1/k,x)$, where $\sin x = k \sin \varphi$.

(b) Show that if $k > 1$, the elliptic integral of the second kind $E(k,\varphi) = (1/k - k)F(1/k,x) + kE(1/k,x)$, where $\sin x = k \sin \varphi$.

10. Find the length of one arch of $y = \sin x$.

11. Through what angle does a pendulum swing if it is 9 in. long and its period is one second?

12. Find the length of the lemniscate with equation $\rho^2 = a^2 \cos 2\theta$. (*Hint:* In polar coordinates, $ds = \sqrt{d\rho^2 + \rho^2 \, d\theta^2}$.)

13. A force moving a particle toward another particle (source of attraction) varies inversely as the distance between particles. If initially both particles are at rest at a distance L apart, how long will it take the particles to meet?

14. Given $J_n(x)$ as one solution of Bessel's equation [Eq. (5-39)], show that the general solution is $c_1 J_n(x) + c_2 Y_n(x)$, where

$$Y_n(x) = J_n(x) \int \frac{dx}{x J_n^2(x)}$$

[*Hint:* Assume that the general solution is $X J_n$, substitute in Eq. (5-39), and solve for X.]

15. (a) Express Eq. (5-53) as an integration formula.

(b) Express Eq. (5-54) as an integration formula.

(c) Using the solution to Prob. 15b, determine $\int x J_0(x) \, dx$.

(d) Using the solution to Prob. 15a, determine $\int J_1(x) \, dx$.

(e) Find $\int J_3(x) \, dx$, using the solution to Prob. 15a.

16. (a) Express Eq. (5-51) as an integration formula.

(b) Using the solution to Prob. 16a, show that $\int J_0(x) \, dx = 2[J_1(x) + J_3(x) + J_5(x) + \cdots]$.

17. Compute to four decimal places:

(a) $J_0(1)$.

(b) $J_1(1)$.

(c) $J_{1/2}(\pi/3)$.

(d) $J_{-1/2}(\pi/2)$.

(e) $J_{5/2}(\pi/2)$.

ANSWERS

1. (a) $f(0) = \cosh 0 = 1$. $f'(0) = f^{(2m+1)}(0) = 0$, where m is an integer. $f''(0) = f^{(2m)}(0) = \cosh 0 = 1$. Hence,

$$\cosh x = 1 + \frac{x^2}{2!} + \frac{x^4}{4!} + \cdots + \frac{x^{2n-2}}{(2n-2)!} + \cdots$$

See Art. 5-1.

(b) $f(1) = \log_e 1 = 0$. $f'(x) = 1/x$ and $f'(1) = 1$. $f''(x) = -1/x^2$ and $f''(1) = -1$. $f^{(n)}(x) = (-1)^{n-1}(n-1)!/x^n$ and $f^{(n)}(1) = (-1)^{n-1}(n-1)!$ Therefore,

$$\log_e x = (x - 1) - \frac{1}{2!}(x - 1)^2 + \frac{2!}{3!}(x - 1)^3 - \cdots$$

$$+ \frac{(-1)^{n-1}(n - 1)!}{n!}(x - 1)^n - \cdots$$

$$= (x - 1) - \frac{1}{2}(x - 1)^2 + \frac{1}{3}(x - 1)^3 - \cdots + \frac{(-1)^{n-1}}{n}(x - 1)^n - \cdots$$

$$0 < x < 2$$

See Art. 5-1.

2. Let $y = \Sigma a_n x^n$. Then, $y' = \Sigma n a_n x^{n-1}$ and $y'' = \Sigma n(n - 1)a_n x^{n-2}$. Substitution in the differential equation yields

$$\Sigma n^2 a_n x^{n-1} + \Sigma a_n x^{n+1} = 0$$

Equating coefficients of like powers of x to zero gives $a_1 = 0$, $a_2 = -a_0/4$, $a_3 = -a_1/9 = 0$, $a_4 = -a_2/16 = a_0/4 \cdot 16$, $a_5 = -a_3/25 = 0$, and $a_6 = -a_4/36 = a_0/4 \cdot 16 \cdot 36 \cdots$. Therefore,

$$y = a_0 \left(1 - \frac{x^2}{2^2} + \frac{x^4}{2^2 4^2} - \frac{x^6}{2^2 4^2 6^2} + \cdots \right)$$

See Art. 5-1.

3. The equation requires that $y' = 1$ when $x = y = 0$. Differentiation of the equation yields $y'' = 1 + y'$, $y^{(3)} = y''$, $y^{(n)} = y^{(n-1)} = y''$. Since $y'(0) = 1$, $y''(0) = 1 + 1 = 2$. Therefore, by Eq. (5-1),

$$y = 0 + x + \frac{2}{2!}x^2 + \frac{2}{3!}x^3 + \frac{2}{4!}x^4 + \cdots - 2e^x - x - 2$$

4. Let $y = \Sigma a_n x^{n+c}$. Then, differentiation gives $y' = \Sigma(n + c)a_n x^{n+c-1}$ and $y'' = \Sigma(n + c - 1)(n + x)a_n x^{n+c-2}$. Substitution in the differential equation yields

$$\Sigma[2(n + c)^2 - (n + c)]a_n x^{n+c-1} - \Sigma[2(n + c)^2 - 3(n + c) + 1]a_n x^{n+c} = 0$$

Equating the coefficients of x^{c-1} to zero yields $c = 0, \frac{1}{2}$. For $c = 0$, the equation becomes

$$\Sigma(2n^2 - n)a_n x^{n-1} - \Sigma(2n^2 - 3n + 1)a_n x^n = 0$$

Equating coefficients of like powers of x to zero gives $a_1 = a_0$ and $a_2 = a_3 = a_4 = \cdots = 0$. Therefore, one solution is $y = a_0 + a_0 x$. For $c = \frac{1}{2}$, the equation becomes

$$\Sigma \left[2 \left(n + \frac{1}{2} \right)^2 - \left(n + \frac{1}{2} \right) \right] a_n x^{n-\frac{1}{2}}$$

$$- \Sigma \left[2 \left(n + \frac{1}{2} \right)^2 - 3 \left(n + \frac{1}{2} \right) + 1 \right] a_n x^{n+\frac{1}{2}} = 0$$

Equating coefficients of like powers of x to zero indicates that $a_n = 0$ for $n > 1$. Hence, a second solution is $y = a_0 x^{\frac{1}{2}}$. And the general solution is $y = c_1(1 + x) + c_2 \sqrt{x}$. See Art. 5-3.

5. Let $y = \Sigma a_n x^{n+c}$. Then, differentiation gives $y' = \Sigma (n + c) a_n x^{n+c-1}$ and $y'' = \Sigma (n + c - 1)(n + c) a_n x^{n+c-2}$. Substitution in the differential equation gives

$$\Sigma[(n + c)^2 - 2(n + c)] a_n x^{n+c-1} + \Sigma a_n x^{n+c+3} = 0$$

Equating the coefficient of x^{c-1} to zero yields $c = 0, 2$. For $c = 0$, the equation becomes

$$\Sigma (n^2 - 2n) a_n x^{n-1} + \Sigma a_n x^{n+3} = 0$$

When we equate coefficients of like powers of x to zero, we find that $a_{1+4m} = a_{2+4m} = a_{3+4m} = 0$, where $m \geq 0$ and is an integer. Also, $a_4 = -a_0/2 \cdot 4 = -a_0/2^2 2!$; $a_8 = a_0/2^4 4!$; $a_{12} = -a_0/2^6 6!$ .... Hence, one solution is

$$y = a_0 \left(1 - \frac{x^4}{2^2 2!} + \frac{x^8}{2^4 4!} - \frac{x^{12}}{2^6 6!} + \cdots \right)$$

For $c = 2$, the equation converts to

$$\Sigma n(n + 2) a_n x^{n+1} + \Sigma a_n x^{n+5} = 0$$

When we equate coefficients of like powers of x to zero, we get $a_{1+4m} = a_{2+4m} = a_{3+4m} = 0$, where $m \geq 0$ and is an integer. Also, $a_4 = -a_0/2^2 3!$; $a_8 = a_0/2^4 5!$; $a_{12} = -a_0/2^6 7!$ .... Therefore, a second solution is

$$y = a_0 \left(x^2 - \frac{x^6}{2^2 3!} + \frac{x^{10}}{2^4 5!} - \frac{x^{14}}{2^6 7!} + \cdots \right)$$

The general solution is the sum of the two solutions with a_0 in each case replaced by different constants. See Art. 5-3.

6. (a) $a_0 = \dfrac{1}{\pi} \displaystyle\int_0^\pi dt = 1$ $a_n = 0$

$$b_n = \frac{1}{\pi} \int_0^\pi \sin nt \, dt = \frac{(-1)^{n+1} + 1}{n\pi}$$

where n is an integer. Hence,

$$f(t) = \frac{1}{2} + \frac{2}{\pi}\left(\sin t + \frac{1}{3}\sin 3t + \frac{1}{5}\sin 5t + \cdots\right)$$

See Art. 5-4.

(b) Expand $f(x)$ in a half-range sine series:

$$b_n = \frac{2}{10}\int_0^5 0.2x \sin\frac{n\pi x}{10}\,dx + \frac{2}{10}\int_5^{10} 0.2(10-x)\sin\frac{n\pi x}{10}\,dx$$

$$= \frac{8}{n^2\pi^2}\sin\frac{n\pi}{2}$$

$$f(x) = \frac{8}{\pi^2}\left(\sin\frac{\pi x}{10} - \frac{1}{3^2}\sin\frac{3\pi x}{10} + \frac{1}{5^2}\sin\frac{5\pi x}{10} - \cdots\right)$$

Since $f(x) = 1$ when $x = 5$, $1 = (8/\pi^2)(1 + 1/3^2 + 1/5^2 + \cdots)$. Therefore

$$\pi^2 = 8(1 + 1/3^2 + 1/5^2 + \cdots)$$

(c) $a_0 = \dfrac{1}{\pi}\displaystyle\int_0^\pi \sin x\,dx = \dfrac{2}{\pi}$

$a_1 = \dfrac{1}{\pi}\displaystyle\int_0^\pi \sin x \cos x\,dx = 0$

$a_n = \dfrac{1}{\pi}\displaystyle\int_0^\pi \sin x \cos nx\,dx = \dfrac{1 + \cos n\pi}{\pi(1 - n^2)}\qquad a_n \neq 1$

$b_1 = \dfrac{1}{\pi}\displaystyle\int_0^\pi \sin^2 x\,dx = \dfrac{1}{2}$

$b_n = \dfrac{1}{\pi}\displaystyle\int_0^\pi \sin x \sin nx\,dx = 0\qquad n > 1$

$$f(x) = \frac{1}{\pi} + \frac{1}{2}\sin x - \frac{2}{\pi}\left(\frac{\cos 2x}{3} + \frac{\cos 4x}{15} + \frac{\cos 6x}{35} + \cdots\right)$$

See Art. 5-4.

7. The bending moment $M = M_P + M_Q$, where M_P is the moment determined for P alone, as given in Art. 5-4, and M_Q is the bending moment for Q with deflection y; that is, $M_Q = -Qy$. The differential equation for the elastic curve then is

$$y'' = -\frac{M}{EI} = -\frac{M_P + M_Q}{EI} = -\frac{M_P + Qy}{EI}$$

On transposing M_Q to the left-hand side of the equation, we get

$$\left(D^2 + \frac{Q}{EI}\right)y = -\frac{M_P}{EI}$$

Assume $y = \Sigma y_n \sin n\pi x$ and substitute in the equation.   With M_P as given in Art. 5-4, this gives

$$\sum \left(\frac{Q}{EI} - \frac{n^2\pi^2}{L^2} \right) y_n \sin n\pi x = -\frac{2PL}{\pi^2 EI} \sum \frac{1}{n^2} \sin n\pi k \sin n\pi x$$

Equating coefficients of $\sin n\pi x$, we obtain

$$y_n = \frac{2PL^3 \sin n\pi k}{n^2\pi^4 EI (n^2 - QL^2/\pi^2 EI)}$$

Let $Q_c = \pi^2 EI/L^2$, so that $QL^2/\pi^2 EI = Q/Q_c = \alpha$.   (*Note:* Q_c is the smallest load under which the beam, as a column, will buckle when P is not present.)   Then, $y_n = (2PL^3 \sin n\pi k)/\pi^4 EI n^2(n^2 - \alpha)$ and

$$y = \frac{2PL^3}{\pi^4 EI} \left[\frac{1}{1 - \alpha} \sin \pi k \sin \pi x + \frac{1}{4(4 - \alpha)} \sin 2\pi k \sin 2\pi x + \cdots \right]$$

Let y_P be the maximum deflection of the beam for P alone and assume that the first term of the series gives the deflection approximately.   Then, for both P and Q acting, the maximum deflection is $y_{\max} = y_P/(1 - \alpha)$.

8. (a) Substitute $y = a_0/2 + \Sigma(a_n \cos nx + b_n \sin nx)$ in

$$\bar{y}^2 = \frac{1}{2\pi} \int_c^{c+2\pi} y^2 \, dx$$

Resulting integrals equal zero, except

$$\int_c^{c+2\pi} \frac{a_0^2}{4} \, dx = \frac{\pi a_0^2}{2}$$

$$\int_c^{c+2\pi} a_n^2 \cos^2 nx \, dx = \pi a_n^2$$

$$\int_c^{c+2\pi} b_n^2 \sin^2 nx \, dx = \pi b_n^2$$

Hence, for $n > 0$

$$\bar{y}^2 = \frac{1}{2\pi} \left(\frac{\pi a_0^2}{2} + \sum \pi a_n^2 + \sum \pi b_n^2 \right) = \frac{a_0^2}{4} + \frac{1}{2} \sum (a_n^2 + b_n^2)$$

See Art. 5-4.

(b) $I_n \sin (n\omega t + \theta_n) = I_n \cos \theta_n \sin n\omega t + I_n \sin \theta_n \cos n\omega t$
$$= A_n \cos n\omega t + B_n \sin n\omega t$$

where $A_n = I_n \sin \theta_n$ and $B_n = I_n \cos \theta_n$.   Hence,

$$I = \Sigma A_n \cos n\omega t + \Sigma B_n \sin n\omega t$$

From the solution to Prob. 8a, the effective value of the current is

$$I_e = \sqrt{\frac{1}{2}\Sigma A_n^2 + \frac{1}{2}\Sigma B_n^2} = \sqrt{\frac{1}{2}(I_1^2 + I_2^2 + I_3^2 + \cdots)}$$

9. (a) Differentiation of $\sin x = k \sin \varphi$ gives

$$\cos x \, dx = k \cos \varphi \, d\varphi = k \sqrt{1 - \sin^2 \varphi} \, d\varphi = k \sqrt{1 - \frac{1}{k^2} \sin^2 x} \, d\varphi$$

Solving for $d\varphi$ and substituting in $F(k,\varphi)$, we get

$$F(k,\varphi) = \int_0^\varphi \frac{d\varphi}{\sqrt{1 - k^2 \sin^2 \varphi}} = \int_0^x \frac{\cos x \, dx}{k \sqrt{1 - \frac{1}{k^2} \sin^2 x} \cos x}$$

$$= \frac{1}{k} \int_0^x \frac{dx}{\sqrt{1 - (1/k^2) \sin^2 x}} = \frac{1}{k} F\left(\frac{1}{k}, x\right)$$

See Art. 5-5.

(b) Differentiating $\sin x = k \sin \varphi$ gives

$$\cos x \, dx = k \sqrt{1 - \frac{1}{k^2} \sin^2 x} \, d\varphi$$

as in Prob. 9*a*. Then,

$$E(k,\varphi) = \int_0^\varphi \sqrt{1 - k^2 \sin^2 \varphi} \, d\varphi = \int_0^x \cos x \, \frac{\cos x \, dx}{k \sqrt{1 - (1/k^2) \sin^2 x}}$$

$$= \int_0^x \frac{dx}{k \sqrt{1 - (1/k^2) \sin^2 x}} - \int_0^x \frac{\sin^2 x \, dx}{k \sqrt{1 - (1/k^2) \sin^2 x}}$$

Notice that

$$F\left(\frac{1}{k}, x\right) - E\left(\frac{1}{k}, x\right) = \int_0^x \frac{dx}{\sqrt{1 - (1/k^2) \sin^2 x}} - \int_0^x \sqrt{1 - \frac{1}{k^2} \sin^2 x} \, dx$$

$$= \int_0^x \frac{\sin^2 x \, dx}{k^2 \sqrt{1 - (1/k^2) \sin^2 x}}$$

Consequently,

$$E(k,\varphi) = \frac{1}{k} F\left(\frac{1}{k}, x\right) - k \left[F\left(\frac{1}{k}, x\right) - E\left(\frac{1}{k}, x\right) \right]$$

$$= \left(\frac{1}{k} - k\right) F\left(\frac{1}{k}, x\right) + kE\left(\frac{1}{k}, x\right)$$

See Art. 5-5.

10. Substitute $dy = \cos x \, dx$ in the equation for differential length of arc, $ds^2 = dx^2 + dy^2$, and integrate.

$$L = 2 \int_0^{\pi/2} \sqrt{dx^2 + \cos^2 x \, dx^2} = 2 \int_0^{\pi/2} \sqrt{1 + \cos^2 x} \, dx$$

$$= 2 \sqrt{2} \int_0^{\pi/2} \sqrt{1 - \frac{1}{2} \sin^2 x} \, dx = 2 \sqrt{2} \, E\left(\sqrt{\frac{1}{2}}\right) = 2 \sqrt{2} \times 1.35$$

$$= 3.82$$

See Art. 5-5.

11. Since the period $T = 1 = 4K \sqrt{L/g}$, $K = \frac{1}{4}\sqrt{(32.2 \times 12)/9} = 1.638$. For this value of K, $\alpha = 23.2°$, or $k = \sin \alpha = 0.395$. (See Art. 5-5.) From Fig. 5-3, $\cos \psi = 1 - 2k^2 = 1 - 2(0.395)^2 = 0.688$ and $\psi = 46.5°$. Hence, the pendulum swings through an angle of $2 \times 46.5 = 93°$.

12. $\rho = \pm a \sqrt{\cos 2\theta}$. Differentiating, we get

$$d\rho = \pm \frac{a \sin 2\theta}{\sqrt{\cos 2\theta}} d\theta$$

and $d\rho^2 = [(a^2 \sin^2 2\theta)/(\cos 2\theta)] d\theta^2$. Hence,

$$ds = \sqrt{\frac{a^2 \sin^2 2\theta}{\cos 2\theta} + a^2 \cos 2\theta}\, d\theta = \frac{a\, d\theta}{\sqrt{\cos 2\theta}}$$

Then, the length of the lemniscate is

$$L = 4 \int_0^{\pi/4} \frac{a\, d\theta}{\sqrt{\cos 2\theta}}$$

Two methods of solution are possible. One treats the integral as an elliptic integral. The other uses gamma functions and Eq. (5-35b). By the former method, and Prob. 9a,

$$L = 4a \int_0^{\pi/4} \frac{d\theta}{\sqrt{1 - 2\sin^2\theta}} = \frac{4a}{\sqrt{2}} K\left(\frac{1}{\sqrt{2}}\right) = 5.24a$$

Alternatively, let $\alpha = 2\theta$, so that $L = 2a \int_0^{\pi/2} \cos^{-1/2}\alpha\, d\alpha$. Then, by Eq. (5-35b), with $n = -\frac{1}{2}$, and Eq. (5-33),

$$L = 2a \frac{\Gamma(\frac{1}{4})\sqrt{\pi}}{\Gamma(\frac{3}{4})\, 2} = a\sqrt{\pi}\, \frac{4\Gamma(\frac{5}{4})}{\frac{4}{3}\Gamma(\frac{7}{4})} = 5.24a$$

13. At time t, from $F = Ma$, $-kM/x = M\, dx^2/dt^2$, where k is a positive constant and x is measured from the source of attraction. Now, let $v =$ velocity. Then,

$$\frac{d^2x}{dt^2} = \frac{dv}{dt} = \frac{dv}{dx}\frac{dx}{dt} = \frac{dv}{dx} v$$

Hence, we can write the differential equation as $v\, dv/dx = -k/x$ after dividing both sides by M. The variables are separable. Integration yields $v^2/2 = -k \log_e x + c$. Since $v = 0$ when $x = L$, we find $c = k \log_e L$. Thus, $v^2 = (dx/dt)^2 = 2k \log_e L/x$, from which

$$dt = -\frac{1}{\sqrt{2k}} \frac{dx}{\sqrt{\log_e L/x}}$$

And the time it takes the particles to meet is

$$T = -\frac{1}{\sqrt{2k}} \int_L^0 \frac{dx}{\sqrt{\log_e L/x}}$$

Let $y = \log_e L/x$, or $x = Le^{-y}$. Then, $dx = -Le^{-y}\,dy$. Therefore,

$$T = \frac{1}{\sqrt{2k}} \int_0^\infty \frac{Le^{-y}\,dy}{\sqrt{y}} = \frac{L}{\sqrt{2k}} \int_0^\infty y^{-\frac{1}{2}} e^{-y}\,dy = \frac{L\Gamma(\frac{1}{2})}{\sqrt{2k}} = L\sqrt{\frac{\pi}{2k}}$$

See Art. 5-6.

14. Let $y = XJ_n$. Then, $y' = XJ_n' + X'J_n$ and $y'' = XJ_n'' + 2X'J_n' + J_nX''$. Substituting in Eq. (5-39) and setting $[x^2J_n'' + xJ_n' + (x^2 - n^2)J_n]X = 0$, since J_n satisfies Eq. (5-39), yields $x^2J_nX'' + (2x^2J_n' + xJ_n)X' = 0$. Since X is not present in this equation, we can reduce its order by letting $X' = v$ and $X'' = v'$. The result is $x^2J_nv' + (2x^2J_n' + xJ_n)v = 0$. The variables are separable: $v'/v = -2J_n'/J_n - 1/x$. Integration gives $\log_e v = -2\log_e J_n - \log_e x + \log_e k = \log_e(k/xJ_n^2)$. Hence, $v = k/xJ_n^2 = X'$ and $X = k\int dx/xJ_n^2$. Therefore, the general solution is $y = c_1J_n(x) + c_2Y_n(x)$, where $Y_n(x) = J_n(x)\int dx/xJ_n^2$.

15. (a)

$$\int x^{-n}J_{n+1}(x)\,dx = -x^{-n}J_n(x) + c$$

(b)

$$\int x^nJ_{n-1}(x)\,dx = x^nJ_n(x) + c$$

(c) With $n = 1$,

$$\int xJ_0(x)\,dx = xJ_1(x) + c$$

(d) With $n = 0$,

$$\int J_1(x)\,dx = -J_0(x) + c$$

(e) Integrate by parts: Let $u = x^2$ and $dv = x^{-2}J_3(x)\,dx$. Then, $du = 2x\,dx$, and $v = -x^{-2}J_2(x)$ when we use the solution to Prob. 15a. Hence, $\int J_3(x)\,dx = -J_2(x) + 2\int x^{-1}J_2(x)\,dx$. Using the solution to Prob. 15a again to evaluate the integral, we get

$$\int J_3(x)\,dx = -J_2(x) - 2x^{-1}J_1(x) + c$$

16. (a) $\int J_{n-1}(x)\,dx = 2J_n(x) + \int J_{n+1}(x)\,dx$

(b) Take $n = 1$. $\int J_0(x)\,dx = 2J_1(x) + \int J_2(x)\,dx$. With $n = 3$, $\int J_2(x)\,dx = 2J_3(x) + \int J_4(x)\,dx$ . . . . Hence, $\int J_0(x)\,dx = 2[J_1(x) + J_3(x) + J_5(x) + \cdots]$.

17. (a) Using Eq. (5-41), we get

$$J_0(x) = 1 - \left(\frac{x}{2}\right)^2 + \left(\frac{1}{2!}\right)^2\left(\frac{x}{2}\right)^4 - \left(\frac{1}{3!}\right)^2\left(\frac{x}{2}\right)^6 + \cdots$$

Setting $x = 1$, we find that

$$J_0(1) = 1 - \frac{1}{4} + \frac{1}{64} - \frac{1}{2,304} + \cdots = 0.7652$$

(b) Using Eq. (5-41), we obtain

$$J_1(x) = \frac{x}{2} - \frac{1}{2!}\left(\frac{x}{2}\right)^3 + \frac{1}{2!3!}\left(\frac{x}{2}\right)^5 - \frac{1}{3!4!}\left(\frac{x}{2}\right)^7 + \cdots$$

With $x = 1$, we get

$$J_1(1) = \frac{1}{2} - \frac{1}{16} + \frac{1}{384} - \frac{1}{18,432} + \cdots = 0.4401$$

(c) Using Eq. (5-43), we get $J_{1/2}(\pi/3) = \sqrt{6/\pi^2} \sin \pi/3 = 0.675$.

(d) By Eq. (5-44), $J_{-1/2}(\pi/2) = 0$.

(e) From Eq. (5-49), we obtain $J_{n+1}(x) = \dfrac{2n}{x} J_n(x) - J_{n-1}(x)$.   Hence,

$$J_{5/2}(x) = \frac{3}{x} J_{3/2}(x) - J_{1/2}(x) = \frac{3}{x}\sqrt{\frac{2}{\pi x}}\left(\frac{\sin x}{x} - \cos x\right) - \sqrt{\frac{2}{\pi x}} \sin x$$

(See example in Art. 5-7.)   Therefore,

$$J_{5/2}(\pi/2) = \sqrt{\frac{4}{\pi^2}}\left(\frac{12}{\pi^2} - 1\right) = 0.1374$$

SIX

Numerical Integration of Ordinary Differential Equations

Chapters 3 through 5 presented methods for obtaining solutions of differential equations in terms of elementary and nonelementary functions. But there are many differential equations for which such solutions do not exist or are too difficult to get. These equations often may be solved by numerical methods. These methods yield solutions in the form of tables of data relating the variables. Usually, such tables are not only an acceptable but also a desirable form for solutions to engineering problems.

Numerical methods generally require tedious, lengthy, routine calculations. Many of them, however, are adaptable for easy, rapid solution by high-speed electronic computers. We shall discuss a few of these methods in this chapter.

The subject should not be completely new to you. We considered some of the techniques in previous chapters: Picard's method for first-order differential equations in Art. 3-3 and solution by series in Art. 5-1. Picard's method permits solution by successive approximations. The series methods are applicable when solutions exist that can be expressed in series form.

This chapter deals mainly with numerical .integration of first-order equations.   The reason for this treatment is that higher-order equations can be represented as a system of first-order equations.   Article 6-2 gives a technique for obtaining this result.

In this chapter we will deal mainly with initial-value problems. Because boundary-value and characteristic-value problems are much more complicated, you will find it desirable to study them from specialized texts.   Similarly, you will find it desirable to investigate the effects of errors in approximate methods by reading texts specializing in numerical methods.

6-1   *Finite Differences.* We define a finite difference as a small change in a variable.   As you will observe later, there are three main types of finite differences.

Because they can be used to approximate derivatives, finite differences can help solve differential equations.   Remember that we defined dy/dx as the limit of the ratio $\Delta y/\Delta x$ as Δx approaches zero. Δx is a small increment in the independent variable x, and Δy is the corresponding increment in the dependent variable y.   Thus, it appears logical to use $\Delta y/\Delta x$ as an approximation of dy/dx.

Many methods for numerical integration of differential equations are based on finite differences.   For convenience of solution, however, formulas often are given in terms of the coordinates, eliminating the need for computing differences.   Nevertheless, you will find a knowledge of finite differences helpful, if only for background purposes.   Furthermore, finite differences are also applicable to interpolation, curve fitting, and smoothing of data.

One type of finite differences is called **forward differences.**   To define this type, let us consider $y = f(x)$ and observe the behavior of y when x is given small successive increments equal to a positive constant h. (Methods of numerical integration have been developed to permit use of nonconstant h, but we will not discuss them in this book.)   Let $y = y_n$ when $x = x_n$, and $y = y_{n+1}$ when $x = x_n + h$.   Then, the first forward difference of y at x_n is defined by

$$\Delta y_n = y_{n+1} - y_n \tag{6-1}$$

Thus, the first forward difference of y is the change in y due to an increase of h in x.

In general, first forward differences of y also change with increments in x.   We account for these changes by defining the second forward difference of y at x_n by

$$\Delta(\Delta y_n) = \Delta^2 y_n = \Delta y_{n+1} - \Delta y_n = y_{n+2} - 2y_{n+1} + y_n \tag{6-2}$$

We can define higher-order forward differences similarly. For example, the kth forward difference is given by

$$\Delta^k y_n = y_{n+k} + \sum_{r=1}^{k} (-1)^r C(k,r) y_{n+k-r} \qquad (6\text{-}3)$$

where $C(k,r) = [k(k-1)(k-2) \cdots (k-r+1)]/r!$

Δ may be used as an operator in the same manner as the differential operator $D = d/dx$ (Art. 3-4). For example, we can represent y at any point $x = x_n + mh$ by

$$y_{n+m} = (1+\Delta)^m y_n = y_n + m\,\Delta y_n + \frac{m(m-1)}{2!}\,\Delta^2 y_n + \cdots \qquad (6\text{-}4)$$

D is related to the first forward difference Δ by

$$D = \frac{1}{h}\left(\Delta - \frac{\Delta^2}{2} + \frac{\Delta^3}{3} - \frac{\Delta^4}{4} + \cdots\right) \qquad (6\text{-}5)$$

Higher derivatives can be obtained by taking successive powers of D.

$$D^2 = \frac{1}{h^2}\left(\Delta^2 - \Delta^3 + \frac{11}{12}\Delta^4 - \frac{5}{6}\Delta^5 + \cdots\right) \qquad (6\text{-}6)$$

$$D^3 = \frac{1}{h^3}\left(\Delta^3 - \frac{3}{2}\Delta^4 + \frac{7}{4}\Delta^5 - \cdots\right) \qquad (6\text{-}7)$$

$$D^4 = \frac{1}{h^4}\left(\Delta^4 - 2\Delta^5 + \frac{17}{6}\Delta^6 - \cdots\right) \qquad (6\text{-}8)$$

Hence, Dy_n at $x = x_n$ is given approximately by $(1/h)\,\Delta y_n$. Also, $D^2 y_n$ is approximated by $(1/h^2)\,\Delta^2 y_n$, and $D^m y_n$ by $(1/h^m)\,\Delta^m y_n$.

As a simple example of the use of the preceding formulas, let us compute $\sin 33°$ to five significant figures, given $\sin 30° = 0.500\,000$; $\sin 31° = 0.515\,038$; $\sin 32° - 0.529\,919$. Assume that third forward differences of $\sin 30°$ are small enough to be ignored. Also, from these data, let us determine the first and second derivatives of $\sin x$ when $x = 30°$, noting that $1° = 0.017\,453$ rad.

The first step is to tabulate the given values of $\sin x$ and their forward differences:

Angle	Sine	Δ	Δ^2
30°	0.500 000		
		0.015 038	
31°	0.515 038		-0.000 157
		0.014 881	
32°	0.529 919		

As indicated in the table, forward differences are written in columns and midway between the values differenced. Consequently, the forward differences for a specific value of the independent variable lie along a diagonal. For example, in the table, $\sin 30°$, $\Delta \sin 30° = 0.015\ 038$, and $\Delta^2 \sin 30° = -0.000\ 157$ lie along a diagonal. Note also that each column of differences is obtained by subtracting pairs of values in the column on the left.

To obtain $\sin 33°$, we use Eq. (6-4):

$$\begin{aligned}
\sin 33° &= \sin (30° + 3°) = (1 + \Delta)^3 \sin 30° \\
&= (1 + 3\Delta + 3\Delta^2 + \Delta^3) \sin 30° \\
&= 0.500\ 000 + 3 \times 0.015\ 038 - 3 \times 0.000\ 157 + 0 \\
&= 0.54464
\end{aligned}$$

Comparison with a table of sines indicates this result is exact.

To determine the derivative of $\sin x$ when $x = 30°$, we use the first two terms of Eq. (6-5), with $h = 1° = 0.017\ 453$ rad:

$$\frac{d}{dx} \sin x \bigg]_{30°} = \frac{0.015\ 038 + 0.000\ 157/2}{0.017\ 453} = 0.8661$$

The exact derivative is $\cos 30° = 0.866\ 025$.

The second derivative may be approximated by the first term of Eq. (6-6):

$$\frac{d^2}{dx^2} \sin x \bigg]_{30°} = -\frac{0.000\ 157}{0.000\ 305} = -0.515$$

The correct answer is $-\sin 30° = -0.500\ 000$. The decrease in accuracy is due to the decrease in significant figures in the higher-order differences and the increasing importance of higher-order differences ignored in the calculations.

A second type of finite differences, called **backward differences,** is evaluated for successive decrements h in x. Let $y = y_{n-1}$ when $x = x_n - h$. Then, the first backward difference of y at x_n is defined by

$$\nabla y_n = y_n - y_{n-1} \tag{6-9}$$

The second backward difference of y at x_n is defined by

$$\nabla(\nabla y_n) = \nabla^2 y_n = \nabla y_n - \nabla y_{n-1} = y_n - 2y_{n-1} + y_{n-2} \tag{6-10}$$

And similar to the kth forward difference, the kth backward difference is

$$\nabla^k y_n = y_n + \sum_{r=1}^{k} (-1)^r C(k,r) y_{n-r} \tag{6-11}$$

∇ (del) may be used as an operator in the same manner as Δ and D. For example, we can represent y at any point $x = x_n - mh$ by

$$y_{n-m} = (1 - \nabla)^m y_n \tag{6-12}$$

D is related to ∇ by

$$D = \frac{1}{h}\left(\nabla + \frac{\nabla^2}{2} + \frac{\nabla^3}{3} + \frac{\nabla^4}{4} + \cdots\right) \tag{6-13}$$

Higher derivatives can be obtained by taking successive powers of D.

$$D^2 = \frac{1}{h^2}\left(\nabla^2 + \nabla^3 + \frac{11}{12}\nabla^4 + \frac{5}{6}\nabla^5 + \cdots\right) \tag{6-14}$$

$$D^3 = \frac{1}{h^3}\left(\nabla^3 + \frac{3}{2}\nabla^4 + \frac{7}{4}\nabla^5 + \cdots\right) \tag{6-15}$$

$$D^4 = \frac{1}{h^4}\left(\nabla^4 + 2\nabla^5 + \frac{17}{6}\nabla^6 + \cdots\right) \tag{6-16}$$

Hence, Dy_n at $x = x_n$ is given approximately by $(1/h)\nabla y_n$. Also, $D^2 y_n$ is approximated by $(1/h^2)\nabla^2 y_n$, and $D^m y_n$ by $(1/h^m)\nabla^m y_n$.

Using the preceding formulas, let us compute $\log_e 2.700$, given $\log_e 2.800 = 1.02962$, $\log_e 2.900 = 1.06471$, and $\log_e 3.000 = 1.09861$. Assume that third backward differences of $\log_e 3.000$ are small enough to be ignored. Also, from these data, let us calculate the first and second derivatives of $\log_e x$ when $x = 3.000$.

The first step is to tabulate the given values of $\log_e x$ and their backward differences:

x	$\log_e x$	∇	∇^2
2.800	1.02962	0.03509	
2.900	1.06471	0.03390	-0.00119
3.000	1.09861		

As for forward differences, backward differences are entered in the table in columns and midway between the values differenced. Hence, the backward differences for a specific value of the independent variable lie along a diagonal. For example, in the table, $\log_e 3.000$, $\nabla \log_e 3.000 = 0.03390$, and $\nabla^2 \log_e 3.000 = -0.00119$ lie along a diagonal.

To obtain $\log_e 2.700$, we use Eq. (6-12):

$$\begin{aligned}
\log_e 2.700 &= \log_e (3.000 - 0.300) = (1 - \nabla)^3 \log_e 3.000 \\
&= (1 - 3\nabla + 3\nabla^2 - \nabla^3) \log_e 3.000 \\
&= 1.09861 - 3 \times 0.03390 - 3 \times 0.00119 - 0. \\
&= 0.9933
\end{aligned}$$

A table of logarithms gives the answer to five significant figures as 0.99325.

To determine the derivative of $\log_e x$ when $x = 3.000$, we use the first two terms of Eq. (6-13):

$$\frac{d}{dx} \log_e x \bigg]_3 = \frac{0.03390 - 0.00119/2}{0.1} = 0.3331$$

The exact derivative is $\frac{1}{3} = 0.3333$.

For the second derivative, let us use the first term of Eq. (6-14):

$$\frac{d^2}{dx^2} \log_e x \bigg]_3 = -\frac{0.00119}{0.01} = -0.119$$

The exact second derivative is $-\frac{1}{9} = -0.111$. The decrease in accuracy occurs because there are fewer significant figures in the higher-order differences and because higher-order differences ignored in the calculations are of greater relative importance.

A third type of finite differences, called **central differences,** is evaluated for successive increments and decrements of $h/2$ in x. Let $y = y_{n\pm1}$ when $x = x_n \pm h$, and $y = y_{n\pm\frac{1}{2}}$ when $x = x_n \pm h/2$. Then, the first central difference of y at x_n is defined by

$$\delta y_n = y_{n+\frac{1}{2}} - y_{n-\frac{1}{2}} \tag{6-17}$$

The second central difference is defined by

$$\delta(\delta y_n) = \delta^2 y_n = \delta y_{n+\frac{1}{2}} - \delta y_{n-\frac{1}{2}} = y_{n+1} - 2y_n + y_{n-1} \tag{6-18}$$

And the kth central difference is given by

$$\delta^k y_n = y_{n+k/2} + \sum_{r=1}^{k} (-1)^r C(k,r) y_{n-r+k/2} \tag{6-19}$$

Use of Eqs. (6-17) to (6-19) requires that y be known at $x_n \pm rh/2$ for every value of r. The need for determining y where r is an odd number can be eliminated, however. One way, when the values of y at $x_n \pm rh$ are known, is to compute odd-order central differences at $x_n \pm rh/2$, and even-order central differences at $x_n \pm rh$. If this is done, $\delta y_{r+1/2} = \Delta y_r$, where Δy_r is the forward difference at x_r. Also, $\delta^2 y_r = \Delta^2 y_{r-1}$. Another way is to use average differences for odd central differences; for example, $\delta y_n = \frac{1}{2}(y_{n+1} - y_{n-1})$.

When averaged odd central differences are used, we can relate δ to D by

$$D = \frac{1}{h}\left(\delta - \frac{\delta^3}{6} + \frac{\delta^5}{30} - \cdots \right) \tag{6-20}$$

$$D^2 = \frac{1}{h^2}\left(\delta^2 - \frac{\delta^4}{12} + \frac{\delta^6}{90} - \cdots \right) \tag{6-21}$$

$$D^3 = \frac{1}{h^3}\left(\delta^3 - \frac{\delta^5}{4} + \frac{7\delta^7}{120} - \cdots \right) \tag{6-22}$$

$$D^4 = \frac{1}{h^4}\left(\delta^4 - \frac{\delta^6}{6} + \frac{7\delta^8}{240} - \cdots \right) \tag{6-23}$$

Hence, Dy_n at $x = x_n$ is given approximately by $(1/2h)(y_{n+1} - y_{n-1})$. When m is an odd number, $D^m y_n$ is approximately $(1/2h^m)\delta^{m-1}(y_{n+1} - y_{n-1})$ and when m is an even number, approximately

$$D^m y_n = \frac{\delta^m}{h^m}\, y_n = \frac{1}{h^m}\left[y_{n+m/2} + \sum_{r=1}^{m} (-1)^r C(m,r) y_{n-r+m/2} \right]$$

6-2   Conversion of Higher-order Equations to First-order. In this chapter, we will examine mainly methods for solving numerically first-order equations.   The reason for doing this is that higher-order equations can be converted to a system of first-order equations, which we can then solve by the methods of this chapter.

The conversion technique is simple.   Suppose, for example, that the given differential equation is $ay'' + by' = f(x)$.   Then, let y' be a new variable p.   So $y'' = p'$.   Thus, we can replace the second-order equation by the system of first-order equations

$$y' = p$$
$$ap' + bp = f(x)$$

Suppose, instead, that the given differential equation is third order: $ay^{(3)} + by'' + cy' + dy = f(x)$.   Then, let y'' be a new variable p. Hence, $y^{(3)} = p'$.   Also, let y' be a new variable q; so $p = q'$.   Then, we can replace the third-order equation by the first-order system

$$y' = q$$
$$q' = p$$
$$ap' = f(x) - bp - cq - dy$$

Thus, you replace an nth-order equation by a system of n first-order equations.   You set each of the derivatives below the nth equal to a new variable to obtain $n - 1$ equations.   You get the nth equation by substituting the new variables in the given equation. We will examine a method for solving these equations in Art. 6-9.

6-3 *Methods for Solving First-order Equations Numerically.* In considering the main methods in use for numerical solution of first-order linear equations, we can divide the equations into two classes. One is of the form $y' = f(x)$, or simply integration. We discuss methods for numerical integration in Art. 6-5. The second class has the form $y' = f(x,y)$, with the given information that $y = y_0$ at $x = x_0$.

To solve the latter numerically, we obtain values y_k for y for specific values of x. We will select values of x at intervals h, where h is a positive constant, called step size. The general procedure is first to establish a range of values of x for which values of y are to be found. (It is often convenient to transform the variables so that they are dimensionless.) You then decide on the number of steps, or values of y, to be calculated. This determines the step size. For if N is the number of steps and $a \le x \le b$, then $h = (a - b)/N$. Within limits, the smaller h is, the more accurate the numerical solution is likely to be. Next, we take advantage of the given initial condition, $y = y_0$ at $x = x_0$. We use this information to find y_1 at $x_0 + h$ (or $x_0 - h$). With y_1 known, we then determine y_2, and after that $y_3, y_4, \ldots, y_N$.

To illustrate, let us apply a simple method, attributed to Euler, to the numerical solution of $y' = x + y$, given $(x_0,y_0) = (0,1)$. Suppose that we wish to find y_k when $0 \le x < 0.5$.

We select a step size of $h = 0.10$. Then, we determine a relation between the value y_n of y at any step and y_{n+1}, the value of y at the next step. For Euler's method, we derive this relation from the definition of a definite integral:

$$\int_{x_n}^{x_{n+1}} f(x,y)\, dx = y(x_{n+1}) - y(x_n) \tag{6-24}$$

We assume that $f(x,y)$ is nearly constant between (x_n,y_n) and $(x_n + h, y_{n+1})$, where $y_n = y(x_n)$, $y_{n+1} = y(x_{n+1})$, and $x_n + h = x_{n+1}$. Thus, the integral equals $(x_{n+1} - x_n)f(x_n,y_n) = hf(x_n,y_n)$. Therefore, we can rewrite Eq. (6-24) to get the recursion formula:

$$y_{n+1} = y_n + hf(x_n,y_n) \tag{6-25}$$

To obtain the solution of the given differential equation, $y' = x + y$, we apply Eq. (6-25) repeatedly, starting with $(x_0,y_0) = (0,1)$.

$$y_1 = y_0 + h(x_0 + y_0) = 1 + 0.10(0 + 1) = 1.10$$
$$y_2 = y_1 + h(x_1 + y_1) = 1.10 + 0.10(0.1 + 1.10) = 1.22$$
$$y_3 = y_2 + h(x_2 + y_2) = 1.22 + 0.10(0.2 + 1.22) = 1.36$$
$$y_4 = y_3 + h(x_3 + y_3) = 1.36 + 0.10(0.3 + 1.36) = 1.53$$
$$y_5 = y_4 + h(x_4 + y_4) = 1.53 + 0.10(0.4 + 1.53) = 1.72$$

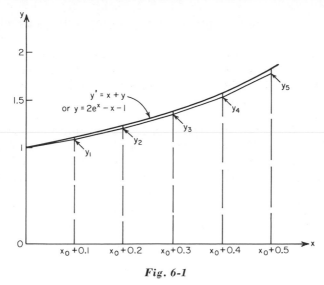

Fig. 6-1

These values are plotted in Fig. 6-1 with the exact solution $y = 2e^x - x - 1$, for comparison.

Other methods strive for greater accuracy, especially when high-speed electronic computers may be used. These methods, in general, are of three types: direct self-starting, iterated self-starting, and aided-starting. The last two usually give greater accuracy for the same step size but may require more calculations. Some methods of the aided-starting type are unstable; that is, errors due to approximations may increase rapidly from step to step.   Also, the influence of step size on accuracy is greater for the aided-starting type than for the iterated self-starting type.   The methods presented in this chapter, however, offer an excellent combination of accuracy, efficiency, and stability and are recommended for automatic computation.

In all cases, the methods require routine calculations, or algorithms, based on recursion formulas, such as Eq. (6-25). These formulas are either explicit or implicit.   Equation (6-25) is an example of an explicit formula.   It gives the quantity to be computed in terms of known quantities only.   Implicit formulas give the quantity to be computed in terms of known quantities and one or more quantities unknown or not accurately determined.   These undetermined quantities must be estimated before the formulas can be used.   Equation (3-11) on which Picard's method is based is an example of an implicit formula.   The integral on the right-hand side includes the variable y, which is to be determined. We might consider the value of y substituted on the right-hand side a predicted value and the result obtained from the formula a corrected value of y.   Some methods provide formulas for getting predicted values also.

Algorithms that require use of both prediction and correction formulas are called predictor-corrector methods.   Usually, the corrected quantity can be substituted in the implicit corrector formula to obtain a more accurate corrected quantity.

6-4   Direct Self-starting Methods. Let us examine first algorithms for solving $y' = f(x,y)$ that can compute y_{n+1} when only x_n and y_n are given. Accuracy of results depends on the step size h chosen.   Within limits, the smaller h, the greater the accuracy; but the greater the amount of calculation required.

Direct self-starting methods are usually derived from a Taylor series expansion of the solution $y(x)$ about (x_n, y_n):

$$y_{n+1} = y_n + hy_n' + \frac{h^2}{2!} y_n'' + \cdots + \frac{h^r}{r!} y_n^{(r)} + T_{n+1} \qquad (6\text{-}26)$$

where T_{n+1} = truncation error resulting from terms ignored.   Or the methods may be derived from the definition of the definite integral, as in Euler's method (Art. 6-3).

When a Taylor series is used, y' is given by the differential equation. Higher derivatives may be obtained by differentiating the equation. Note that y' is a function of both x and y and that the total derivatives are required:

$$\frac{df}{dx} = \frac{\partial f}{\partial x} + \frac{\partial f}{\partial y}\frac{dy}{dx} = \frac{\partial f}{\partial x} + f\frac{\partial f}{\partial y} \qquad (6\text{-}27)$$

$$\frac{d^2f}{dx^2} = \frac{\partial^2 f}{\partial x^2} + 2f\frac{\partial^2 f}{\partial x\,\partial y} + f^2\frac{\partial^2 f}{\partial y^2} + \frac{\partial f}{\partial y}\left(\frac{\partial f}{\partial x} + f\frac{\partial f}{\partial y}\right) \qquad (6\text{-}28)$$

The amount of calculation and complexity of formulas may increase rapidly for higher-order derivatives.   Hence, direct use of Taylor series with higher-order derivatives can be very difficult.   But some self-starting methods use a Taylor series indirectly without computing higher-order derivatives.

Equation (6-25), on which Euler's method is based, is given by the first two terms of Eq. (6-26).   Another method, the Runge-Kutta type, substitutes a polynomial for the first few terms of a Taylor series.   The general procedure for this and other direct self-starting methods is much like that for Euler's method illustrated in Art. 6-3.   Article 6-6 explains the Runge-Kutta method.

6-5   Numerical Integration of Definite Integrals. Before we tackle the solution of the more general first-order equations in detail, let us investigate evaluation of definite integrals, the case where y' is a function of x only. We might consider this a problem in approximating the area under a curve $f(x)$ and between given limits $x = a$ and $x = b$.

Numerical integration methods generally obtain an approximate solution by representing the integral $\int_a^b f(x)\,dx = y$ by

$$y = \sum_i X_i f(x_i) \tag{6-29}$$

In each method, X_i is a coefficient determined to make the error in the calculations zero under certain conditions specified for $f(x)$. Usually, the accuracy of the methods depends on the size h of step into which the interval $b - a$ is divided.

The trapezoidal rule for integration, for example, approximates $f(x)$ in each step by a straight line (Fig. 6-2). So Eq. (6-29) becomes

$$y = \sum \frac{h}{2}\,(f_{i+1} + f_i) \tag{6-30}$$

where $f_i = f(x_i)$ is the value of $f(x)$ at x_i, and $f_{i+1} = f(x_{i+1})$ is the value of $f(x)$ at $x_i + h$. We can expand Eq. (6-30) to give a first approximation to the integral as

$$y_1 = h(\tfrac{1}{2}f_0 + f_1 + f_2 + \cdots + f_{n-1} + \tfrac{1}{2}f_n) \tag{6-31}$$

where $f_0 = f(a)$ and $f_n = f(b)$.

The accuracy of the approximation sometimes may be improved by repeating the calculation with a smaller value of the step to obtain another

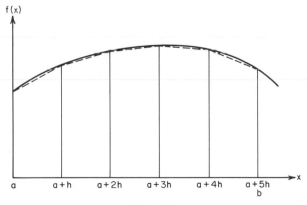

Fig. 6-2

approximation of y, say y_2. Then, a better approximation is usually

$$y = \frac{m^2}{m^2 - n^2} y_2 - \frac{n^2}{m^2 - n^2} y_1 = y_2 + \frac{n^2}{m^2 - n^2} (y_2 - y_1) \qquad (6\text{-}32)$$

where m = number of steps used in determining y_2

$\qquad n$ = number of steps used in determining y_1

For example, let us evaluate

$$y = \int_0^{\pi/2} \sqrt{1 - 0.25 \sin^2 x} \; dx$$

Let us first try $n = 2$, $h = \pi/4$. Thus, if $f = \sqrt{1 - 0.25 \sin^2 x}$, $f_0 = f(0) = 1$, $f_1 = f(\pi/4) = 0.9354$, and $f_2 = f(\pi/2) = 0.8660$. Substitution in Eq. (6-31) gives

$$y_1 = h \left(\frac{1}{2} f_0 + f_1 + \frac{1}{2} f_2 \right)$$

$$= \frac{\pi}{4} \left(\frac{1}{2} \times 1 + 0.9354 + \frac{1}{2} \times 0.8660 \right) = 1.4674$$

Next, let us try $n = 4$, $h = \pi/8$. Then, $f_0 = 1$, $f_1 = 0.9815$, $f_2 = 0.9354$, $f_3 = 0.8868$, and $f_4 = 0.8660$. This time Eq. (6-31) yields

$$y_2 = \frac{\pi}{8} \left(\frac{1}{2} \times 1 + 0.9815 + 0.9354 + 0.8868 + \frac{1}{2} \times 0.8660 \right) = 1.4674$$

Since we obtained the same result in both trials, we can assume that 1.4674 is a close approximation to the integral.

Simpson's rule for integration is another method of evaluating a definite integral. For this method, when the number of steps n is an even number, Eq. (6-29) becomes

$$y = \sum \frac{h}{3} (f_{2i} + 4f_{2i+1} + f_{2i+2}) \qquad (6\text{-}33)$$

We can rewrite this to give a first approximation to the integral as

$$y_1 = \frac{h}{3} (f_0 + 4f_1 + 2f_2 + 4f_3 + \cdots + 2f_{n-2} + 4f_{n-1} + f_n) \qquad (6\text{-}34)$$

When n is an odd number, apply this formula to the first $n - 1$ parts and apply the trapezoidal rule to the last part. Equation (6-33) gives the sum of the areas under second-degree parabolas passed through the points (x_{2i}, f_{2i}), (x_{2i+1}, f_{2i+1}), and (x_{2i+2}, f_{2i+2}), where $i = 0, 1, 2, \ldots,$ $(n - 2)/2$.

Accuracy may be improved sometimes by repeating the calculation with more steps, say m, to obtain another approximation of y, say y_2. Then, you can usually obtain a closer approximation from

$$y = \frac{m^4}{m^4 - n^4} y_2 - \frac{n^4}{m^4 - n^4} y_1 = y_2 + \frac{n^4}{m^4 - n^4} (y_2 - y_1) \qquad (6\text{-}35)$$

As an example, let us evaluate $y = \int_0^1 x^2 e^x \, dx$. Let us first try $n = 4$, $h = \frac{1}{4}$. Equation (6-34) gives, for $f = x^2 e^x$,

$$
\begin{aligned}
y_1 &= \frac{h}{3} (f_0 + 4f_1 + 2f_2 + 4f_3 + f_4) \\
&= \frac{1}{12}(0 + 4 \times 0.080\ 252 + 2 \times 0.412\ 180 \\
&\qquad\qquad\qquad + 4 \times 1.190\ 813 + 2.718\ 282) \\
&= 0.718\ 909
\end{aligned}
$$

Next, let us try $n = 8$, $h = \frac{1}{8}$. This time Eq. (6-34) yields

$$
\begin{aligned}
y_2 &= \frac{1}{24}(0 + 4 \times 0.017\ 705 + 2 \times 0.080\ 252 + 4 \times 0.204\ 608 \\
&\qquad + 2 \times 0.412\ 180 + 4 \times 0.729\ 784 + 2 \times 1.190\ 813 \\
&\qquad\qquad\qquad\qquad + 4 \times 1.836\ 639 + 2.718\ 282) \\
&= 0.718\ 313
\end{aligned}
$$

Finally, Eq. (6-35) gives us the closer approximation

$$y = \frac{8^4}{8^4 - 4^4} 0.718\ 313 - \frac{4^4}{8^4 - 4^4} 0.718\ 909 = 0.71827$$

The exact answer is $e - 2 = 0.718\ 282$.

Romberg's method improves on the trapezoidal rule. The method evaluates a definite integral by successive approximations. In this method, you use Eq. (6-31) repeatedly, doubling the number of steps in successive calculations. Let us call the results $y_1, y_2, \ldots, y_r$, where there are r sets of computations. If any result y_{i+1} is acceptably close to a preceding value y_i, you have obtained the solution and need continue no further. When you consider the step size sufficiently small and still have not obtained a solution, you halt this procedure. Then, to each pair of results, you apply Eq. (6-36) to get improved values $y_1', y_2', \ldots, y_{r-1}'$.

$$y_i' = y_{i+1} - \frac{1}{3} \Delta \qquad (6\text{-}36)$$

Equation (6-36) is obtained from Eq. (6-32) by setting $m = 2n$ and $\Delta = y_i - y_{i+1}$. If any result y_{i+1}' is acceptably close to a preceding result

y_i' or y_{i+2}', you have obtained the solution and need continue no further. Otherwise, to each pair of the last set of results, you apply Eq. (6-37) to get improved values y_1'', y_2'', . . . , y_{r-2}''.

$$y_i'' = y_{i+1}' - \frac{1}{15} \Delta' \qquad\qquad (6\text{-}37)$$

Equation (6-37) is obtained from Eq. (6-35) by setting $m = 2n$ and $\Delta' = y_i' - y_{i+1}'$. If any result y_{i+1}'' is acceptably close to a preceding result y_i'' or y_{i+2}'', you have obtained the solution and need continue no further. If more iteration is required, use an equation of the same form as Eqs. (6-36) and (6-37) with $1/(4^k - 1)$ as the coefficient of Δ for the kth set of improvement calculations.

To illustrate this method, let us solve the preceding problem, which we solved by Simpson's rule. First, we try $n = 1$, $h = 1$. Equation (6-31) gives $y_1 = 1.359\,141$. For $n = 2$, $h = \frac{1}{2}$, we get $y_2 = 0.885\,661$. For $n = 4$, $h = \frac{1}{4}$, we obtain $y_3 = 0.760\,596$. And for $n = 8$, $h = \frac{1}{8}$, we find $y_4 = 0.728\,890$. (See table below.)

Suppose at this stage that we consider h sufficiently small. Yet, there are still substantial changes in successive values of y. Hence, we apply Eq. (6-36) to these results. For the first pair, we get $\Delta = 1.359\,141 - 0.885\,661 = 0.473\,480$, and $\Delta/3 = 0.157\,827$. This yields an improved value $y_1' = 0.885\,661 - 0.157\,827 = 0.727\,834$. Similarly, for $0.885\,661$ and $0.760\,596$, Eq. (6-36) gives $y_2' = 0.718\,908$. And for $0.760\,596$ and $0.728\,890$, we get $y_3' = 0.718\,321$. Successive results are getting closer to each other, but they still are not acceptably close. So now we apply Eq. (6-37) to the last set of results. For the first pair, $\Delta' = 0.727\,834 - 0.718\,908 = 0.008\,926$, and $\Delta'/15 = 0.000\,595$. Thus, we get the improved value $y_1'' = 0.718\,908 - 0.000\,595 = 0.718\,313$. Similarly, for $0.718\,908$ and $0.718\,321$, Eq. (6-37) yields $0.718\,282$. We are getting very close now. Using the last pair of results, we have $\Delta'' = 0.718\,313 - 0.718\,282 = 0.000\,031$, and $\Delta''/(4^3 - 1) = 0.000\,000\,5$, too small to change our last result. Therefore, the solution is $0.718\,282$. Summarizing the calculations, we have

Number of steps	Trapezoid sums	Repeated linear interpolations		
		$k = 1$	$k = 2$	$k = 3$
1	1.359 141			
2	0.885 661	0.727 834		
4	0.760 596	0.718 908	0.718 313	
8	0.728 890	0.718 321	0.718 282	0.718 282

6-6 Runge-Kutta Method for First-order Equations. Let us now examine a direct self-starting method of the Runge-Kutta type for solving numerically $y' = f(x,y)$. Formulas for the Runge-Kutta methods are derived by approximating a Taylor series by a polynomial. The polynomial most frequently used is based on a Taylor series up to and including the fourth derivative of $f(x,y)$.

Fourth-order Runge-Kutta methods are often used to start aided-starting methods of solving differential equations, by methods providing the first few terms of the solution. Runge-Kutta methods, however, can be used for a complete solution. But they are not as efficient for continuing after the first few terms as some of the aided-starting methods. Runge-Kutta methods require determination of the value of $f(x,y)$ at four points for each step of the solution.

With the initial condition y_0 given at x_0, values of y for other values of x may be computed from the fourth-order Runge-Kutta recursion formula:

$$y_{n+1} = y_n + \frac{k_1 + 2k_2 + 2k_3 + k_4}{6} \tag{6-38}$$

where $k_1 = hf(x_n, y_n)$
$\quad\quad k_2 = hf(x_n + h/2,\ y_n + k_1/2)$
$\quad\quad k_3 = hf(x_n + h/2,\ y_n + k_2/2)$
$\quad\quad k_4 = hf(x_n + h,\ y_n + k_3)$
$\quad\quad h =$ step size

Notice that h need not have the same value in successive steps. Also, k_1 must be computed before k_2, k_2 before k_3, and k_3 before k_4.

As an example, let us solve $y' = x + y$, for $0 \leq x \leq 0.5$, given $y_0 = 1$ when $x_0 = 0$. Assume a step size $h = 0.1$. For the first step, $n = 0$, and we seek y_1. We compute k_1, k_2, k_3, and k_4 for substitution in Eq. (6-38). Thus, $k_1 = 0.1f(x_0, y_0) = 0.1(x_0 + y_0) = 0.1(0 + 1) = 0.1000$. $k_2 = 0.1f(x_0 + 0.05,\ y_0 + k_1/2) = 0.1(0.05 + 1 + 0.1000/2) - 0.1100$. $k_3 = 0.1f(x_0 + 0.05,\ y_0 + k_2/2) = 0.1(0.05 + 1 + 0.1100/2) = 0.1105$. $k_4 = 0.1f(x_0 + 0.1, y_0 + k_3) = 0.1(0.1 + 1 + 0.1105) = 0.1211$. Substitution of these values in Eq. (6-38) gives

$$y_1 = \frac{0.1000 + 2 \times 0.1100 + 2 \times 0.1105 + 0.1211}{6} + 1 = 1.1104$$

These results are entered on the first line of Table 6-1.

For the second step, $n = 1$. Using the value of y_1 just obtained, we calculate k_1, k_2, k_3, and k_4, and obtain $y_2 = 1.2429$ from Eq. (6-38). You can continue in the same manner to get y_3, y_4, and y_5. The results are

TABLE 6-1   Solution of $y' = x + y$ by the Runge-Kutta Method

x	k_1	k_2	k_3	k_4	y	Exact y
0.1	0.1000	0.1100	0.1105	0.1211	1.1104	1.110 342
0.2	0.1210	0.1321	0.1326	0.1443	1.2429	1.242 806
0.3	0.1443	0.1565	0.1571	0.1700	1.3998	1.399 718
0.4	0.1700	0.1835	0.1842	0.1984	1.5838	1.583 650
0.5	0.1984	0.2133	0.2140	0.2298	1.7976	1.797 442

shown in Table 6-1. For comparison, the exact solution to seven significant figures, obtained from $y = 2e^x - x - 1$, is included.

6-7   *Iterated Self-starting Method.* Accuracy of some direct self-
starting methods can be improved considerably by including a procedure
of converging approximations, or iteration. The Romberg-Euler iterated
self-starting method for solving $y' = f(x,y)$ is a good example. It is
based on the simple Euler recursion formula

$$y_{n+1} = y_n + hf(x_n,y_n) \tag{6-25}$$

discussed in Art. 6-3. But iteration is applied to obtain very efficiently
an accuracy rivaling the best predictor-corrector methods. The Romberg-Euler method, in addition, has the advantage that accuracy does
not depend as much as in the other methods on a choice of step size h.

The iteration procedure resembles Romberg's improvement of the
trapezoidal rule for numerical integration. In that method, the integral
is evaluated approximately for several step sizes, each step being half the
size of the preceding one. Then, a formula is used to predict the value
toward which the results are converging. The Romberg-Euler method
applies the halving and prediction concepts to each step.

The method starts with the selection of a convenient step size h.
Then, given y_0 at x_0, apply Eq. (6-25) to the first step and obtain a first
approximation to y_1. Let us call this result Y_1, where the subscript
indicates first approximation. Divide the step into two equal intervals,
and apply Eq. (6-25) to them to obtain Y_2, a second approximation to
y_1. If the results are acceptably close, you need proceed no further.
If not, apply Eq. (6-39) to Y_1 and Y_2 to get a first predicted value for
the solution Y_1'.

$$Y_i' = Y_{i+1} - \Delta \tag{6-39}$$

where $\Delta = Y_i - Y_{i+1}$.

Next, divide the step into four equal intervals, and apply Eq. (6-25)
to them to obtain Y_3, a third approximation to y_1. If Y_3 is acceptably

close to Y_2, you need proceed no further; for Y_3 is the solution.   If not, apply Eq. (6-39) to Y_2 and Y_3 to get a second predicted value for the solution Y'_2.   If Y'_2 and Y'_1 are acceptably close, you need not continue; for Y'_2 is the solution.   If not, apply Eq. (6-40) to get Y''_1, the first of a second generation of predicted values of the solution.

$$Y''_i = Y'_{i+1} - \frac{1}{3}\Delta' \tag{6-40}$$

where $\Delta' = Y'_i - Y'_{i+1}$.

Now, divide the step into eight equal parts.   Apply Eq. (6-25) to each part to get Y_4.   If the result is acceptably close to Y_3, you need not continue; Y_4 is the answer.   If not, apply Eq. (6-39) to Y_3 and Y_4 to obtain Y'_3.   If Y'_3 and Y'_2 are acceptably close, you need proceed no further; Y'_3 is the solution.   If not, apply Eq. (6-40) to Y'_2 and Y'_3 to get Y''_2.   If Y''_2 is acceptably close to Y''_1, then Y''_2 is the answer.

If additional generations of predicted values of the solution are necessary, you can use an equation of the form of Eqs. (6-39) and (6-40) with $1/(2^k - 1)$ as the coefficient of Δ, for the kth generation.

After you obtain y_1, repeat the procedure for the following steps to obtain y_2, y_3, ...

As an example, let us solve $y' = x + y$ in the range $0 \le x \le 0.3$, given $y_0 = 1$ at $x_0 = 0$.   Assume a step size $h = 0.1$.   First, we will compute y_1 at $x = 0.1$.   We start with an interval $h_1 = 0.1$.   Using Eq. (6-25), we find $Y_1 = 1 + 0.1(0 + 1) = 1.1000$.   Then, we cut the interval in half and apply Eq. (6-25) to each in succession: $Y_2 = [1 + 0.05(0 + 1) = 1.0500] + 0.05(0.05 + 1.0500) = 1.1050$.   Since Y_1 and Y_2 differ considerably, we compute $\Delta = 1.1000 - 1.1050 = -0.0050$, and use Eq. (6-39) to get a first prediction of the solution: $Y'_1 = 1.1050 + 0.0050 = 1.1100$.   These results are entered in Table 6-2.

Now, let us cut the interval in half again and apply Eq. (6-25) to each in succession.   This calculation yields $Y_3 = 1.1076$.   Since this value is not sufficiently close to Y_2, we find $\Delta = 1.1050 - 1.1076 = -0.0026$. With this value of Δ, we determine from Eq. (6-39), $Y'_2 = 1.1076 + 0.0026 = 1.1102$.   We could consider this close enough to Y'_1 and stop at this stage.

Or we could start a second generation of predicted values.   In that case, we compute from Eq. (6-40), with $\Delta' = 1.1100 - 1.1102 = 0.0002$, $Y''_1 = 1.1102 + \frac{1}{3} \times 0.0002 = 1.1103$.   The preceding results are entered in Table 6-2.   Next, we repeat the calculations with the step divided into eight parts.   As soon as two values in any column are close enough for our purposes, we have the desired value of y_1.

We then follow the same procedure to determine y_2 and y_3.   The

TABLE 6-2   **Solution of** $y' = x + y$ **by the Romberg-Euler Method**

Interval	Y	Y'	Y''	Exact y_i
		Step 1. $x = 0.1$		
0.100	1.1000			
0.050	1.1050	1.1100		
0.0250	1.1076	1.1102	1.1103	1.110 342
		Step 2. $x = 0.2$		
0.100	1.2313			
0.050	1.2368	1.2423		
0.025	1.2398	1.2428	1.2430	1.242 806
		Step 3. $x = 0.3$		
0.100	1.3871			
0.050	1.3931	1.3991		
0.025	1.3965	1.3999	1.4002	1.399 718

results of these calculations are also given in Table 6-2, along with the exact answers to seven significant figures, for comparison.

6-8 *Aided-starting Methods.* Several predictor-corrector methods have been developed for solving $y' = f(x,y)$ with y_0 given at x_0. They generally require fewer calculations than do self-starting methods for the same accuracy. But the recursion formulas used for prediction require two or more values of y, at successive steps, to be known. Some other method must be used to determine these values so a predictor-corrector method can start.

Predictor formulas are often developed by using backward differences of known values of y' to approximate $f(x,y)$ by a polynomial. Integration of the polynomial leads to a recursion formula. When ordinates y_i are substituted for the backward differences, the result is an explicit formula giving y_{n+1} in terms of y_n, y'_n, y'_{n-1}, . . . , and step size h.

Corrector formulas similarly are based on backward differences of known values of y', except that $f(x_{n+1},y_{n+1})$ is included. With ordinates y_i substituted for the backward differences, these implicit formulas give y_{n+1} in terms of y_n, y'_{n+1}, y'_n, y'_{n-1}, . . . , and step size h. If the resulting value of y_{n+1} differs considerably from the predicted value, the corrected value can be substituted in the corrector formula to obtain an improved value. The procedure can be repeated until convergence of two successive values of y_{n+1}.

One example of the predictor-corrector type is the **Adams-Moulton method.** It offers an excellent combination of accuracy and stability with relatively little computation. Predictor and corrector formulas have been derived to include the first backward differences, second backward differences, and much higher-order differences. For example, with the first and second backward differences included, the predictor is

$$y_{n+1}^p = y_n + \frac{h}{12} (23f_n - 16f_{n-1} + 5f_{n-2}) \tag{6-41}$$

where $f = f(x,y)$. And the corrector is

$$y_{n+1}^c = y_n + \frac{h}{12} (5f_{n+1} + 8f_n - f_{n-1}) \tag{6-42}$$

As an example, let us solve $y' = x + y$ in the range $0 \le x \le 0.5$, given $y_0 = 1$ at $x_0 = 0$. Assume $h = 0.1$.

Before we can use Eq. (6-41), we need values for $f(x,y)$ at x_0, $x_1 = x_0 + 0.1$, and $x_2 = x_0 + 0.2$. From the given initial condition, we find at once $y' = f(x_0,y_0) = x_0 + y_0 = 0 + 1 = 1$. But to determine $f_1 = f(x_1,y_1)$ and $f_2 = f(x_2,y_2)$, we have to obtain y_1 and y_2 by a self-starting method. The fourth-order Runge-Kutta method (Art. 6-6) is suitable for the purpose. Let us assume that this method yields $y_1 = 1.1103$ and $y_2 = 1.2428$. Substitution in the given differential equation then gives $f_1 = 0.1 + 1.1103 = 1.2103$ and $f_2 = 0.2 + 1.2428 = 1.4428$. These values are entered in Table 6-3.

Now, we can use predictor formula (6-41) to obtain a first approximation of y_3:

$$y_3^p = y_2 + \frac{0.1}{12} (23f_2 - 16f_1 + 5f_0)$$

$$= 1.2428 + \frac{0.1}{12} (23 \times 1.4428 - 16 \times 1.2103 + 5 \times 1) - 1.3996$$

With this result, we get a first approximation to $f_3: f_3^p = 0.3 + 1.3996 = 1.6996$. Using this value in the corrector formula (6-42), we get

$$y_3^c = y_2 + \frac{0.1}{12} (5f_3 + 8f_2 - f_1)$$

$$= 1.2428 + \frac{0.1}{12} (5 \times 1.6996 + 8 \times 1.4428 - 1.2103)$$

$$= 1.3997$$

This result is so close to the predicted value that we can accept it as y_3. Substitution in the corrector formula would not give greater accuracy.

TABLE 6-3 Solution of $y' = x + y$ by
the Adams-Moulton Method

x	y^p	f^p	y	$y' = f(x,y)$	Exact y
0			1.0000	1.0000	1.000 000
0.1			1.1103	1.2103	1.110 342
0.2			1.2428	1.4428	1.242 806
0.3	1.3996	1.6996	1.3997	1.6997	1.399 718
0.4	1.5835	1.9835	1.5836	1.9836	1.583 650
0.5	1.7973	2.2973	1.7974		1.797 442

With y_1, y_2, and y_3 known, we can obtain a predicted value for y_4 from Eq. (6-41). With this value, we get a more accurate result from Eq. (6-42). Then, with y_4 known, we can use the same procedure to obtain y_5. The results are summarized in Table 6-3. For comparison, the exact values of y to seven significant figures, obtained from the solution $y = 2e^x - x - 1$, also are given in the table.

6-9 *Numerical Solution of Higher-order Equations and Simultaneous Equations.* Article 6-2 showed how an nth-order differential equation $y^{(n)} = f(x,y,y', \ldots ,y^{(n-1)})$ can be transformed into a system of n first-order equations. This article will show how to solve such a system of equations and thus how to solve an nth-order differential equation. The method can also be used to solve the more general case of simultaneous differential equations.

Let us assume that in the general case we can express the given equations in the form of a system of first-order equations:

$$\frac{dx_1}{dt} = f_1(t,x_1,x_2, \ldots ,x_N)$$

$$\frac{dx_2}{dt} = f_2(t,x_1,x_2, \ldots ,x_N) \tag{6-43}$$

$$\cdots \cdots \cdots \cdots \cdots \cdots \cdots \cdots$$

$$\frac{dx_N}{dt} = f_n(t,x_1,x_2, \ldots ,x_N)$$

Let us assume also that the initial conditions give us x_1, x_2, $\ldots$, x_N at $t = t_0$. We can then select a step size h for t and find values for the variables by a direct self-starting method, such as the fourth-order Runge-Kutta method (Art. 6-6). After we have found values for the variables for a few steps, we can continue with an aided-starting method, such as the Adams-Moulton method (Art. 6-8).

Suppose that you decide to use the Runge-Kutta method for starting.

Then, you will compute values for the variables from Eq. (6-38). But now there will be a set k_1, k_2, k_3, and k_4 for each $f_i = f_i(t, x_1, x_2, \ldots, x_N)$ in Eq. (6-43). And the members of each set must be calculated in the proper order, as was done for a single first-order equation in Art. 6-6. For example, for each variable x_i, you first compute $k_{1i} = hf_i(t_n, x_{1n}, \ldots, x_{Nn})$. Then, you can calculate $k_{2i} = hf_i(t_n + h/2, x_{1n} + k_{11}/2, \ldots, x_{Nn} + k_{1N}/2)$; after that, k_{3i} and k_{4i}.

To illustrate the method, let us solve, with step size $h = 0.02$, the second-order equation $y'' = -25y - y^3$, given $y = 1$ and $y' = 0$ when $t = 0$. First, we transform the given equation into two first-order equations. We introduce a new variable z and set

$$y' = z \tag{6-44}$$
$$z' = -25y - y^3 \tag{6-45}$$

Note that according to the initial conditions, $y_0 = 1$ and $z_0 = 0$. We now apply the Runge-Kutta method to find y_1, z_1, y_2, and z_2.

To obtain y_1 and z_1, we start by computing k_{11} and k_{12}. For the purpose, we set $f_1 = z$ and $f_2 = -25y - y^3$. We then extend the definition of k_1 of Eq. (6-38) to several variables:

$$k_{11} = hf_1(t_0, y_0, z_0) = hz_0 = 0$$
$$k_{12} = hf_2(t_0, y_0, z_0) = h(-25y_0 - y_0{}^3) = 0.02(-25 - 1) = -0.5200$$

With k_{1i} known, we can now compute k_{2i}. We extend the definition of k_2 of Eq. (6-38) to several variables to get

$$k_{21} = hf_1\left(t_0 + \frac{h}{2}, y_0 + \frac{k_{11}}{2}, z_0 + \frac{k_{12}}{2}\right)$$
$$- 0.02\left(z_0 + \frac{k_{12}}{2}\right) - 0.02\left(0 - \frac{0.5200}{2}\right) = -0.0052$$

$$k_{22} = hf_2\left(t_0 + \frac{h}{2}, y_0 + \frac{k_{11}}{2}, z_0 + \frac{k_{12}}{2}\right)$$
$$= 0.02[-25(y_0 + 0) - (y_0 + 0)^3] = 0.02(-25 - 1) = -0.5200$$

Similarly, with these values we can compute k_{3i}:

$$k_{31} = hf_1\left(t_0 + \frac{h}{2}, y_0 + \frac{k_{21}}{2}, z_0 + \frac{k_{22}}{2}\right)$$
$$= 0.02\left(z_0 + \frac{k_{22}}{2}\right) = 0.02\left(0 - \frac{0.5200}{2}\right) = -0.0052$$

$$k_{32} = hf_2\left(t_0 + \frac{h}{2}, y_0 + \frac{k_{21}}{2}, z_0 + \frac{k_{22}}{2}\right)$$
$$= 0.02\left[-25\left(1 - \frac{0.0052}{2}\right) - \left(1 - \frac{0.0052}{2}\right)^3\right] = -0.5185$$

And with these values we can calculate k_{4i}:

$$k_{41} = hf_1(t_0 + h, y_0 + k_{31}, z_0 + k_{32}) = 0.02(0 - 0.5185) = -0.0104$$
$$k_{42} = hf_2(t_0 + h, y_0 + k_{31}, z_0 + k_{32})$$
$$= 0.02[-25(1 - 0.0052) - (1 - 0.0052)^3] = -0.5171$$

Finally, substitution of these results in Eq. (6-38) yields

$$y_1 = y_0 + \frac{k_{11} + 2k_{21} + 2k_{31} + k_{41}}{6}$$

$$= 1 + \frac{0 + 2(-0.0052) + 2(-0.0052) - 0.0104}{6} = 0.9948$$

$$z_1 = z_0 + \frac{k_{12} + 2k_{22} + 2k_{32} + k_{42}}{6}$$

$$= 0 + \frac{-0.5200 + 2(-0.5200) + 2(-0.5185) - 0.5171}{6}$$

$$= -0.5190$$

With the known values of y_1 and z_1, we continue with the Runge-Kutta method in the same manner to get y_2 and z_2. As in the first step, we start with

$$k_{11} = hz_1 = 0.02(-0.5190) = -0.0104$$
$$k_{12} = h(-25y_1 - y_1^3) = 0.02(-25 \times 0.9948 - 0.9948^3) = -0.5171$$

Now, we can compute

$$k_{21} = h\left(z_1 + \frac{k_{12}}{2}\right) = 0.02\left(-0.5190 - \frac{0.5171}{2}\right) = -0.0156$$

$$k_{22} = h\left[-25\left(y_1 + \frac{k_{11}}{2}\right) - \left(y_1 + \frac{k_{11}}{2}\right)^3\right]$$

$$= 0.02\left[-25\left(0.9948 - \frac{0.0104}{2}\right) - \left(0.9948 - \frac{0.0104}{2}\right)^3\right]$$

$$= -0.5142$$

With these values, we calculate

$$k_{31} = h\left(z_1 + \frac{k_{22}}{2}\right) = 0.02\left(-0.5190 - \frac{0.5142}{2}\right) = -0.0155$$

$$k_{32} = h\left[-25\left(y_1 + \frac{k_{21}}{2}\right) - \left(y_1 + \frac{k_{21}}{2}\right)^3\right]$$

$$= 0.02\left[-25\left(0.9948 - \frac{0.0156}{2}\right) - \left(0.9948 - \frac{0.0156}{2}\right)^3\right]$$

$$= -0.5127$$

And then we obtain

$$k_{41} = h(z_1 + k_{32}) = 0.02(-0.5190 - 0.5127) = -0.0206$$
$$k_{42} = h[-25(y_1 + k_{31}) - (y_1 + k_{31})^3]$$
$$= 0.02[-25(0.9948 - 0.0155) - (0.9948 - 0.0155)^3] = -0.5084$$

Finally, substitution in Eq. (6-38) gives

$$y_2 = 0.9948 + \frac{-0.0104 - 2 \times 0.0156 - 2 \times 0.0155 - 0.0206}{6}$$

$$= 0.9793$$

$$z_2 = -0.5190 + \frac{-0.5171 - 2 \times 0.5142 - 2 \times 0.5127 - 0.5084}{6}$$

$$= -1.0322$$

At this stage, we have enough known values of y and z to shift to the Adams-Moulton method. We start with the predictor Eq. (6-41):

$$y_3{}^p = y_2 + \frac{h}{12} (23z_2 - 16z_1 + 5z_0)$$

$$= 0.9793 + \frac{0.02}{12} [23(-1.0322) - 16(-0.5190)]$$

$$= 0.9536$$

Now, we can evaluate $z' = f_2$ at $y_3{}^p$, y_2, and y_1 for use in corrector Eq. (6-42). At $y_3{}^p$, we have $f_2 = -25 \times 0.9536 - 0.9536^3 = -24.7743$. At y_2, $f_2 = 25.4216$, and at y_1, $f_2 = -25.8544$. Substitution in Eq. (6-42) gives

$$z_3 = -1.0322 + \frac{0.02}{12} [5(-24.7743)$$

$$+ 8(-25.4216) + 25.8544] = -1.5345$$

To obtain an improved value of y_3, we substitute this value for z_3 in Eq. (6-42):

$$y_3{}^c = 0.9793 + \frac{0.02}{12} [5(-1.5345) + 8(-1.0322) + 0.5190] = 0.9536$$

Since this value is the same as $y_3{}^p$, we now have the final values of y_3 and z_3. We can continue in the same way to get additional values of y and z.

6-10   *Boundary-value Problems.* In the previous articles of this chapter, we discussed the solution of differential equations with condi-

tions given at only one point.   When conditions are given at more than
one point, you have a boundary-value problem, and the numerical integration is more complex.   In general, you can get the solution by expressing
the derivatives in terms of central differences [Eqs. (6-20) to (6-23) or
the approximations given after Eq. (6-23)].   You then use the initial
conditions to obtain the values of the variables at each step, by trial and
error, by interpolation, or by setting up simultaneous equations.

The procedure can be explained with an example.   Let us solve
$y'' - (18x + 4)y = -(18x + 4)e^x$ in the interval $0 \leq x \leq 1$, given $y = 0$
when $x = 0$, $y = 1$ when $x = 1$.   For illustrative purposes, we will use
a step size $h = 0.25$.   But we would get more accurate results with a
smaller h.   For $h = 0.25$, we have $x_0 = 0$, $x_1 = 0.25$, $x_2 = 0.50$, $x_3 = 0.75$,
and $x_4 - 1$.   Using central differences, we approximate y'' by

$$y'' = \frac{1}{h^2}(y_{m+1} - 2y_m + y_{m-1})$$

We then can express the given differential equation as a difference equation.   After we multiply both sides by $h^2 = 0.0625$, this equation becomes

$$y_{m+1} - 2y_m + y_{m-1} - 0.0625(18x_m + 4)y_m = -0.0625(18x_m + 4)e^{x_m}$$

Collecting like terms of y, we get

$$y_{m+1} - (1.125x_m + 2.25)y_m + y_{m-1} = -(1.125x_m + 0.25)e^{x_m}$$

Now, we make use of the given conditions $(x_0,y_0) = (0,0)$ and $(x_4,y_4) =
(1,1)$.   We set $m = 1$, 2, and 3 and get three equations:

$$y_2 - (1.125 \times 0.25 + 2.25)y_1 + 0 = -(1.125 \times 0.25 + 0.25)e^{0.25}$$
$$y_3 - (1.125 \times 0.50 + 2.25)y_2 + y_1 = -(1.125 \times 0.50 + 0.25)e^{0.50}$$
$$1 - (1.125 \times 0.75 + 2.25)y_3 + y_2 = -(1.125 \times 0.75 + 0.25)e^{0.75}$$

Their simultaneous solution yields $y_1 = 0.7752$, $y_2 = 1.2801$, and
$y_3 = 1.4854$.

6-11   *Characteristic-value Problems.* Suppose that you are solving a
boundary-value problem by the method given in Art. 6-10 and the given
differential equation contains a parameter.   You may find that the
resulting algebraic equations will have a nontrivial solution only for
specific values of the parameter.   In that case, you have a characteristic-
value problem.   Set the determinant of the coefficients of the unknowns
in the equations equal to zero and solve for the parameter.

For example, let us solve for the smallest value of K that will satisfy

$$\frac{d^4y}{dx^4} + K\frac{d^2y}{dx^2} = 0$$

given $y = y' = 0$ when $x = 0$, and $y = y'' = 0$ when $x = 1$. For illustrative purposes, we will use a step size $h = 0.25$. But more accurate results will be obtained with smaller h. Using central differences, we approximate the derivatives by

$$\frac{d^4y}{dx^4} = \frac{1}{h^4}(y_{m-2} - 4y_{m-1} + 6y_m - 4y_{m+1} + y_{m+2})$$

$$\frac{d^2y}{dx^2} = \frac{1}{h^2}(y_{m-1} - 2y_m + y_{m+1})$$

We can then express the differential equation as a difference equation. After we multiply both sides of the equation by h^4 and set $k = Kh^2$, we get

$$y_{m-2} - 4y_{m-1} + 6y_m - 4y_{m+1} + y_{m+2} + k(y_{m-1} - 2y_m + y_{m+1}) = 0$$

Collecting like terms of y, we obtain

$$y_{m-2} + (k-4)y_{m-1} - (2k-6)y_m + (k-4)y_{m+1} + y_{m+2} = 0$$

The given conditions require that $y_0 = 0$; $y_1 - y_{-1} = 0$; $y_4 = 0$; $y_5 - 0 + y_3 = 0$. Using these conditions, we set $m = 1$, 2, and 3 to obtain three equations:

$$y_1 + 0 - (2k-6)y_1 + (k-4)y_2 + y_3 = 0$$
$$0 + (k-4)y_1 - (2k-6)y_2 + (k-4)y_3 + 0 = 0$$
$$y_1 + (k-4)y_2 - (2k-6)y_3 + 0 - y_3 = 0$$

If these equations are not to have the trivial solution $y_m = 0$, the determinant of the coefficients of y_m must be zero.

$$\begin{vmatrix} -(2k-7) & (k-4) & 1 \\ (k-4) & -(2k-6) & (k-4) \\ 1 & (k-4) & -(2k-5) \end{vmatrix} = 0$$

Evaluation of the determinant yields $-4k^3 + 30k^2 - 68k + 44 = 0$. The smallest root of this equation is $k = 1.111$. Hence, $K = k/h^2 = 17.776$. On substitution of k in the equations, you can solve for y_1, y_2, and y_3.

6-12 *Bibliography*

R. P. AGNEW, "Differential Equations," McGraw-Hill Book Company, New York.
F. CESCHINO and J. KUNTZMAN, "Numerical Solution of Initial-value Problems," Prentice-Hall, Inc., Englewood Cliffs, N.J.
S. D. CONTE, "Elementary Numerical Analysis," McGraw-Hill Book Company, New York.

A. H. Fox, "Fundamentals of Numerical Analysis," The Ronald Press Company, New York.

R. W. Hamming, "Numerical Methods for Scientists and Engineers," McGraw-Hill Book Company, New York.

L. D. Harris, "Numerical Methods Using FORTRAN," Charles E. Merrill Books, Inc., Columbus, Ohio.

T. R. McCalla, "Introduction to Numerical Methods and FORTRAN Programming," John Wiley & Sons, Inc., New York.

J. M. McCormick and M. G. Salvadori, "Numerical Methods in FORTRAN," Prentice-Hall, Inc., Englewood Cliffs, N.J.

A. Ralston, "A First Course in Numerical Analysis," McGraw-Hill Book Company, New York.

M. G. Salvadori and M. L. Baron, "Numerical Methods in Engineering," Prentice-Hall, Inc., Englewood Cliffs, N.J.

PROBLEMS

1. Given the following values of the Bessel function of zero order: $J_0(0) = 1.000\,000$, $J_0(0.5) = 0.938\,470$, $J_0(1.0) = 0.765\,198$, and $J_0(1.5) = 0.511\,828$, use forward differences to predict to four significant figures:

    (a) $J_0(2.0)$.

    (b) The derivative $J_0'(0.5)$.

    (c) $J_0(0.3)$, with the forward Gregory-Newton interpolation formula,

$$f(x) = f(x_0 + rh) = f(x_0) + r\,\Delta f(x_0) + \frac{r(r-1)}{2!}\,\Delta^2 f(x_0)$$

$$+ \frac{r(r-1)(r-2)}{3!}\,\Delta^3 f(x_0) + \cdots$$

where h is the step size.

2. Given the following values of the Bessel function of zero order: $J_0(2.4) = 0.002\,508$, $J_0(2.3) = 0.055\,540$, $J_0(2.2) = 0.110\,362$, and $J_0(2.1) = 0.166\,607$, use backward differences to predict to four significant figures:

    (a) $J_0(2.0)$.

    (b) The derivative $J_0'(2.3)$.

3. Given the following values of the Bessel function of zero order: $J_0(0) = 1$, $J_0(0.1) = 0.997\,502$, $J_0(0.2) = 0.990\,025$, $J_0(0.3) = 0.977\,626$, $J_0(0.4) = 0.960\,398$, and $J_0(0.5) = 0.938\,470$, use central differences to compute to four significant figures the derivative $J_0'(0.3)$. (Use average differences for odd central differences.)

4. Compute to three significant figures by the trapezoidal rule $\int_0^x e^x\,dx$ for $x = 0.2, 0.4, 0.6, 0.8,$ and 1.0.

5. (a) Use Simpson's rule to compute to four significant figures $\int_0^\pi \sin x\,dx$, with step size $h = \pi/2$.

    (b) Solve Prob. 5a with $h = \pi/4$.

    (c) Use Eq. (6-35) to improve the solutions to Prob. 5a and b.

6. Calculate $\int_0^\pi \sin x \, dx$ to five significant figures by Romberg's method.

7. (a) Given $y = -1$ when $x = 0$, solve by the Runge-Kutta method $y' = x^2 + y$, with step size $h = 0.1$, for $0 \le x \le 0.2$.   Use five significant figures.

   (b) Check by the Romberg-Euler method.

   (c) Continue with the Adams-Moulton method to find y at $x = 0.3$ and 0.4.

8. (a) Given $y = 1$ when $x = 0$, solve by the Runge-Kutta method $y' = xy(y - 2)$, with step size $h = 0.1$, for $0 \le x \le 0.2$.   Use six significant figures.

   (b) Continue with the Adams-Moulton method to find y at $x = 0.3$.

9. (a) Given $y = -\frac{1}{2}$ and $y' = 0$ when $x = 0$, solve by the Runge-Kutta method $(1 - x^2)y'' - 2xy' + 6y = 0$, with step size $h = 0.1$, for $0 \le x \le 0.2$.   Use six significant figures.

   (b) Continue with the Adams-Moulton method to find y at $x = 0.3$.

10. (a) Given $x = -1$ and $y = 1$ when $t = 0$, solve by the Runge-Kutta method, with step size $h = 0.01$,

$$x' = \frac{dx}{dt} = x(1 - y)$$

$$y' = \frac{dy}{dt} = y(x - 1)$$

for x and y when $t = 0.01$ and 0.02.   Use six significant figures.

   (b) Continue with the Adams-Moulton method to find x and y when $t = 0.03$.

ANSWERS

1. (a) Forward differences are:

x	$J_0(x)$	Δ	Δ^2	Δ^3
0	1.000 000			
		$-0.061\ 530$		
0.5	0.938 470		$-0.111\ 742$	
		$-0.173\ 272$		0.031 644
1.0	0.765 198		$-0.080\ 098$	
		$-0.253\ 370$		
1.5	0.511 828			

From Eq. (6-4):

$$J_0(2.0) = J_0(0 + 4 \times 0.5) = (1 + \Delta)^4 J_0(0)$$
$$= 1.000\ 000 + 4(-0.061\ 530) + 6(-0.111\ 742)$$
$$+ 4(0.031\ 644) + \cdots$$
$$= 0.2100$$

Exact answer: $J_0(2.0) = 0.223\ 891$.

(b) From Eq. (6-5):

$$J_0'(0.5) = \frac{1}{0.5}\left(-0.173\ 272 + \frac{0.080\ 098}{2} + \cdots\right) = -0.2664$$

Exact answer: $J_0'(0.5) = -0.242\ 268$.

(c) The Gregory-Newton formula yields

$$J_0(0.3) = J_0(0 + 0.6 \times 0.5) = 1.000\ 000 + 0.6(-0.061\ 530)$$

$$+ \frac{0.6(-0.4)}{2!}(-0.111\ 742) + \frac{0.6(-0.4)(-1.4)}{3!}(0.031\ 644) + \cdots$$

$$= 0.9783$$

Exact answer: $J_0(0.3) = 0.977\ 626$.

2. (a) Backward differences are:

x	$J_0(x)$	∇	∇^2	∇^3
2.1	0.166 607			
		−0.056 245		
2.2	0.110 362		0.001 423	
		−0.054 822		0.000 367
2.3	0.055 540		0.001 790	
		−0.053 032		
2.4	0.002 508			

From Eq. (6-12):

$$J_0(2.0) = J_0(2.4 - 4 \times 0.1) = (1 - \nabla)^4 J_0(2.4)$$
$$= 0.002\ 508 - 4(-0.053\ 032) + 6(0.001\ 790) - 4(0.000\ 367) + \cdots$$
$$= 0.2239$$

Exact answer: $J_0(2.0) = 0.223\ 891$.

(b) From Eq. (6-13):

$$J_0'(2.3) = \frac{1}{0.1}\left(-0.054\ 822 + \frac{0.001\ 423}{2} + \cdots\right) = -0.5411$$

Exact answer: $J_0'(2.3) = -0.539\ 873$.

3. Central differences are:

x	$J_0(x)$	δ	δ^2	δ^3
0	1.000 000			
0.1	0.997 502	−0.004 988	−0.004 979	
0.2	0.990 025	−0.009 938	−0.004 922	
0.3	0.977 626	−0.014 814	−0.004 829	0.000 111
0.4	0.960 398	−0.019 578	−0.004 700	
0.5	0.938 470			

$\delta J_0(0.3) = \frac{1}{2}(0.960\ 398 - 0.990\ 025) = -0.014\ 814$.   By Eq. (13-18), $\delta^2 J_0(0.3)$
$= 0.960\ 398 - 2(0.977\ 626) + 0.990\ 025 = -0.004\ 829$.
$\delta^3 J_0(0.3) = \frac{1}{2}(-0.004\ 700 + 0.004\ 922) = 0.000\ 111$.   From Eq. (6-20), $J_0'(0.3)$
$= (1/0.1)(-0.014\ 814 - 0.000\ 111/6) = -0.1483$.
Exact answer: $J_0'(0.3) = -0.148\ 319$.

4. For a compact, tabulated solution, rewrite Eq. (6-31) as

$$y = h\left(\frac{f_0 + f_1}{2} + \frac{f_1 + f_2}{2} + \frac{f_2 + f_3}{2} + \cdots + \frac{f_{n-1} + f_n}{2}\right)$$

Let $f = e^x$.   Then, organize the calculations in a table:

x	e^x	$\dfrac{f_n + f_{n+1}}{2}$	$\sum \dfrac{f_n + f_{n+1}}{2}$	$y = 0.2 \sum \dfrac{f_n + f_{n+1}}{2}$	$e^x - 1$
0	1.000	.....	0	0	0
0.2	1.221	1.111	1.111	0.222	0.221
0.4	1.492	1.357	2.468	0.494	0.492
0.6	1.822	1.657	4.125	0.825	0.822
0.8	2.226	2.024	6.149	1.23	1.23
1.0	2.718	2.472	8.621	1.72	1.72

5. (a) $\sin 0 = 0$, $\sin \pi/2 = 1$, $\sin \pi = 0$.

$$\int_0^\pi \sin x\ dx = \frac{\pi}{6}[0 + 4(1) + 0] = 2.094$$

(b) $\sin \pi/4 = 0.7071$, $\sin 3\pi/4 = 0.7071$.

$$\int_0^\pi \sin x\ dx = \frac{\pi}{12}[0 + 4(0.7071) + 2(1) + 4(0.7071) + 0] - 2.005$$

(c)

$$\int_0^\pi \sin x\ dx = 2.005 + \frac{2^4}{4^4 - 2^4}(2.005 - 2.094) = 1.999$$

6. Results from Romberg's method are:

Number of steps	Trapezoid sums	Repeated linear interpolations		
		$k = 1$	$k = 2$	$k = 3$
1	0			
2	1.57080	2.09440		
4	1.89611	2.00455	1.99856	
8	1.97423	2.00027	1.99998	2.00000

$$\int_0^{\pi} \sin x \, dx = \frac{\pi}{8} [0 + 2(0.382\ 683) + 2(0.707\ 101) + 2(923\ 880) + 1] = 1.97423$$

For $k = 1$: $1.97423 - \dfrac{1.89611 - 1.97423}{3} = 2.00027$

For $k = 2$: $2.00027 - \dfrac{2.00455 - 2.00027}{15} = 1.99998$

For $k = 3$: $1.99998 - \dfrac{1.99856 - 1.99998}{63} = 2.00000$

7. (a) For use in Eq. (6-38), compute for y:

$k_1 = -0.10000$ $k_2 = -0.10475$ $k_3 = -0.10499$ $k_4 = -0.10950$

$$y_1 = -1 + \frac{-0.10000 + 2(-0.10475) + 2(-0.10499) - 0.10950}{6}$$

$$= -1.1048$$

With this value, compute for y_2:

$k_1 = -0.10948$ $k_2 = -0.11370$ $k_3 = -0.11392$ $k_4 = -0.11787$

$$y_2 = -1.1048 + \frac{-0.10948 + 2(-0.11370) + 2(-0.11392) - 0.11787}{6}$$

$$= -1.2186$$

(b) The Romberg-Euler method yields:

Interval	Y	Y'	Y''
		Step 1. $x = 0.1$	
0.100	−1.1000		
0.050	−1.1024	−1.1048	
0.025	−1.1036	−1.1048	$y_1 = -1.1048$
		Step 2. $x = 0.2$	
0.100	−1.2143		
0.050	−1.2163	−1.2183	
0.025	−1.2174	−1.2185	−1.2186 $y_2 = 1.2186$

(c) From Eq. (6-41) with $n = 2$:

$$y_3{}^p = -1.2186 + \frac{0.1}{12} [23(-1.1786) - 16(-1.0948) + 5(-1)] = -1.3401$$

$$f_3{}^p = -1.3401 + 0.3^2 = -1.2501$$

From Eq. (6-42) with $n = 2$:

$$y_3{}^c = -1.2186 + \frac{0.1}{12} [5(-1.2501) + 8(-1.1786) + 1.0948] = -1.3401$$

$$f_3{}^c = -1.3401 + 0.3^2 = -1.2501$$

x	y^p	f^p	y	$y' = f(x,y)$
0			-1.0000	-1.0000
0.1			-1.1048	-1.0948
0.2			-1.2186	-1.1786
0.3	-1.3401	-1.2501	-1.3401	-1.2501
0.4	-1.4682	-1.3082	-1.4681	

Using these values, we get

$$y_4{}^p = -1.3401 + \frac{0.1}{12} [23(-1.2501) - 16(-1.1786) + 5(-1.0948)]$$

$$= -1.4682$$

$$f_4{}^p = -1.4682 - 0.4^2 = -1.3082$$

$$y_4{}^c = -1.3401 + \frac{0.1}{12} [5(-1.3082) + 8(-1.2501) + 1.1786] = -1.4681$$

8. (a) For use in Eq. (6-38), compute:

$$k_1 = 0 \qquad k_2 = -0.005 \qquad k_3 = -0.005\ 000 \qquad k_4 = -0.010\ 000$$

$$y_1 = 1 + \frac{0 + 2(-0.005) + 2(-0.005\ 000) - 0.010\ 000}{6} = 0.995\ 000$$

With this value, compute for y_2:

$$k_1 = -0.010\ 000 \qquad k_2 = -0.014\ 999 \qquad k_3 = -0.014\ 998$$

$$k_4 = -0.019\ 992$$

$$y_2 = 0.995\ 000 + \frac{-0.010\ 000 + 2(-0.014\ 999) + 2(-0.014\ 998) - 0.019\ 992}{6}$$

$$= 0.980\ 002$$

(b) From Eq. (6-41):

$$y_3{}^p = 0.980\ 002 + \frac{0.1}{12} [23(-0.199\ 920) - 16(-0.099\ 998)] = 0.955\ 010$$

$$f_3{}^p = 0.3(0.955\ 010)(0.955\ 010 - 2) = -0.299\ 393$$

$$y_3{}^c = 0.980\ 002 + \frac{0.1}{12} [5(-0.299\ 393) + 8(-0.199\ 920) + 0.099\ 998]$$

$$= 0.955\ 033$$

x	y^p	f^p	y	$y' = f(x,y)$
0			1.000 000	0
0.1			0.995 000	−0.099 998
0.2			0.980 002	−0.199 920
0.3	0.995 910	−0.299 393	0.955 033	

9. (a) Let $y' = z$, $y'' = z'$. Then, $(1 - x^2)y'' - 2xy' + 6y = 0$ can be replaced by the system $y' = z$, $z' = 2(xz - 3y)/(1 - x^2)$. For use in Eq. (6-38) extended to three variables, compute:

$$k_{11} = hz_0 = 0.1(0) = 0$$

$$k_{12} = 2h\frac{x_0z_0 - 3y_0}{1 - x_0^2} = 0.2\frac{0 + 1.5}{1 - 0} = 0.3$$

$$k_{21} = h\left(z_0 + \frac{k_{12}}{2}\right) = 0.1(0 + 0.15) = 0.015$$

$$k_{22} = 2h\frac{(x_0 + h/2)(z_0 + k_{12}/2) - 3(y_0 + k_{11}/2)}{1 - (x_0 + h/2)^2}$$

$$= 0.2\frac{0.05(0 + 0.15) - 3(-0.5 + 0)}{1 - 0.0025} = 0.302\ 255$$

$$k_{31} = h\left(z_0 + \frac{k_{22}}{2}\right) = 0.1(0 + 0.151\ 128) = 0.015\ 113$$

$$k_{32} = 2h\frac{(x_0 + h/2)(z_0 + k_{22}/2) - 3(y_0 + k_{21}/2)}{1 - (x_0 + h/2)^2}$$

$$= 0.2\frac{0.05(0 + 0.151\ 128) - 3(-0.5 + 0.0075)}{1 - 0.0025} = 0.297\ 755$$

$$k_{41} = h(z_0 + k_{32}) = 0.1(0 + 0.297\ 755) = 0.029\ 776$$

$$k_{42} = 2h\frac{(x_0 + h)(z_0 + k_{32}) - 3(y_0 + k_{31})}{1 - (x_0 + h)^2}$$

$$= 0.2\frac{0.1(0 + 0.297\ 755) - 3(-0.5 + 0.015\ 113)}{1 - 0.01} = 0.299\ 886$$

Substitution in Eq. (6-38) yields

$$y_1 = -0.5 + \frac{0 + 2(0.015) + 2(0.015\ 113) + 0.029\ 776}{6} = -0.485\ 000$$

$$z_1 = 0 + \frac{0.3 + 2(0.302\ 255) + 2(0.297\ 755) + 0.299\ 886}{6} = 0.299\ 984$$

With these values, compute for y_2 and z_2:

$$
\begin{aligned}
k_{11} &= 0.029\ 998 & k_{12} &= 0.299\ 970 \\
k_{21} &= 0.044\ 997 & k_{22} &= 0.302\ 301 \\
k_{31} &= 0.045\ 113 & k_{32} &= 0.297\ 698 \\
k_{41} &= 0.059\ 768 & k_{42} &= 0.299\ 833
\end{aligned}
$$

Substitution in Eq. (6-38) gives $y_2 = -0.440\ 002$ and $z_2 = 0.599\ 951$.

(b) By Eq. (6-41),

$$y_3{}^p = -0.440\ 002 + (0.1/12)[23(0.599\ 951) - 16(0.299\ 984)] = -0.365\ 009$$

Compute

$$z_0' = 2[0 - 3(-0.5)]/(1 - 0) = 3;\ z_1' = 2.999\ 996;\ z_2' = 2.999\ 983$$

Then, $z_3{}^p = 0.599\ 951 + (0.1/12)[23(2.999\ 983) - 16(2.999\ 996) + 5(3)] = 0.899\ 948$.   From Eq. (6-42), $y_3{}^c = -0.365\ 016$.

10. (a) For use in Eq. (6-38), compute:

$$k_{11} = hx_0(1 - y_0) = 0.01(-1)(1 - 1) = 0$$
$$k_{12} = hy_0(x_0 - 1) = 0.01(1)(-1 - 1) = -0.02$$
$$k_{21} = h\left(x_0 + \frac{k_{11}}{2}\right)\left(1 - y_0 - \frac{k_{12}}{2}\right) = 0.01(-1 + 0)(1 - 1 + 0.01)$$
$$= -0.0001$$
$$k_{22} = h\left(y_0 + \frac{k_{12}}{2}\right)\left(x_0 + \frac{k_{11}}{2} - 1\right) = 0.01(1 - 0.01)(-1 + 0 - 1)$$
$$= -0.019\ 800$$
$$k_{31} = h\left(x_0 + \frac{k_{21}}{2}\right)\left(1 - y_0 - \frac{k_{22}}{2}\right) = -0.000\ 099$$
$$k_{32} = h\left(y_0 + \frac{k_{22}}{2}\right)\left(x_0 + \frac{k_{21}}{2} - 1\right) = -0.019\ 803$$
$$k_{41} = h(x_0 + k_{31})(1 - y_0 - k_{32}) = -0.000\ 198$$
$$k_{42} = h(y_0 - k_{32})(x_0 + k_{31} - 1) = -0.019\ 605$$

Substitution in Eq. (6-38) gives

$$x_1 = -1 + \frac{0 + 2(-0.0001) + 2(-0.000\ 099) - 0.000\ 198}{6}$$
$$= -1.000\ 083$$
$$y_1 = 1 + \frac{-0.02 + 2(-0.019\ 800) + 2(-0.019\ 803) - 0.019\ 605}{6}$$
$$= 0.980\ 198$$

With these values, compute for x_2 and y_2:

$$k_{11} = -0.000\ 198 \qquad k_{12} = -0.019\ 605$$
$$k_{21} = -0.000\ 296 \qquad k_{22} = -0.019\ 410$$
$$k_{31} = -0.000\ 295 \qquad k_{32} = -0.019\ 410$$
$$k_{41} = -0.000\ 393 \qquad k_{42} = -0.019\ 216$$

Substitution in Eq. (6-38) produces $x_2 = -1.000\ 379$ and $y_2 = 0.960\ 788$.
(b)

$$x_0' = 0 \qquad x_1' = -0.019\ 802 \qquad x_2' = -0.039\ 227$$
$$y_0' = -2 \qquad y_1' = -1.960\ 483 \qquad y_2' = -1.921\ 931$$

By Eq. (6-41),

$$x_3{}^p = -1.000\ 379 + \frac{0.01}{12}\ [23(-0.039\ 227) - 16(-0.019\ 802)] = -1.000\ 861$$

Similarly, $y_3{}^p = 0.941\ 757$. From these values, compute $x_3' = -0.058\ 293$ and $y_3' = -1.884\ 323$. Substitution in Eq. (6-42) gives $x_3{}^c = -1.000\ 867$ and $y_3{}^c = 0.941\ 758$.

SEVEN

Partial Differential Equations

When engineering problems involve rates of change of more than two variables, they often require solution of partial differential equations. In general, such problems lead to a system of equations containing partial derivatives of several dependent variables with respect to several independent variables. Solving these equations usually is far more difficult than solving ordinary differential equations, which deal with only two variables.

To solve a system of partial differential equations, the first step generally is to convert the system into a canonical, or standard, form. This consists of one or more partial differential equations, each containing the partial derivatives of one dependent variable with respect to the independent variables. Much more is known about this type of equation than the more general type. So the chances of finding a solution are considerably improved.

In addition to dealing with more variables, partial differential equations are also more difficult to solve for other reasons. Their general solution, for instance, contains arbitrary functions, whereas the general solution of an ordinary differential equation contains arbitrary constants. Suppose, for example, that $z = f(x,y) + g_1(y)$. When you take the partial derivative of z with respect to x, you treat y as a constant. The result is $\partial z / \partial x = f_x(x,y)$, for the partial derivative of g_1 with respect to x is zero. Suppose, instead that $z = f(x,y) + g_2(y)$, where $g_2 \neq g_1$. Again, $\partial z / \partial x = f_x(x,y)$, because $\partial g_2 / \partial x = 0$. Consequently, when you integrate $\partial z / \partial x = f_x(x,y)$ with respect to x, the result is $z = f(x,y) + g(y)$, where $g(y)$ is an arbitrary function of y.

Having obtained a general solution of a partial differential equation in terms of arbitrary functions, you then are faced with the problem of determining them to satisfy given initial and boundary conditions. This problem may be difficult or impossible to solve. Sometimes you may have to settle for a particular solution that meets only some of the boundary values.

As with ordinary differential equations, linear partial differential equations usually are easier to solve than the more general types. In a **linear equation,** the dependent variable and all its derivatives are of the first degree. Consequently, when you set up partial differential equations, use simplifying assumptions, if possible, so that the resulting equations are linear.

You will find in your studies of partial differential equations that many methods of solution have at least two characteristics in common. One is an attempt to convert a given equation into one or more ordinary differential equations, whose solution will lead to the solution of the given equation. The second characteristic is an attempt, guided by the initial and boundary values, to guess at the form of the final solution. This technique often expedites solution of linear equations and may work with nonlinear equations after a suitable transformation of variables.

In this chapter, we can touch on only a few of the many types of partial differential equations and methods for solving them. If your work involves more complex problems, you will find it well worth while to continue your studies of this subject with the aid of specialized texts, such as those listed in Art. 7-7.

7-1 *Exact Partial Differential Equations.* As an illustration of how you can use boundary conditions to determine arbitrary functions in the general solution of a partial differential equation, let us solve

$$\frac{\partial^2 z}{\partial x\, \partial y} = \frac{\partial z}{\partial x}$$

Assume that the boundary conditions are $z = 0$ when $x = 0$, for all values of y; and $z = x$ when $y = 0$, for all values of x.

If we write the given equation as

$$\frac{\partial}{\partial x}\left(\frac{\partial z}{\partial y} - z\right) = 0$$

we can see that it is exact; we can integrate directly with respect to x, keeping y constant.   The result of integration is

$$\frac{\partial z}{\partial y} - z = f(y)$$

where $f(y)$ is an arbitrary function of y.   This is a linear first-order equation with e^{-y} as an integrating factor.   After multiplying both sides of the equation by e^{-y} and integrating, we get

$$e^{-y}z = \int e^{-y}f(y)\, dy + g(x)$$

where $g(x)$ is an arbitrary function of x.   This is the general solution.

From the boundary condition $z(0,y) = 0$, we find that $\int e^{-y}f(y)\, dy = -g(0)$.   Now, we can write that $z = e^{y}g(x) - e^{y}g(0)$.   Then, from the boundary condition, $z(x,0) = x$, we obtain $x = g(x) - g(0)$, which indicates that $g(x) = x + g(0)$.   This gives us the particular solution we seek:

$$z = xe^{y}$$

Substitution in the given equation verifies that this result is indeed the solution.

7-2   Linear Partial Differential Equations of the First Order. As for ordinary differential equations, the order of a partial differential equation is the order of the highest-order derivative in the equation. Thus, a first-order partial differential equation contains no derivatives higher than the first order.   In canonical form, a first-order linear equation can be written

$$P_1\frac{\partial z}{\partial x} + P_2\frac{\partial z}{\partial y} + \cdots + P_n\frac{\partial z}{\partial t} = f \tag{7-1}$$

where z, P_1, P_2, ... , P_n, and f are functions of the independent variables x, y, ... , t.

Solution of Eq. (7-1) can be converted into the problem of solving the system of linear ordinary differential equations

$$\frac{dx}{P_1} = \frac{dy}{P_2} = \cdots = \frac{dt}{P_n} = \frac{dz}{f} \tag{7-2}$$

For if $Z(x,y, \ldots ,t) = c$, a constant, satisfies Eqs. (7-2), then $z = Z$ satisfies Eq. (7-1). Consequently, if the general solution of Eqs. (7-2) is $Z_1 = c_1$, $Z_2 = c_2$, $\ldots$, $Z_n = c_n$, where the c's are constants, the general solution of Eq. (7-1) is $\phi(Z_1, Z_2, \ldots ,Z_n) = 0$, where ϕ is an arbitrary function of Z_1, Z_2, $\ldots$, Z_n.

For example, let us solve $yz_x - xz_y = x^2 - y^2$, given $z = 0$ when $x = -y$. ($z_x = \partial z/\partial x$ and $z_y = \partial z/\partial y$.) The given equation is equivalent to the system of ordinary differential equations

$$\frac{dx}{y} = -\frac{dy}{x} = \frac{dz}{x^2 - y^2}$$

The first pair of terms, when integrated, yields $x^2 + y^2 = c_1{}^2$. Hence, we can set $Z_1 = x^2 + y^2$. The first and third terms, when multiplied by $x^2 - y^2$, form the differential equation $x^2\, dx/y - y\, dx = dz$. To solve it, we obtain $dx = -y\, dy/x$ from the first pair of terms and substitute it in the first term of the equation. The result is $-x\, dy - y\, dx = dz$. This is an exact differential equation. Integration yields $-xy + c_2 = z$, or $z + xy = c_2$. Therefore, we can set $Z_2 = z + xy$. And the general solution of the partial differential equation is

$$\phi(x^2 + y^2, z + xy) = 0$$

The initial condition requires that $z = 0$ when $x = -y$. Consequently, we must have $\phi(y^2 + y^2, 0 - y^2) = 0$, for all values of y. This can be satisfied by $\phi = 2y^2 - 2y^2$. So a particular solution of the given partial differential equation is $x^2 + y^2 + 2(z + xy) = 0$, or $z = -(x + y)^2/2$. But this is not the complete solution. See if you can find additional particular solutions.

7-3   Uniform Higher-order Linear Equations.
There is a special case of higher-order linear partial differential equations that is relatively easy to solve. In this type of equation, the coefficients of the derivatives are constants and the order of all the terms are equal. The equations have the form

$$(a_0 D_x{}^n + a_1 D_x{}^{n-1} D_y + \cdots + a_{n-1} D_x D_y{}^{n-1} + a_n D_y{}^n)z = 0 \qquad (7\text{-}3)$$

where the a's are constants and D_x and D_y indicate partial differentiation with respect to x and y, respectively.

Solution of Eq. (7-3) parallels that of linear ordinary differential equations with constant coefficients (Art. 3-7). First, you solve the auxiliary equation

$$a_0 m^n + a_1 m^{n-1} + \cdots + a_{n-1} m + a_n = 0 \qquad (7\text{-}4)$$

If the roots $m_1, m_2, \ldots, m_n$ are distinct, the general solution of Eq. (7-3) is

$$z = \phi_1(y + m_1 x) + \phi_2(y + m_2 x) + \cdots + \phi_n(y + m_n x) \qquad (7\text{-}5)$$

where ϕ_i represents arbitrary functions. If m is an r-fold root, however, not only is $\phi_1(y + mx)$ an integral but so also are $x\phi_2(y + mx)$, $x^2\phi_3(y + mx), \ldots, x^{r-1}\phi_r(y + mx)$, and $y\theta_2(y + mx), y^2\theta_3(y + mx)$, $\ldots, y^{r-1}\theta_r(y + mx)$. When the right-hand side of Eq. (7-3) is not zero, a particular solution can be found by methods analogous to those for linear ordinary differential equations.

As an example, let us solve

$$\frac{\partial^2 z}{\partial x^2} + \frac{\partial^2 z}{\partial x\, \partial y} - 2\frac{\partial^2 z}{\partial y^2} = (D_x{}^2 + D_x D_y - 2D_y{}^2)z = \sin(x + y)$$

The auxiliary equation is

$$m^2 + m - 2 = 0$$

Its roots are $m_1 = -2$ and $m_2 = 1$. Hence, the complementary function is

$$z = \phi_1(y - 2x) + \phi_2(y + x)$$

For a particular integral, let us try $z = cx \cos(x + y)$, which includes x as a factor because $f(x + y)$ is already a part of the complementary function. When we substitute this value of z in the given equation and equate coefficients of $\sin(x + y)$, we find $c = -\frac{1}{3}$. The general solution then is

$$z = \phi_1(y - 2x) + \phi_2(y + x) - \frac{1}{3}x \cos(x + y)$$

Suppose now that the initial conditions are $z = 0$ when $x = 0$ and $x = -y$. These give the system of equations

$$\phi_1(y - 0) + \phi_2(y + 0) = 0$$

$$\phi_1(y + 2y) + \phi_2(y - y) + \frac{1}{3}y \cos(-y + y) = 0$$

The first equation indicates that $\phi_2(y,0) = -\phi_1(y,0)$. Then, from the second equation, you can conclude that we can get a particular solution from $f(3y) + y/3 = 0$. Thus, one particular solution is $z = (y + x)/9 - (y - 2x)/9 - [x \cos(x + y)]/3$, as you can verify by substitution in the given partial differential equation. Can you find other particular solutions?

You will find it instructive to set up pairs of boundary conditions and try to determine the arbitrary functions from them.

7-4   *Solution by Separation of Variables.* Problems in different fields of engineering often lead to the same partial differential equation. The symbols may be different or, if they are the same, they may represent different physical phenomena; but mathematically, the equations may be identical. If they are, the solution developed for one field of engineering also applies to others when the symbols are interpreted appropriately. Therefore, when you encounter partial differential equations in your work, you can save yourself considerable time and effort if you find a solution already developed.

There are several types of partial differential equations, especially second-order linear ones, that have recurred in several engineering fields and that consequently have been thoroughly analyzed. We will examine some important ones in this chapter.

Such equations are often amenable to solution by the powerful method of separation of variables. This method attempts to convert a given partial differential equation into a system of ordinary differential equations. The method is based on the observation that for many engineering problems, the particular solution can be expressed as a product of functions, and each function contains only one variable. So you start by assuming a solution consisting of unknown functions in this form. On substitution of these functions in the given equation, you can often obtain a system of equivalent ordinary differential equations. (Sometimes, especially for nonlinear partial differential equations, you may have to try a transformation of variables to succeed.) Solution of the ordinary equations and use of initial and boundary conditions lead to solution of the given equation.

As an example of the use of this method, and also to illustrate a typical technique for setting up a partial differential equation, let us investigate the motion of a flexible string vibrating with small amplitude in a plane. Let us choose coordinate axes as indicated in Fig. 7-1a. If we select and examine a tiny length Δs of the string (Fig. 7-1b), we find that the end at x is subjected to a tangential tension T and the end at $x + \Delta x$, to a tangential tension $T + \Delta T$. If the motion was initiated by a force parallel to the y axis, then the components H of T and $T + \Delta T$ parallel to the x axis are equal but opposite in direction. Hence, the component V of T parallel to the y axis equals $H \tan \alpha_1$. And the component $V + \Delta V$ of $T + \Delta T$ parallel to the y axis equals $H \tan \alpha_2$. The net force parallel to the y axis is consequently $H \tan \alpha_2 - H \tan \alpha_1 - w \Delta x$, where w is the weight per unit length of string. Let us assume that w is constant and that we can approximate $w \Delta s$ by $w \Delta x$.

The net force parallel to the y axis causes an acceleration $\partial^2 y / \partial t^2$ of the string segment parallel to the y axis. (A partial derivative is used here because y is a function of both x and t, and we are concerned with the rate of change of y while x is held constant.) By Newton's law, $F = Ma$, we can relate the force and acceleration:

$$H(\tan \alpha_2 - \tan \alpha_1) - w\,\Delta x = \frac{w\,\Delta x}{g} \frac{\partial^2 y}{\partial t^2}$$

Transposing and dividing both sides of the equation by $H\,\Delta x$, we get

$$\frac{\tan \alpha_2 - \tan \alpha_1}{\Delta x} - \frac{w}{Hg} \frac{\partial^2 y}{\partial t^2} = \frac{w}{H}$$

Notice now that $\tan \alpha_1$, the slope of the string at x, equals $\partial y / \partial x$ at x. Also, $\tan \alpha_2 = \partial y / \partial x$ at $x + \Delta x$. So if we let Δx approach zero, and set $Hg/w = a^2$, the equation becomes

$$a^2 \frac{\partial^2 y}{\partial x^2} - \frac{\partial^2 y}{\partial t^2} = g \tag{7-6}$$

Equation (7-6) is often given with the right-hand side taken as zero. In that form, it is called the **one-dimensional wave equation,** because it represents, among other phenomena, the motion of electromagnetic waves propagated with velocity a.

Comparison of Eq. (7-6) with Eq. (7-3) indicates that $a^2 y_{xx} - y_{tt} = 0$ is a uniform equation and can be solved by the method of Art. 7-3. The auxiliary equation is $a^2 - m^2 = 0$. Its roots are $m = \pm a$. Hence, the general solution is $y = \phi_1(x + at) + \phi_2(x - at)$.

The problem still remains, however, of determining the arbitrary functions ϕ_1 and ϕ_2 so that given initial and boundary conditions are satisfied. This part of the solution may be difficult or impossible, depending on the given conditions.

In the case of a vibrating string, and many other engineering problems, however, we can anticipate that the solution will be harmonic. Thus,

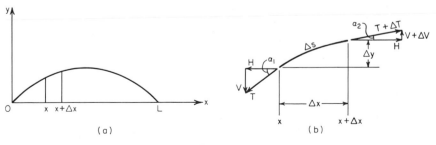

Fig. 7-1

the arbitrary functions will have the form sin $(x \pm at)$ and cos $(x \pm at)$. But sin $(x \pm at)$ = sin x cos $at \pm$ cos x sin at, and cos $(x \pm at)$ = cos x cos $at \mp$ sin x sin at. In each case, the terms on the right of the equal sign are the products of a function of x and a function of t. Consequently, we can replace the arbitrary harmonic functions of $x \pm at$ by the product XT, where X is an arbitrary function of x and T is an arbitrary function of t, or by the sum of such products.

For example, let us solve the equation of a vibrating string $y_{tt} - a^2 y_{xx} = 0$. Boundary conditions are $y = 0$ where $x = 0$ or L. Initial conditions are $y = y_0$ sin $(\pi x/L)$, and $y_t = \partial y/\partial t = 0$, when $t = 0$. From knowledge of the physical conditions and the given harmonic initial condition, we can anticipate that the solution will consist of harmonic functions. So we assume that the solution will have the form $y(x,t) = XT$. Differentiation then gives us $y_{tt} = XT''$, where $T'' = d^2T/dt^2$, and $y_{xx} = TX''$, where $X'' = d^2X/dx^2$. Substitution in the partial differential equation yields the ordinary differential equation $XT'' - a^2TX'' = 0$. Notice that the variables are separable. Thus, we can write the equation as

$$\frac{T''}{a^2 T} = \frac{X''}{X} = K \tag{7-7}$$

where K is a function to be determined. The ratios in the equation must be equal for all values of X and T. But T'' and T are functions only of t; so T''/a^2T must be independent of x. Hence, X''/X must be independent of x also. Furthermore, since X'' and X are functions only of x, X''/X must be independent of t. Consequently, T''/a^2T must also be independent of t. Since both ratios in the equation are independent of x and t, they must be equal to a constant. Let K be this constant.

This analysis enables us to replace the given partial differential equation by the system of linear ordinary differential equations

$$T'' - a^2KT = 0 \qquad X'' - KX = 0$$

Now, we started with the assumption that the solution will consist of harmonic functions. We can get such a result only if K is negative. Let $k = -K$. Then, solution of these equations yields

$$X = c_1 \sin \sqrt{k}\, x + c_2 \cos \sqrt{k}\, x$$
$$T = c_3 \sin at \sqrt{k} + c_4 \cos at \sqrt{k}$$

And the general solution of $y_{tt} - a^2 y_{xx}$, therefore, is

$$y = XT = (c_1 \sin \sqrt{k}\, x + c_2 \cos \sqrt{k}\, x)(c_3 \sin at \sqrt{k} + c_4 \cos at \sqrt{k})$$

For the particular solution, we have first $x = 0$, $y = 0$. Hence, $c_2 = 0$. Next, we have $y_t = 0$ when $t = 0$. Consequently, after differentiating with respect to t, we find $c_3 = 0$. Also, $y = 0$ for all values of t when $x = L$. We cannot take $c_1 = 0$ to satisfy this requirement, because we would be left with the trivial solution $y = 0$. We can conclude, therefore, that $\sin \sqrt{k}\, L = 0$, or $\sqrt{k}\, L = n\pi$, where n is an integer. So $\sqrt{k} = n\pi/L$. At this stage, then, our particular solution is

$$y = c_1 \sin \frac{n\pi x}{L}\, c_4 \cos \frac{n a \pi t}{L}$$

We have one more condition to use: $y = y_0 \sin (\pi x/L)$ when $t = 0$, for all values of x. This requires that

$$y_0 \sin \frac{\pi x}{L} = c_1 c_4 \sin \frac{n\pi x}{L}$$

for all values of x. Hence, $c_1 c_4 = y_0$ and $n = 1$. And the particular solution of $y_{tt} - a^2 y_{xx} = 0$ therefore is

$$y = y_0 \sin \frac{\pi x}{L} \cos \frac{a\pi t}{L}$$

For other boundary conditions, infinite series, such as Fourier series, may have to be used to find the particular solution. Suppose, for example, that instead of $y = y_0 \sin (\pi x/L)$ at $t = 0$, you are given as an initial condition

$$y = \begin{cases} \dfrac{2x}{L} & 0 \le x \le \dfrac{L}{2} \\[2ex] 2\dfrac{L - x}{L} & \dfrac{L}{2} \le x \le L \end{cases}$$

We can not, as before, set this equal to $c_1 c_4 \sin (n\pi x/L)$ and determine c_1, c_4, and n directly. But since the given partial differential equation is linear, we can expect that the solution can be expressed as the sum of terms of the form $c_n \sin (n\pi x/L) \cos (na\pi t/L)$. Therefore, we expand y at $t = 0$ in a half-range Fourier sine series in the interval $0 \le x \le L$ and determine c_n by equating that series to $c_n \sin (n\pi x/L)$. From Eq. (5-18), we obtain

$$b_n = \frac{2}{L} \int_0^{L/2} \frac{2x}{L} \sin \frac{n\pi x}{L}\, dx + \frac{2}{L} \int_{L/2}^L \frac{2(L-x)}{L} \sin \frac{n\pi x}{L}\, dx = \frac{8}{\pi^2} \left(\frac{1}{n^2} \sin \frac{n\pi}{2} \right)$$

Now, on setting $c_n = b_n$ we have the particular solution

$$y = \frac{8}{\pi^2} \left(\sin \frac{\pi x}{L} \cos \frac{a\pi t}{L} - \frac{1}{3^2} \sin \frac{3\pi x}{L} \cos \frac{3a\pi t}{L} \right.$$

$$\left. + \frac{1}{5^2} \sin \frac{5\pi x}{L} \cos \frac{5a\pi t}{L} - \cdots \right)$$

You can verify that this is the solution by substitution in the given partial differential equation and checking the initial and boundary conditions.

7-5 *More General Partial Differential Equations.* When you set up partial differential equations to solve an engineering problem, you may not initially have them in canonical form, with the partial derivatives of only one dependent variable with respect to one or more independent variables. To use the methods in preceding articles, however, you have to convert the equations to canonical form.

There is no set procedure for doing this. In general, you should operate on the equations much as you would do to solve a system of ordinary differential equations (Art. 3-9). In addition, you sometimes may be able to employ another technique: express two or more, or even all the dependent variables as derivatives of a single function. These procedures can be demonstrated by an example, which also can serve again as a demonstration of how to set up a partial differential equation.

The irregular shape in Fig. 7-2a represents an elastic body loaded by forces in a plane and, except for the weight of the body, acting along the boundary. Let us select coordinate axes x_1 and x_2 in the plane of the loads, as indicated in Fig. 7-2a. Also, let us take the dimension of the body normal to the plane as dx_3. We will assume that the loads produce stresses in the body that can be resolved only into normal stresses, such as tension and compression, f_1 parallel to the x_1 axis; normal stresses f_2 parallel to the x_2 axis; and shears v acting on the same surfaces as f_1 and f_2 and in the plane of the loads.

Let us examine a rectangular piece of the body with sides dx_1 and dx_2. We will take its weight as $\rho g \, dx_1 \, dx_2 \, dx_3$, where ρ is the mass per unit

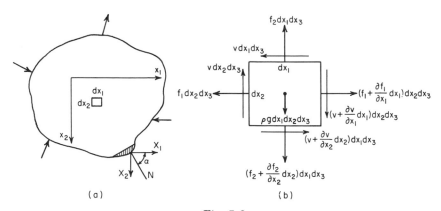

Fig. 7-2

volume and g is the acceleration due to gravity. In general, such a block will have normal forces and shears acting on its sides, as shown in Fig. 7-2b. For example, one face parallel to the x_2 axis is subjected to a normal stress f_1 and a shear v. These act on an area $dx_2\, dx_3$. Hence, the total forces exerted by these stresses are $f_1\, dx_2\, dx_3$ and $v\, dx_2\, dx_3$, respectively.

Since stresses are generally not constant throughout a body, the stresses on the opposite face will differ from f_1 and v by a small amount. We can represent the change in f_1 by $(\partial f_1/\partial x_1)\, dx_1$ and change in v by $(\partial v/\partial x_1)\, dx_1$. Then, the total forces acting on the face are $[f_1 + (\partial f_1/\partial x_1)\, dx_1]\, dx_2\, dx_3$ and $[v + (\partial v/\partial x_1)\, dx_1]\, dx_2\, dx_3$.

Similarly, stresses f_2 and v act on one face parallel to the x_1 axis. Also, $f_2 + (\partial f_2/\partial x_2)\, dx_2$ and $v + (\partial v/\partial x_2)\, dx_2$ act on the opposite face.

The block is in equilibrium under the action of these forces. Hence, the sum of the forces parallel to the x_1 axis must be zero, and the sum of the forces parallel to the x_2 axis must be zero. Summation of the forces parallel to the x_1 axis yields

$$\left(f_1 + \frac{\partial f_1}{\partial x_1}\, dx_1\right) dx_2\, dx_3 - f_1\, dx_2\, dx_3$$
$$+ \left(v + \frac{\partial v}{\partial x_2}\, dx_2\right) dx_1\, dx_3 - v\, dx_1\, dx_3 = 0$$

On combining like terms and dividing through by $dx_1\, dx_2\, dx_3$, we get

$$\frac{\partial f_1}{\partial x_1} + \frac{\partial v}{\partial x_2} = 0 \tag{7-8}$$

Similarly, by summing the forces parallel to the x_2 axis, we obtain

$$\frac{\partial f_2}{\partial x_2} + \frac{\partial v}{\partial x_1} + \rho g = 0 \tag{7-9}$$

Equations (7-8) and (7-9) are the differential equations of equilibrium for two-dimensional stress problems. They must be satisfied throughout the entire body.

At the boundaries, the stress components must be in equilibrium with the loads. We can treat the loads as a continuation of the stress distribution determined by Eqs. (7-8) and (7-9). Let us resolve the surface forces per unit area along the boundary into components X_1 parallel to the x_1 axis and X_2 parallel to the x_2 axis. Then, for equilibrium at any point on the boundary (Fig. 7-2a):

$$X_1 = f_1 \cos \alpha + v \sin \alpha \tag{7-10}$$
$$X_2 = f_2 \sin \alpha + v \cos \alpha \tag{7-11}$$

where α is the angle the normal to the boundary makes with the x_1 axis.

Suppose, for example, that the body is rectangular, with sides parallel to the coordinate axes. Then, at a point on a side parallel to the x_1 axis, $X_1 = \pm v$ and $X_2 = \pm f_2$, since $\alpha = \pi/2$. (The positive sign should be taken if the normal extends outward in the positive direction of the x_2 axis.) Thus, the stress components equal the components of the boundary loads per unit area along the boundary.

The equilibrium equations and boundary equations are not sufficient to determine the state of stress in a body. Additional information must be obtained from a knowledge of the deformations produced in the body.

Unit strains at a point in a two-dimensional body can be resolved, like the stresses, into three components: elongations ϵ_1 parallel to the x_1 axis, elongations ϵ_2 parallel to the x_2 axis, and shearing strains γ. These strains are related to continuous functions of displacement u and w in the x_1 and x_2 directions, respectively, by

$$\epsilon_1 = \frac{\partial u}{\partial x_1} \qquad \epsilon_2 = \frac{\partial w}{\partial x_2} \qquad \gamma = \frac{\partial u}{\partial x_2} + \frac{\partial w}{\partial x_1} \tag{7-12}$$

Inspection of these equations indicates a relation between the strain components. We can obtain it by differentiating Eqs. (7-12):

$$\frac{\partial^2 \epsilon_1}{\partial x_2{}^2} + \frac{\partial^2 \epsilon_2}{\partial x_1{}^2} = \frac{\partial^2 \gamma}{\partial x_1\,\partial x_2} \tag{7-13}$$

Since the body is elastic, we can relate the strains to the stresses by

$$\epsilon_1 = \frac{1}{E}\,(f_1 - \mu f_2) \tag{7-14}$$

$$\epsilon_2 = \frac{1}{E}\,(f_2 - \mu f_1) \tag{7-15}$$

$$\gamma = \frac{2(1 + \mu)}{E}\,v \tag{7-16}$$

where E is the modulus of elasticity and μ is Poisson's ratio. Substitution in Eq. (7-13) gives

$$\frac{\partial^2}{\partial x_2{}^2}\,(f_1 - \mu f_2) + \frac{\partial^2}{\partial x_1{}^2}\,(f_2 - \mu f_1) = 2(1 + \mu)\,\frac{\partial^2 v}{\partial x_1\,\partial x_2} \tag{7-17}$$

Thus, we have expressed the strain-compatibility condition in terms of the stresses. But we can simplify this equation. We can use Eqs.

(7-8) and (7-9) to eliminate v from it. Differentiate Eq. (7-8) with respect to x_1 and Eq. (7-9) with respect to x_2 and add the results to obtain

$$2 \frac{\partial^2 v}{\partial x_1 \, \partial x_2} = - \frac{\partial^2 f_1}{\partial x_1{}^2} - \frac{\partial^2 f_2}{\partial x_2{}^2}$$

Substitution in Eq. (7-17) produces the simpler equation

$$\left(\frac{\partial^2}{\partial x_1{}^2} + \frac{\partial^2}{\partial x_2{}^2} \right) (f_1 + f_2) = 0 \tag{7-18}$$

Equation (7-18) in conjunction with Eqs. (7-8) and (7-9) and boundary conditions are usually sufficient for determination of two-dimensional stress distributions. But they contain the derivatives of three dependent variables with respect to two independent variables. They are not in canonical form. To solve them, we require an equation with the derivatives of only one dependent variable with respect to x_1 and x_2.

As suggested at the start of this article, let us introduce the so-called **Airy stress function** ϕ. An arbitrary function of x_1 and x_2, ϕ is defined by

$$f_1 = \frac{\partial^2 \phi}{\partial x_2{}^2} - \rho g x_2 \tag{7-19}$$

$$f_2 = \frac{\partial^2 \phi}{\partial x_1{}^2} - \rho g x_2 \tag{7-20}$$

$$v = - \frac{\partial^2 \phi}{\partial x_1 \, \partial x_2} \tag{7-21}$$

Consequently, ϕ satisfies the equilibrium equations (7-8) and (7-9). To determine the conditions under which ϕ also meets the compatibility requirement, substitute Eqs. (7-19) to (7-21) in Eq. (7-18). We then find that ϕ must satisfy

$$\frac{\partial^4 \phi}{\partial x_1{}^4} + 2 \frac{\partial^4 \phi}{\partial x_1{}^2 \, \partial x_2{}^2} + \frac{\partial^4 \phi}{\partial x_2{}^4} = 0 \tag{7-22}$$

We have now expressed the two-dimensional stress distribution as a partial differential equation in canonical form.

This equation has been widely studied. Many different methods have been developed to solve it for a wide variety of boundary conditions. (S. Timoshenko and J. N. Goodier, "Theory of Elasticity," and I. S. Sokolnikoff, "Mathematical Theory of Elasticity," both McGraw-Hill Book Company, New York, give comprehensive treatments.)

One method applicable to bodies with rectangular boundaries attempts to find a solution in terms of polynomials. First, a polynomial of second degree in x_1 and x_2 with undetermined, constant coefficients is chosen as ϕ. The stresses given by Eqs. (7-19) to (7-21) are then examined. Next, the procedure is repeated for a third-degree polynomial. After that, the procedure is repeated for a fourth-degree, fifth-degree, sixth-degree polynomial in succession, and relations are determined between the coefficients of each polynomial so that in each case ϕ satisfies Eq. (7-22). To solve a specific boundary-value problem, you have to combine the polynomials to satisfy given conditions.

For example, let us choose $\phi_2 = (a_2/2)x_1{}^2 + b_2 x_1 x_2 + (c_2/2)x_2{}^2$. It satisfies Eq. (7-22). Let us also assume that $\rho g = 0$. Then, Eqs. (7-19) to (7-21) give $f_1 = c_2$, $f_2 - a_2$, and $v = -b_2$. Since the coefficients are constants, all three stress components are constant throughout the body. The forces per unit area acting on the boundaries must equal these stresses.

For use in a problem to follow, let us now choose

$$\phi_4 = \frac{a_4}{4 \cdot 3} x_1{}^4 + \frac{b_4}{3 \cdot 2} x_1{}^3 x_2 + \frac{c_4}{2} x_1{}^2 x_2{}^2 + \frac{d_4}{3 \cdot 2} x_1 x_2{}^3 + \frac{e_4}{4 \cdot 3} x_2{}^4$$

This will satisfy Eq. (7-22) only if $e_4 = -(2c_4 + a_4)$. For our later use, let us assume that all coefficients except d_4 are zero. In that case, Eqs. (7-19) to (7-21) yield $f_1 = d_4 x_1 x_2$, $f_2 = 0$, and $v = -d_4 x_2{}^2/2$. Figure 7-3b shows the boundary forces on a rectangular body for this stress distribution.

As an example of the use of the polynomials that satisfy Eq. (7-22), let us determine the stresses in a prismatic, rectangular cantilever (Fig. 7-3a). Length of the cantilever is L, depth D, and thickness 1. The beam carries a concentrated load P at its free end. We will not evaluate stresses due to the weight of the beam in this problem. Let us select coordinate axes x_1 and x_2 as shown in Fig. 7-3a.

Boundary conditions require shears and normal stresses at the fixed end to prevent movement. At the free end, there must be shears adding up to P. And top and bottom surfaces of the cantilever, being free of external forces, must be free of stresses normal to the surfaces and of shears.

Let us examine polynomial ϕ_4 as a possible solution. As a start, we take all coefficients but d_4 as zero, to get the stress distribution $f_1 = d_4 x_1 x_2$, $f_2 = 0$, and $v = -d_4 x_2{}^2/2$. This provides acceptable normal stresses $f_1 = d_4 L x_2$ and shears at the fixed end (Fig. 7-3b). At the free end, stresses are also acceptable: $f_1 = 0$ and there are shears to counteract

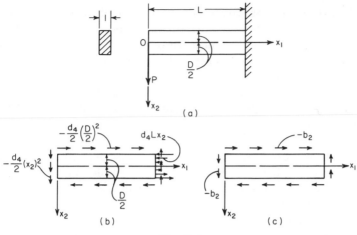

Fig. 7-3

P.   On the top and bottom surfaces, however, conditions are not acceptable.   While $f_2 = 0$, as desired, there are constant shears $v = -d_4D^2/2$.

To remove these shears, let us include polynomial ϕ_2 in the solution.
If we take coefficients a_2 and c_2 as zero, the stress distribution becomes
pure shear, $v = -b_2$.   Thus, the boundaries will carry constant shears
$-b_2$, as indicated in Fig. 7-3c.   Then, if we subtract these shears from
those provided by the ϕ_4 solution, we will have the boundary conditions
we seek.

So let us choose $\phi = d_4x_1x_2{}^3/6 - b_2x_1x_2$.   It satisfies Eq. (7-22).   We
have only to determine d_4 and b_2 to satisfy the boundary conditions.

The stresses are $f_1 = d_4x_1x_2$, $f_2 = 0$, and $v = b_2 - d_4x_2{}^2/2$.   For the
top and bottom surfaces of the beam to be free of stresses, $v = 0$ when
$x_2 = \pm D/2$.   Hence, we must have $b_2 = d_4D^2/8$.   Substitution of this
in the equation for v gives $v = (D^2 - 4x_2{}^2)d_4/8$.   We can determine d_4
from the boundary condition at the free end, where the shears must
resist the load P.

$$- \int_{-D/2}^{D/2} v \, dx_2 = -\frac{d_4}{8} \int_{-D/2}^{D/2} (D^2 - 4x_2{}^2) \, dx_2 = P$$

This gives $d_4 = -12P/D^3$.   Since $D^3/12 = I$, the moment of inertia of
the beam cross section, we can write $d_4 = -P/I$.   Therefore, the stresses
in the cantilever are $f_1 = -Px_1x_2/I$, $f_2 = 0$, and $v = -(D^2 - 4x_2{}^2)P/8I$.

7-6   Numerical Integration of Partial Differential Equations.   The
preceding articles demonstrated methods of obtaining particular solu-

tions of partial differential equations in closed form. Not all such equations, however, are amenable to that type of solution. You may sometimes have to resort to numerical integration to obtain answers.

For the purpose, you generally will find it expedient to convert the given equation into a finite-difference equation. Often, transforming derivatives into central differences will give good results. But sometimes you may have to use combinations of central and forward or backward differences.

The formulas used for partial derivatives are extensions of those for ordinary derivatives, as given in Art. 6-1. For one thing, you may use a different step size for each variable. For example, if z is the dependent variable and x and y the independent variables, you might select a step size h for x and a step size k for y.

Geometrically, you are laying a mesh with spacings h and k over the xy plane. Your objective is to approximate the partial derivatives of z with respect to x and y at the mesh intersections, which are called nodes, or pivotal points. With the derivatives, you set up difference equations that you solve to find values of z at the nodes.

Figure 7-4 shows part of a mesh in the xy plane. At (x_m, y_m), the node is labeled m. At $(x_m - h, y_m)$ and $(x_m + h, y_m)$, the nodes are marked l and r, respectively. At $(x_m, y_m - k)$ and $(x, y_m + k)$, the nodes are marked d and u, respectively. Similarly, related symbols are assigned to nodes more distant from m. The symbols are used as subscripts to identify the nodes at which values of z and its derivatives are determined.

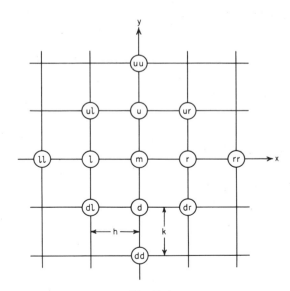

Fig. 7-4

Using central differences, we can approximate the partial derivatives of z with respect to x and y by

$$D_x z_m = \frac{1}{2h}(z_r - z_l)$$

$$D_y z_m = \frac{1}{2k}(z_u - z_d)$$

$$D_x{}^2 z_m = \frac{1}{h^2}(z_r - 2z_m + z_l)$$

$$D_y{}^2 z_m = \frac{1}{k^2}(z_u - 2z_m + z_d)$$

$$D_x{}^3 z_m = \frac{1}{2h^3}(z_{rr} - 2z_r + 2z_l - z_{ll})$$

$$D_y{}^3 z_m = \frac{1}{2h^3}(z_{uu} - 2z_u + 2z_d - z_{dd})$$

$$D_x{}^4 z_m = \frac{1}{h^4}(z_{rr} - 4z_r + 6z_m - 4z_l + z_{ll})$$

$$D_y{}^4 z_m = \frac{1}{k^4}(z_{uu} - 4z_u + 6z_m - 4z_d + z_{dd})$$

$D_{xy} z_m$ can be obtained by operating on D_y with D_x. When $h = k$, for example,

$$D_{xy} z_m = \frac{1}{4h^2}[(z_r - z_l)_u - (z_r - z_l)_d] = \frac{1}{4h^2}(z_{ur} - z_{ul} - z_{dr} + z_{dl})$$

You can set up similar approximations for partial derivatives with respect to additional independent variables.

Analysis of the effects of errors in these approximations on solutions has to be left to your continuing study of this subject with the aid of books listed in Art. 7-7. Also, left to your continuing study are discussions of stability and convergence of methods of solution and explanations of a wide variety of methods of solution not covered in this book. But we will examine some characteristics of partial differential equations that are important in numerical integration.

To illustrate these characteristics, let us refer to three types of linear second-order equations. In analogy with second-degree algebraic equations representing conic sections, partial differential equations of the type

$$af_{xx} + bf_{xy} + cf_{yy} = F(x,y,f,f_x,f_y) \tag{7-23}$$

are classified as elliptic, parabolic, and hyperbolic. In Eq. (7-23), a, b, and c are real, continuous functions of x and y, and F also is a continuous function.   The equation is called **elliptic** when $b^2 - 4ac < 0$, **parabolic** when $b^2 - 4ac = 0$, and **hyperbolic** when $b^2 - 4ac > 0$.

When you are dealing with an equation that is elliptic in a region R, you usually will find that boundary conditions specify the function f, or its normal derivative, or a linear combination of these at every point of the closed boundary of R within which you are seeking a solution. For elliptic equations, these boundary conditions completely determine the solution within the boundary.   To solve such equations, you can set up a system of linear, algebraic equations involving values of f at interior and boundary nodes.   **Laplace's equation**

$$\nabla^2 U = \frac{\partial^2 U}{\partial x^2} + \frac{\partial^2 U}{\partial y^2} = 0 \qquad (7\text{-}24)$$

is an example of an elliptic equation.

With an equation that is parabolic in a region R, you generally will be given the initial value of f at some time t_0.   You will also need the value of f, or of its normal derivative, or of a linear combination of these on the boundary.   These conditions do not define a solution within a closed domain.   So, starting with initial values and given points on an open boundary, you determine the solution at adjacent nodes and then at nodes farther and farther away.   The **one-dimensional heat-flow equation**

$$K \frac{\partial^2 U}{\partial x^2} = \frac{\partial U}{\partial t} \qquad (7\text{-}25)$$

is an example of a parabolic equation.

With an equation that is hyperbolic in a region R, you often will be given the values of f and its first derivative with respect to time at some time t_0.   Also, you will need the value of f, or of its normal derivative, or a linear combination of these on the boundary.   These conditions do not define a solution within a closed domain.   You therefore have to find the solution in the same way as for parabolic equations.   The **one-dimensional wave equation**

$$a^2 \frac{\partial^2 U}{\partial x^2} = \frac{\partial^2 U}{\partial t^2} \qquad (7\text{-}26)$$

treated in Art. 7-4 is an example of a hyperbolic equation.

To illustrate, let us start with a simple problem requiring solution of an elliptic equation.   Given the boundary temperatures shown in Fig. 7-5, let us find the temperature distribution in an insulated, thin, metal

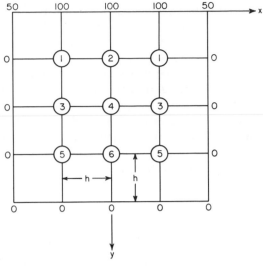

Fig. 7-5

plate. With the distribution independent of time, we are required to find the particular solution of the Laplace equation $U_{xx} + U_{yy} = 0$, where $U(x,y)$ is the temperature of the plate that satisfies the boundary conditions.

Let us select a mesh with spacing h in both the x and y directions, as indicated in Fig. 7-5. Let us also number the internal nodes from 1 to 6, giving the same number to pivotal points symmetrically located with respect to the y axis. Now, using central differences, we can approximate U_{xx} at any point (x_m,y_m) by $(U_r - 2U_m + U_l)/h^2$ and U_{yy} at that point by $(U_u - 2U_m + U_d)/h^2$. Substitution in the Laplace equation gives the difference equation

$$U_r + U_l + U_u + U_d - 4U_m = 0 \qquad (7\text{-}27)$$

When we apply Eq. (7-27) to nodes 1 to 6 in succession, we get the following system of six linear, algebraic equations:

Node	U_1	U_2	U_3	U_4	U_5	U_6	= Constant
1	−4	1	1				−100
2	2	−4		1			−100
3	1		−4	1	1		0
4		1	2	−4		1	0
5			1		−4	1	0
6				1	2	−4	0

Solution of these equations yields the temperatures at the nodes: $U_1 = 42.86$, $U_2 = 52.68$, $U_3 = 18.75$, $U_4 = 25.00$, $U_5 = 7.14$, and $U_6 = 9.82$. Notice how the boundary conditions completely determined the temperature distribution throughout the plate.

Now, let us try a problem requiring solution of a parabolic equation. Suppose a metal bar of length L and constant cross section is heated to 100°C. Suppose also that at time $T = 0$, the bar is enclosed in insulation except at one end, $X = L$, which is cooled rapidly to 10°C and held at that temperature. What is the temperature distribution throughout the length of the bar for $T > 0$?

The temperature variation is governed by the one-dimensional heat-flow equation

$$K\frac{\partial^2 U}{\partial X^2} = \frac{\partial U}{\partial T} \tag{7-28}$$

where $U(X,T)$ is the temperature at any point X in the bar at any time T, and K is the thermal diffusivity of the metal, or ratio of thermal conductivity to the product of specific heat and mass density. U must satisfy the initial condition $U(X,0) = 100$, and the boundary conditions $U(L,T) = 10$, and $\partial U/\partial X = 0$ when $X = 0$, for all T. Thus, to find the temperature distribution for $T > 0$, we must start with known initial conditions at $T = 0$ and determine values of U that satisfy Eq. (7-28) and the boundary conditions at a later time. From the new values of U, we can find the temperature distribution at a still later time. Since the boundary for T is open, we can continue in this fashion indefinitely.

Before we start this procedure, however, we should transform the variables to make them nondimensional. Usually, such a step simplifies the calculations and makes selection of mesh spacing easier. So let us set $X = xL$, $T = L^2 t/K$, and $U = u(100 - 10) + 10 = 90u + 10$. Then, Eq. (7-28) becomes $u_{xx} = u_t$. The initial condition is now $u(x,0) = 1$. And the boundary conditions are $u(1,t) = 0$ and $u_x(0,t) = 0$.

Let us select a mesh with spacing h in the x direction and k in the t direction. We can approximate u_{xx} at any point (x_m,t_m), with central differences, by $(u_r - 2u_m + u_l)/h^2$. We can use central differences because the initial conditions supply sufficient information for a start. But it is desirable to approximate u_t at (x_m,t_m), with forward differences, by $(u_u - u_m)/k$, because there is insufficient information for central differences. Substitution in the differential equation leads to

$$u_u = \alpha u_r + (1 - 2\alpha)u_m + \alpha u_l \tag{7-29}$$

where $\alpha = k/h^2$. Thus, when u is known at (x_m,t_m) and adjoining nodes at the same time t_m, Eq. (7-29) gives u at the node $(x_m, t_m + k)$.

Inspection of Eq. (7-29) indicates that if $\alpha = \frac{1}{2}$, the coefficient of u_m becomes 0. Then, the equation simplifies to

$$u_u = \frac{1}{2}(u_r + u_l) \qquad\qquad (7\text{-}30)$$

For illustrative purposes, let us select a rather coarse mesh, with spacing $h = \frac{1}{2}$ and $k = \frac{1}{8}$, $\alpha = \frac{1}{2}$. Using these values, we can approximate the boundary condition $u_x(0,t) = 0$, with central differences, by $(u_r - u_l)/2h = 0$. This gives $u_r = u_l$ at $x = 0$ for all t.

The xt plane is represented in Fig. 7-6 for $-\frac{1}{2} < x < 1$ and $0 < t < 1$. Values of u for $x = -\frac{1}{2}$ are determined by $u_r = u_l$ at $x = 0$. They are needed for finding u_u at $x = 0$ from Eq. (7-30). Values of u at $x = 1$ are determined by the boundary condition $u(1,t) = 0$. And at $t = 0$, we are given $u = 1$ for all x. There is a conflict in these conditions, however, We cannot have $u = 1$ at $x = 1$ when $t = 0$ and also $u = 0$ at $x = 1$ for all t. So let us arbitrarily take $u = \frac{1}{2}$ at $(1,0)$.

Now, using Eq. (7-30) and $u_r = u_l$ at $x = 0$, we can find the temperature distribution in the bar at $t = \frac{1}{8}$. For example, $u(\frac{1}{2}, \frac{1}{8}) = \frac{1}{2}(1 + \frac{1}{2}) = \frac{3}{4}$. The result is entered on the node in Fig. 7-6. From this value, we obtain $u(-\frac{1}{2}, \frac{1}{8}) = u(\frac{1}{2}, \frac{1}{8}) = \frac{3}{4}$. This, too, is written on a node. When we obtain all the values of u at $t = \frac{1}{8}$, we follow the same procedure for $t = \frac{1}{4}$, then $t = \frac{3}{8}$, etc. Figure 7-6 gives the values of u on the nodes up to $t = 1$. Then, the transformation $U = 90u + 10$ gives the particular solution of Eq. (7-28).

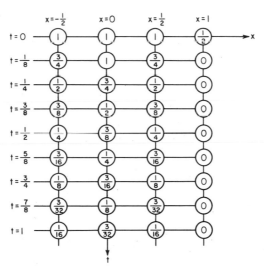

Fig. 7-6

The problems we treated in this chapter dealt with rectangular boundaries. For these, Cartesian coordinates are suitable. For other types of boundaries, other coordinate systems may simplify the computations. For example, for a circular boundary, you may find polar coordinates advantageous. The methods can be extended to solution of problems involving more variables.

7-7 *Bibliography*

I. Babuska, M. Prager, and E. Vitasek, "Numerical Processes in Differential Equations," John Wiley & Sons, Inc., New York.

R. Beckett and J. Hurt, "Numerical Calculations and Algorithms," McGraw-Hill Book Company, New York.

L. Bers, F. John, and M. Schecter, "Partial Differential Equations," John Wiley & Sons, Inc., New York.

R. Dennemeyer, "Introduction to Partial Differential Equations," McGraw-Hill Book Company, New York.

G. E. Forsythe and W. R. Wasow, "Finite Difference Methods for Partial Differential Equations," John Wiley & Sons, Inc., New York.

P. R. Garabedian, "Partial Differential Equations," John Wiley & Sons, Inc., New York.

D. Greenspan, "Introduction to Partial Differential Equations," McGraw-Hill Book Company, New York.

A. G. Hansen, "Similarity Analyses of Boundary-value Problems in Engineering," Prentice-Hall, Inc., Englewood Cliffs, N.J.

M. G. Salvadori and M. L. Baron, "Numerical Methods in Engineering," Prentice-Hall, Inc., Englewood Cliffs, N.J.

PROBLEMS

1. Solve the following equations:
   (a) $\partial z/\partial t = \sin x$, given $z = 0$ when $t = 0$.
   (b) $\partial^2 z/\partial x\, \partial y = 0$, given $z(0,y) = \sin y$ and $z_x = x^2$ for all y.
   (c) $\partial^2 z/\partial x\, \partial y - \partial z/\partial y = y$, given $z(x,0) = x$ and $z_y(0,y) = -y$.

2. Find the general solutions of the following linear equations:
   (a) $xzz_x + yzz_y = xy$.
   (b) $xU_y + (1 + z^2)U_z - yU_x = 0$.

3. Find the general solutions of the following linear equations:
   (a) $z_{xx} - 3z_{xy} + 2z_{yy} = 0$.
   (b) $z_{xx} - 2z_{xy} + z_{yy} = 0$.
   (c) $\partial^3 z/\partial x^2\, \partial y - 2\partial^3 z/\partial x\, \partial y^2 + \partial^3 z/\partial y^3 = 0$.
   (d) $z_{xx} - z_{xy} - 2z_{yy} = x - y$.

4. (a) The amount of heat required to raise the temperature U of a metal body of mass m to $U + \Delta U$ is $ms\, \Delta U$, where s is the specific heat of the metal. The amount of heat Q flowing through a cross section of area A of the body per unit time is given by $Q = -kA\, \Delta t\, \partial U/\partial x$, where Δt is the time during which heat flow occurs, x is the distance normal to the cross section, and k is the thermal conductivity of the metal. Using these assumptions, derive the partial differential equation governing heat flow through a pris-

matic metal bar insulated on its lateral surfaces.   Assume also that no heat is generated internally.

(b) Expand the derivation in Prob. 4a to three-dimensional heat flow.

(c) Solve $U_t = U_{xx}$, given the boundary conditions $U = 0$ when $x = 0$ and $x = 1$ for all t.   The initial condition is $U = \sin \pi x$ when $t = 0$.

(d) Solve $U_t = U_{xx}$, given the boundary conditions $U = 0$ when $x = 0$ and $x = 1$ for all t.   The initial condition is $U = \sin \pi x + \frac{1}{2} \sin 2\pi x$ when $t = 0$.

(e) Solve $U_t = U_{xx}$, given the boundary conditions $U = 0$ when $x = 0$ and $x = 1$ for all t.   The initial condition is $U = 100$ when $t = 0, 0 \le x \le 1$.

5. A 2-ft-long string weighing 0.25 lb is placed under 1-lb tension.   If the center is pulled $\frac{1}{4}$ in. out of horizontal alignment, then released at time $t = 0$, what is the position of the string at any later time?

6. (a) A long transmission line is imperfectly insulated.   It has capacitance C per unit length of cable and current leakage to ground.   Its resistance per unit length is R, its inductance per unit length is L, and conductance to ground is G.   Develop a partial differential equation that determines the voltage $V(x,t)$ at any point x of the cable at any time t in terms of L, C, R, and G.   Do the same for the current $I(x,t)$.

(b) Assume that leakage and inductance are negligible in the transmission line of Prob. 6a.   Determine from the solution of that problem the partial differential equations for voltage and current in the cable.

(c) Assume that there are high-frequency currents and voltages in the transmission line of Prob. 6a.   Hence, V and V_t are negligible compared with V_{tt}.   Also, I and I_t are negligible compared with I_{tt}.   Determine from the solution of Prob. 6a the partial differential equations for voltage and current in the cable.

7. A constant pressure p is applied to a thin, flexible membrane stretched over a square hole with sides L (Fig. 7-7a and b).   This places the membrane

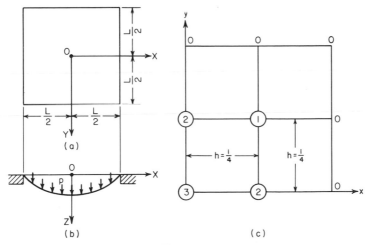

Fig. 7-7

under a constant tension T per unit of length. When the membrane deflection Z is small, it is determined by the **Poisson equation**

$$\frac{\partial^2 Z}{\partial X^2} + \frac{\partial^2 Z}{\partial Y^2} + f = 0$$

where $f = p/T$. Using a mesh in the XY plane with spacing $h = \frac{1}{4}$ in the X and Y directions (Fig. 7-7c), find by numerical integration the deflections to three significant figures at the interior nodes. (*Hint:* Make the variables nondimensional and take advantage of symmetry.)

8. An electrical transmission line 100 miles long has a resistance R of 0.5 ohm per mile and a capacitance $C = 2 \times 10^{-7}$ farad per mile. Inductance L and conductance G to ground are negligible. Hence, the voltage V at any point x along the cable is determined at any time t, sec, by $V_{xx} = RCV_t$ (see Prob. 6b). At time $t = 0$, with 6 volts at $x = 0$ and 2 volts at $x = 100$, the terminal end is suddenly grounded.

    (a) What are the voltages throughout the cable after that? (*Hint:* Assume that the voltage at any point is composed of a steady-state voltage and a transient voltage, decreasing rapidly with time.)

    (b) Solve Prob. 8a to three significant figures by numerical integration at $x = 25$, 50, and 75 miles for $0 < t \leq 1/4,000$ in increments of t of 1/32,000 sec.

9. Solve Prob. 5 by numerical integration to three significant figures at the eighth points of the string for $0 < t \leq \frac{1}{16}$ in increments of t of $\frac{1}{64}$ sec.

ANSWERS

    1. (a) Integration with respect to t gives $z = t \sin x + f(x)$. Since $z(x,0) = 0$, $f(x) = 0$. Hence, $z = t \sin x$. (See Art. 7-1.)

    (b) Integration with respect to y gives $z_x = f(x)$. Since we are given $z_x = x^2$, $f(x) = x^2$. Integration with respect to x produces $z = x^3/3 + g(y)$. Since $z(0,y) = \sin y$, $g = \sin y$. Therefore, $z = x^3/3 + \sin y$. (See Art. 7-1.)

    (c) Consider the equation an ordinary differential equation with z_y as the independent variable and y a constant: $(D - 1)z_y = y$. Integration with respect to x gives $z_y = c_1 e^x - y$. Since we are given $z_y(0,y) = -y$, we have $c_1 = 0$. So, $z_y = -y$. Integration with respect to y gives $z = -y^2/2 + g(x)$. Since $z(x,0) = x$, we have $x = 0 + g$, from which $g = x$. Therefore, $z = x - y^2/2$. (See Art. 7-1.)

    2. (a) Since we have a first-order equation, the solution is given by

$$\frac{dx}{xz} = \frac{dy}{yz} = \frac{dz}{xy}$$

(See Art. 7-2.) Multiplying through by xyz, we get $y\,dx = x\,dy = z\,dz$. From the first pair, we find $dx/x = dy/y$. Integration yields $\log_e y = \log_e x + \log_e c_1$, from which $y/x = c_1$. Adding the first and third pair of terms to the second and third terms, we get $y\,dx + x\,dy = 2z\,dz$. This is an exact equation. Integration gives $xy = z^2 + c_2$, or $xy - z^2 = c_2$.

Hence, $f(xy - z^2, y/x) = 0$ is the general solution.

(b) The solution is given by

$$\frac{dy}{x} = \frac{dz}{1 + z^2} = \frac{dx}{-y} = \frac{dU}{0}$$

(See Art. 7-2.) The fourth term requires $dU = 0$. Hence, $U = c_1$. Integration of the first and third terms gives $x^2 + y^2 = c_2$. Now, multiplying the first pair of terms by x^2 leads to $x \, dy = x^2 \, dz/(1 + z^2)$. Multiplying the second and third terms by y^2 produces $-y \, dx = y^2 \, dz/(1 + z^2)$. Addition of these results yields $x \, dy - y \, dx = (x^2 + y^2) \, dz/(1 + z^2)$, an exact differential equation. After division by $x^2 + y^2$, integration gives $\tan^{-1}(y/x) - \tan^{-1} z = \tan^{-1} c_3$, or $(y - xz)/(x + yz) = c_4$. Therefore, the general solution is

$$f_1\left(U, x^2 + y^2, \frac{y - xz}{x + yz}\right) = 0 \quad \text{or} \quad U = f_2\left(x^2 + y^2, \frac{y - xz}{x + yz}\right)$$

3. (a) This is a uniform equation. The auxiliary equation is $m^2 - 3m + 2 = 0$. (See Art. 7-3.) Its roots are $m = 1, 2$. Hence, the general solution is $z = f_1(y + x) + f_2(y + 2x)$.

(b) The auxiliary equation is $m^2 - 2m + 1 = 0$. The roots are $m = 1, 1$. Therefore, the general solution is $z = f_1(y + x) + xf_2(y + x) + yf_3(y + x)$. See Art. 7-3.

(c) With the given arrangement of the differential equation, the auxiliary equation is $0m^3 + m^2 - 2m + 1 = 0$. Its roots are $m = 1, 1$ and $1/m = 0$. Hence, the general solution is $z = f_1(y + x) + xf_2(y + x) + yf_3(y + x) + f_4(x)$. (See Art. 7-3.) Alternatively, by writing the given equation in reverse order, you can take the auxiliary equation as $m^3 - 2m^2 + m = 0$, from which $m = 0, 1, 1$. It yields the same solution.

(d) The auxiliary equation is $m^2 - m - 2 = 0$. Its roots are $m = 2, -1$. So the complementary function is $z = f_1(y + 2x) + f_2(y - x)$. (See Art. 14-3.) Note that $x \quad y = -(y - x)$ is included in the complementary function and therefore the particular integral must contain $x(y - x)$ or $y(y - x)$ or $xy(y - x)$. So for particular integrals, try $axy^2 + bx^2y$. Substitution in the differential equation yields $2by - 2ay - 2bx - 4ax = x - y$. On collecting terms in x and y, we get $(-2b - 4a)x + (2b - 2a)y = x - y$. This requires $a = 0$, $b = -\frac{1}{2}$. Hence, the solution is $z = f_1(y + 2x) + f_2(y - x) - \frac{1}{2}x^2y$.

4. (a) Consider the portion P of the bar between a normal cross section at x and one at $x + \Delta x$. P has cross-sectional area A and mass $m = (\rho/g)A \, \Delta x$, where ρ is the density and g is acceleration due to gravity. The heat flowing into P at x is $Q_1 = -kA \, \Delta t U_x\big]_x$. The heat flowing out of P at $x + \Delta x$ is $Q_2 = -kA \, \Delta t U_x\big]_{x+\Delta x}$. The heat accumulating in P then is

$$Q_1 - Q_2 = -kA \, \Delta t U_x\big]_x + kA \, \Delta t U_x\big]_{x+\Delta x}$$

$$= kA \, \Delta t \left(U_x\big]_{x+\Delta x} - U_x\big]_x\right)$$

This heat changes the temperature of P by ΔU. Hence,

$$kA \, \Delta t \left(\Big[U_x \Big]_{x+\Delta x} - \Big[U_x \Big]_x \right) = ms \, \Delta U = \frac{\rho}{g} As \, \Delta x \, \Delta U$$

Divide both sides of the equation by $\Delta t \, \Delta x (\rho/g) As$ and set $K = kg/\rho s$. Then, let Δx and Δt approach zero. The result is the one-dimensional heat-flow equation

$$K \frac{\partial^2 U}{\partial x^2} = \frac{\partial U}{\partial t}$$

(b)

$$\frac{\partial U}{\partial t} = K \left(\frac{\partial^2 U}{\partial x^2} + \frac{\partial^2 U}{\partial y^2} + \frac{\partial^2 U}{\partial z^2} \right)$$

(c) Let $U = X(x)T(t)$. (See Art. 7-4.) Substitution in $U_t = U_{xx}$ gives $XT' = TX''$. Dividing through by XT gives $X''/X = T'/T = C$, a constant. Thus, we have $X'' = XC$ and $T' = TC$. Trial will show that we can obtain a solution only if C is negative. So let us take $C = -c^2$. Then, we have the general solutions $X = A_1 \cos cx + B_1 \sin cx$ and $T = C_1 e^{-c^2 t}$. So setting $A = A_1 C_1$ and $B = B_1 C_1$, we get $U = e^{-c^2 t}(A \cos cx + B \sin cx)$. When $x = 0$, $U = 0 = Ae^{-c^2 t}$, which gives $A = 0$. When $x = 1$, $U = 0 = Be^{-c^2 t} \sin c$. Let us take $c = n\pi$, where n is an integer. So far then, $U = Be^{-n^2\pi^2 t} \sin n\pi x$. The initial condition requires that when $t = 0$, $U = \sin \pi x = B \sin n\pi x$. Hence, $B = 1$ and $n = 1$. Therefore, the particular solution is $U = e^{-\pi^2 t} \sin \pi x$.

(d) The solution is the same as in Prob. 4c up to the application of the initial condition; that is, $U = Be^{-n^2\pi^2 t} \sin n\pi x$. But no values that we can choose for B and n will satisfy the initial condition $U = \sin \pi x + (\sin 2\pi x)/2$. Therefore, let us assume a solution in the form $U = B_1 e^{-n^2\pi^2 t} \sin n\pi x + B_2 e^{-m^2\pi^2 t} \sin m\pi x$. When $t = 0$, then, $U = \sin \pi x + (\sin 2\pi x)/2$ requires $B_1 = 1$, $B_2 = \frac{1}{2}$, $n = 1$, and $m = 2$. Hence, the particular solution is

$$U = e^{-\pi^2 t} \sin \pi x + \frac{1}{2} e^{-4\pi^2 t} \sin 2\pi x$$

(e) The solution is the same as in Prob. 4c up to the application of the initial condition; that is, $U = Be^{-n^2\pi^2 t} \sin n\pi x$. To satisfy the initial condition, we assume U in the form $U = \Sigma b_n e^{-n^2\pi^2 t} \sin n\pi x$ and expand 100 in a half-range Fourier sine series: $100 = \Sigma b_n \sin n\pi x$ for $0 \le x \le 1$. Using Eq. (5-18), we get

$$b_n = 2 \int_0^1 100 \sin n\pi x \, dx = \frac{200}{n\pi} (1 - \cos n\pi)$$

Therefore, the particular solution is

$$U = \frac{400}{\pi} \left(e^{-\pi^2 t} \sin \pi x + \frac{1}{3} e^{-9\pi^2 t} \sin 3\pi x + \frac{1}{5} e^{-25\pi^2 t} \sin 5\pi x + \cdots \right)$$

5. Article 7-4 gives the equation of a vibrating string as $a^2 y_{xx} - y_{tt} = 0$, with $a^2 = Hg/w = 1 \times 32.2 \times 2/0.25 = 257.6$. So $a = 16.1$. The boundary conditions are $y = 0$ when $x = 0$ and 2, for all t. Also, $y_t = 0$ when $t = 0$, for all x. The initial condition is $y = x/48$ for $0 \le x \le 1$ and $y = (2 - x)/48$ for $1 \le x \le 2$, when $t = 0$.

Let $y = X(x)T(t)$. Then, as in Art. 7-4, you find that

$$y = XT = (c_1 \sin \sqrt{k}\, x + c_2 \cos \sqrt{k}\, x)(c_3 \sin 16.1t \sqrt{k} + c_4 \cos 16.1t \sqrt{k})$$

From the boundary condition $y = 0$ when $x = 0$, we find $c_2 = 0$. From $y = 0$ when $x = 2$, we get $\sin 2 \sqrt{k} = 0$, since c_1 cannot also be zero. So $\sqrt{k} = n\pi/2$, where n is an integer. And from $y_t = 0$ when $t = 0$, we obtain, on differentiating, $c_3 = 0$. At this stage, we have

$$y = c_1 c_4 \sin \frac{n\pi x}{2} \cos 8.05 n\pi t$$

To satisfy the initial condition, assume that the solution is in the form

$$y = \sum b_n \sin \frac{n\pi x}{2} \cos 8.05 n\pi t$$

Also, expand y when $t = 0$ in a half-range Fourier sine series

$$y = \sum b_n \sin \frac{n\pi x}{2} \qquad \text{for } 0 \le x \le 2$$

From Eq. (5-18),

$$b_n = \frac{2}{2} \int_0^1 \frac{x}{48} \sin \frac{n\pi x}{2} \, dx + \frac{2}{2} \int_1^2 \frac{2 - x}{48} \sin \frac{n\pi x}{2} \, dx = \frac{1}{6n^2\pi^2} \sin \frac{n\pi}{2}$$

Therefore, the particular solution is

$$y = \frac{1}{6\pi^2} \left(\sin \frac{\pi x}{2} \cos 8.05\pi t - \frac{1}{9} \sin \frac{3\pi x}{2} \cos 24.15\pi t + \cdots \right)$$

6. (a) Consider a short length of cable between a point P at x and a point Q at $x + \Delta x$. The voltage V_Q at Q equals the voltage V_P at P minus the loss in PQ, or $V_Q = V_P - IR\,\Delta x - LI_t\,\Delta x$. Let $V_Q - V_P = \Delta V$. Then, $\Delta V = -IR\,\Delta x - LI_t\,\Delta x$. Divide both sides by Δx and let Δx approach zero. The result is $V_x = -IR - LI_t$.

Now, the current I_Q at Q equals the current I_P at P minus the leakage to ground and the charge on PQ, or $I_Q = I_P - VG\,\Delta x - CV_t\,\Delta x$. Let $I_Q - I_P = \Delta I$. Then, $\Delta I = -VG\,\Delta x - CV_t\,\Delta x$. Divide both sides by Δx and let Δx approach zero. The result is $I_x = -VG - CV_t$.

The equations are not in canonical form, because we have two dependent variables in each. (See Art. 7-5.) So differentiate the first with respect to x and the second with respect to t, and eliminate I_{xt} from the result.   On substituting I_x from the second equation, we get the desired form:

$$V_{xx} = LCV_{tt} + (RC + GL)V_t + RGV$$

Similarly, differentiate the first equation with respect to t and the second with respect to x.   Then, eliminate the derivatives of V to get

$$I_{xx} = LCI_{tt} + (RC + GL)I_t + RGI$$

These equations are called the **telephone equations.**

(b) With $G = L = 0$, we get $V_{xx} = RCV_t$ and $I_{xx} = RCI_t$.   These are known as the **telegraph equations.**   Compare them with the one-dimensional heat-flow equation.

(c) With $V = V_t - 0$ and $I = I_t = 0$, we get $V_{xx} = LCV_{tt}$ and $I_{xx} = LCI_{tt}$.   Compare these equations with the one-dimensional wave equation.

7. The boundary conditions are $Z = 0$ when $X = \pm L/2$ and $Y = \pm L/2$. To make the variables nondimensional, let $X = xL$, $Y = yL$, and $Z = fL^2z$. The partial differential equation becomes $z_{xx} + z_{yy} + 1 = 0$.   The boundary conditions then are $z = 0$ when $x = \pm\frac{1}{2}$ and $y = \pm\frac{1}{2}$.   Approximate z_{xx} at any point (x_m, y_m) by $(z_r - 2z_m + z_l)/h^2$ and z_{yy} at that point by $(z_u - 2z_m + z_d)/h^2$, with $h = \frac{1}{4}$.   (See Art. 7-6.)   Substitution in the equation gives

$$4z_m - z_r - z_l - z_u - z_d = \frac{1}{16}$$

Thus, we can establish the following system of linear algebraic equations:

Node	z_1	z_2	z_3	Constant
1	4	−2		$\frac{1}{16}$
2	−2	4	−1	$\frac{1}{16}$
3		−4	4	$\frac{1}{16}$

Solution of these equations yields $z_1 = \frac{9}{128}$, $z_2 = \frac{7}{128}$, and $z_3 = \frac{11}{256}$.   Hence, the deflections are $Z_1 = 0.0703fL^2$, $Z_2 = 0.0547fL^2$, and $Z_3 = 0.0429fL^2$.

8. (a) The governing equation is $V_{xx} = 10^{-7}V_t$.   The initial condition is $V = 6 - (6 - 2)x/100 = 6 - 0.04x$ at $t = 0$.   The boundary conditions are $V(0,t) = 6$ and $V(100,t) = 0$.

For $t > 0$, let $V = V_S + V_T$, where V_S is a steady-state voltage and V_T is a transient voltage. V_S is given by the boundary conditions as $6 - 0.06x$.   Hence, the initial condition for V_T at $t = 0$ is $V_T = 0.02x$, and the boundary conditions are $V_T(0,t) = 0$ and $V_T(100,t) = 0$.

Let $V_T = X(x)T(t)$.   (See Art. 7-4.)   Substitution in $V_{xx} = 10^{-7}V_t$ gives $TX'' = 10^{-7}XT'$.   Dividing through by XT, we get $X''/X = 10^{-7}T'/T = C$, a constant.   Thus, we have $X'' = XC$ and $T' = 10^7TC$.   Trial will show that we can have a solution if C is negative.   Let us take $C = -c^2$.

Then, we have the general solutions $X = A_1 \cos cx + B_1 \sin cx$ and $T = C_1 e^{-10^7 c^2 t}$. Setting $A = A_1 C_1$ and $B = B_1 C_1$, we get

$$V_T = e^{-10^7 c^2 t}(A \cos cx + B \sin cx)$$

From $V_T(0,t) = 0$, we find $A = 0$. From $V_T(100,t) = 0$, we get $c = n\pi/100$, where n is an integer. At this stage, we have

$$V_T = Be^{-1,000n^2\pi^2 t} \sin \frac{n\pi x}{100}$$

To satisfy the initial condition, assume V_T in the form

$$V_T = \sum b_n e^{-1,000n^2\pi^2 t} \sin \frac{n\pi x}{100}$$

and expand $0.02x$ in a half-range Fourier sine series in the range $0 \leq x \leq 100$. Using Eq. (12-18), we get

$$b_n = 0.02 \int_0^{100} 0.02x \sin \frac{n\pi x}{100} \, dx = -\frac{4}{n\pi}(-1)^n$$

Therefore,

$$V = V_s + V_T = 6 - 0.06x + \frac{4}{\pi}\left(e^{-1,000\pi^2 t} \sin \frac{\pi x}{100}\right.$$
$$\left. -\frac{1}{2}e^{-4,000\pi^2 t} \sin \frac{2\pi x}{100} + \cdots\right)$$

(b) Make the variables nondimensional by taking $x = 100y$, $t = 10^{-3}\tau$, and $V = 6E$. The differential equation then becomes $E_{yy} = E_\tau$, with initial condition $E = 1 - \frac{2}{3}y$ at $t = 0$. Boundary conditions are $E(0,\tau) = 1$ and $E(1,\tau) = 0$. We are required to find E at $y = \frac{1}{4}$, $\frac{1}{2}$, and $\frac{3}{4}$ for τ in increments of $\frac{1}{32}$.

Approximate $\partial^2 E/\partial y^2$ by $16(E_r - 2E_m + E_l)$ and $\partial E/\partial \tau$ by $32(E_u - E_m)$. Substitution in the differential equation gives $E_u = (E_r + E_l)/2$. Using this recursion formula, we find for E at the nodes:

	$y = 0$	$y = \frac{1}{4}$	$y = \frac{1}{2}$	$y = \frac{3}{4}$	$y = 1$
$\tau = 0$	1	0.833	0.667	0.500	0.166
$\tau = \frac{1}{32}$	1	0.833	0.667	0.467	0
$\tau = \frac{1}{16}$	1	0.833	0.650	0.333	0
$\tau = \frac{3}{32}$	1	0.825	0.583	0.325	0
$\tau = \frac{1}{8}$	1	0.792	0.575	0.292	0
$\tau = \frac{5}{32}$	1	0.788	0.542	0.278	0
$\tau = \frac{3}{16}$	1	0.771	0.533	0.271	0
$\tau = \frac{7}{32}$	1	0.767	0.521	0.267	0
$\tau = \frac{1}{4}$	1	0.761	0.517	0.261	0

(Note the need for a closer mesh for small values of τ.)   Multiplication of the values of E by 6 give the voltages V.

9. As in Prob. 5, the governing equation is $257.6y_{xx} - y_{tt} = 0$.   Make the the variables nondimensional by setting $x = 2X$, $t = T/\sqrt{2g} = 0.1242T$, and $y = Y/48$.   The equation then becomes $Y_{XX} = Y_{TT}$.   The boundary conditions convert to $Y(0,T) = Y(1,T) = 0$.   The initial condition for $T = 0$ becomes $Y = 2X$ for $0 \leq X \leq \frac{1}{2}$ and $Y = 2(1 - X)$ for $\frac{1}{2} < X \leq 1$.   Also, for $T = 0$, $Y_T = 0$.   We are required to find Y for increments of $\frac{1}{8}$ in X and for $0 \leq T \leq 0.5$ in increments of about $\frac{1}{8}$.

Using central differences, approximate Y_{XX} by $64(Y_r - 2Y_m + Y_l)$ and Y_{TT} by $64(Y_u - 2Y_m + Y_d)$.   Substitution in the differential equation gives the recursion formula

$$Y_u = Y_r + Y_l - Y_d$$

The initial condition $Y_T = 0$ may be approximated by $Y_u - Y_d = 0$ at $T = 0$. Substitution in the recursion formula yields $Y_u = (Y_r + Y_l)/2$ at $T = 0$. Starting with this formula, then using the recursion formula thereafter, we find for Y at the nodes:

	$X = 0$	$X = \frac{1}{8}$	$X = \frac{1}{4}$	$X = \frac{3}{8}$	$X = \frac{1}{2}$
$T = 0$	0	0.250	0.500	0.750	1.000
$T = \frac{1}{8}$	0	0.250	0.500	0.750	0.750
$T = \frac{1}{4}$	0	0.250	0.500	0.500	0.500
$T = \frac{3}{8}$	0	0.250	0.250	0.250	0.250
$T = \frac{1}{2}$	0	0	0	0	0

Dividing Y by 48 yields the positions of the eighth points of the string.

EIGHT

Complex Variables and Conformal Mapping

In previous chapters, we have had occasion to deal with complex numbers and sometimes have used them to solve problems. For example, we employed complex numbers to obtain trigonometric solutions of differential equations. In this chapter, we advance deeper into the complex domain. We examine the theory of functions of a complex variable and applications to solution of engineering problems.

This theory makes available several new techniques. We can, for instance, simplify some problems by treating real variables as special values of complex variables. We can use analytic functions of complex variables for conformal mapping of the points of a plane into points of another plane. Also, we can apply complex variables to integration of real integrals and to the solution of differential equations, in particular, the Laplace partial differential equation.

8-1 *Fundamental Properties of Complex Numbers.* To refresh your memory, here is a brief summary of the more important properties of complex numbers.

A complex number is an ordered pair of real numbers (a,b). The first component a is called a real number. The second component b is called an imaginary number. Complex numbers obey the following rules:

  1. Two complex numbers (a_1,b_1) and (a_2,b_2) are equal only if $a_1 = a_2$ and $b_1 = b_2$. In particular, $(a,b) = 0$ only if $a = 0$ and $b = 0$.
  2. The sum and product of two complex numbers are:

$$(a_1,b_1) + (a_2,b_2) = (a_1 + a_2, b_1 + b_2) \tag{8-1}$$
$$(a_1,b_1)(a_2,b_2) = (a_1a_2 - b_1b_2, a_1b_2 + a_2b_1) \tag{8-2}$$

As usual, subtraction is the inverse of addition, and division is the inverse of multiplication.

These rules are satisfied if the imaginary number is taken as the square root of a negative real number or the product of a real number and i, where

$$i = \sqrt{-1} \tag{8-3}$$

Note that $i^2 = (-i)^2 = -1$.

A complex variable $z = (x,y)$, where x and y are real variables, can be written in the forms

$$z = x + iy \tag{8-4a}$$
$$z = r(\cos \theta + i \sin \theta) \tag{8-4b}$$
$$z = re^{i\theta} \tag{8-4c}$$

The **modulus**, or **absolute value**, of z is

$$|z| = r = \sqrt{x^2 + y^2} = |x + iy| \tag{8-5}$$

The **amplitude**, or **argument**, of z is

$$\theta = \text{amp } z = \arg z = \arctan \frac{y}{x} = \text{amp } (x + iy) \tag{8-6}$$

8-2   *Graphical Representation of Complex Numbers.* A complex number can be represented geometrically in a plane, called the complex (Argand or Gauss) plane. In Cartesian coordinates, the real component is plotted parallel to the x axis (real axis) and the imaginary component is plotted parallel to the y axis (Fig. 8-1). In polar coordinates, the length of the radius vector from the origin to the point representing a complex number is the modulus r. The angle the vector makes with the x axis is the amplitude θ (Fig. 8-1).

8-3   *Complex Algebra.* Two complex numbers can be added (or subtracted) geometrically by vector addition (or subtraction) of their

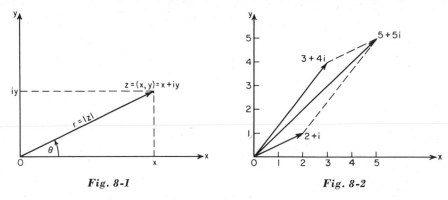

Fig. 8-1   Fig. 8-2

Fig. 8-1   **Fig. 8-2**

radius vectors (parallelogram law). For example, add $z_1 = 3 + 4i$ and $z_2 = 2 + i$. By Eq. (8-1), $z_1 + z_2 = (3 + 2) + i(4 + 1) = 5 + 5i$. Alternatively, you can add the radius vectors, as indicated in Fig. 8-2, to obtain the same result.

The **conjugate** of a complex number $z = (x,y) = x + iy$ is

$$\bar{z} = (x, -y) = x - iy \tag{8-7}$$

Note that the conjugate of the sum of two complex numbers is the sum of their conjugates:

$$\overline{z_1 + z_2} = \bar{z}_1 + \bar{z}_2 \tag{8-8}$$

And the conjugate of the product of two complex numbers is the product of their conjugates:

$$\overline{z_1 z_2} = \bar{z}_1 \bar{z}_2 \tag{8-9}$$

You can also easily verify that the sum of a complex number and its conjugate is a real number: $z + \bar{z} = 2x$. And their difference is an imaginary number: $z - \bar{z} = 2iy$. Furthermore, the product of a complex number and its conjugate is a real number, the square of the modulus: $z\bar{z} = x^2 + y^2 = r^2 = |z|^2$.

Geometrically, a conjugate $\bar{z}$ can be represented by the reflection of point z in the x axis (Fig. 8-3). The polar form is

$$\bar{z} = r[\cos(-\theta) + i \sin(-\theta)] = r(\cos\theta - i\sin\theta) = re^{-i\theta} \tag{8-10}$$

In polar coordinates, the product of two complex numbers, $z_1 = r_1(\cos\theta_1 + i\sin\theta_1)$ and $z_2 = r_2(\cos\theta_2 + i\sin\theta_2)$ is a complex number

$$z_1 z_2 = r_1 r_2 [\cos(\theta_1 + \theta_2) + i \sin(\theta_1 + \theta_2)] = r_1 r_2 e^{i(\theta_1 + \theta_2)} \tag{8-11}$$

Thus, the modulus of the product is the product of the moduli, $r_1 r_2$. The amplitude of the product is the sum of the amplitudes, $\theta_1 + \theta_2$.

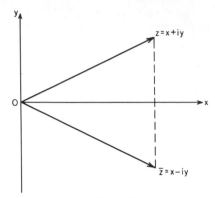

Fig. 8-3

For example, let us multiply $z_1 = 3 + 4i$ by $z_2 = 12 + 5i$. By Eq. (8-2), we get $z_1z_2 = (3 \times 12 - 4 \times 5) + i(3 \times 5 + 4 \times 12) = 16 + 63i$. Alternatively, we have in polar form, $r_1 = \sqrt{3^2 + 4^2} = 5$ and $\theta_1 = \arctan 4\!/\!3 = 53°8'$; so that $z_1 = 5(\cos 53°8' + i \sin 53°8')$. Similarly, $r_2 = \sqrt{12^2 + 5^2} = 13$, $\theta_2 = \arctan 5\!/\!12 = 22°37'$, and $z_2 = 13(\cos 22°37' + i \sin 22°37')$. Hence,

$$z_1z_2 = 5 \times 13[\cos (53°8' + 22°37') + i \sin (53°8' + 22°37')]$$
$$= 65(\cos 75°45' + i \sin 75°45') = 16 + 63i$$

From the rule for multiplication, we get the important result:
Multiplication by i rotates the radius vector of a complex number 90°. Multiplication by $i^2 = -1$ rotates the radius vector 180°.

The quotient of two complex numbers z_1 and z_2 is

$$\frac{z_1}{z_2} = \frac{r_1}{r_2}[\cos (\theta_1 - \theta_2) + i \sin (\theta_1 - \theta_2)] = \frac{r_1}{r_2} e^{i(\theta_1 - \theta_2)} \qquad (8\text{-}12)$$

For example, let us divide $z_1 = 3 + 4i$ by $z_2 = 12 + 5i$. In polar form, $z_1 = 5(\cos 53°8' + i \sin 53°8')$ and $z_2 = 13(\cos 22°37' + i \sin 22°37')$. Therefore,

$$\frac{z_1}{z_2} = \frac{5}{13}[\cos (53°8' - 22°37') + i \sin (53°8' - 22°37')]$$

$$= \frac{5}{13} (\cos 30°31' + i \sin 30°31') = 0.331 + 0.195i$$

Alternatively,

$$\frac{z_1}{z_2} = \frac{3 + 4i}{12 + 5i} \cdot \frac{12 - 5i}{12 - 5i} = \frac{56 + 33i}{169} = 0.331 + 0.195i$$

By repeated application of Eq. (8-11) to any complex number, we can derive **De Moivre's formula** for the power of a number:

$$z^n = r^n(\cos n\theta + i \sin n\theta) = r^n e^{in\theta} \tag{8-13}$$

where n is any integer. But Eq. (8-13) also applies when n is any real number.

As an example of the use of this formula, let us cube $z = 3 + 4i$. In polar form, $z = 5(\cos 53°8' + i \sin 53°8')$. Hence,

$$z^3 = 5^3[\cos 3(53°8') + i \sin 3(53°8')] = 125(\cos 159°24' + i \sin 159°24')$$
$$= 125(-0.936 + 0.352i) = -117 + 44i$$

By taking n as a fraction, we can use Eq. (8-13) to extract roots of complex numbers. For example, let us compute the fourth roots of unity. The amplitude of any real number is given by zero or an integral multiple of 2π. Thus, in polar form, unity can be denoted by $z = \cos 2\pi k + i \sin 2\pi k$, where $k = 0, 1, 2, 3, \ldots$ Use of De Moivre's theorem gives, for the fourth roots,

$$w = z^{1/4} = \cos \frac{2\pi k}{4} + i \sin \frac{2\pi k}{4}$$

Substitution of 0, 1, 2, and 3 for k in succession yields the roots

$$w_1 = \cos 0 + i \sin 0 = +1$$

$$w_2 = \cos \frac{\pi}{2} + i \sin \frac{\pi}{2} = +i$$

$$w_3 = \cos \pi + i \sin \pi = -1$$

$$w_4 = \cos \frac{3\pi}{2} + i \sin \frac{3\pi}{2} = -i$$

The nth roots of unity are given in the complex plane by vectors to the vertices of a regular polygon of n sides inscribed in the circle $|z| = 1$, with one vertex at $z = 1$.

8-4   *Analytic Functions.* Let $w = f(z)$ be a function of a complex variable $z = x + iy$. Then, we also can express w as

$$w = u(x,y) + iv(x,y) \tag{8-14}$$

where u and v are functions of real variables, x and y. For example, take $w = z^2 = x^2 - y^2 + 2ixy$. Then, $u = x^2 - y^2$ and $v = 2xy$.

We can define limits for functions of a complex variable in the same way as for real variables. Thus, a complex constant w_0 is the limit of $f(z)$ as z approaches z_0 if $f(z)$ can be kept arbitrarily close to w_0 by keeping

z very close to but distinct from z_0. Also, as for real variables, $f(z)$ is continuous at z_0 if

$$\lim_{z \to z_0} f(z) = f(z_0)$$

In addition, we can define the **derivative of** $f(z)$ in the same way as for real variables:

$$\frac{dw}{dz} = w' = f'(z) = \lim_{\Delta z \to 0} \frac{f(z + \Delta z) - f(z)}{\Delta z} \tag{8-15}$$

Thus, formulas for differentiation of functions of a complex variable are the same as those for differentiation of functions of real variables, if the functions are appropriately defined and if the derivatives exist.

In many cases, there is no derivative. For example, the simple function $w = \bar{z} = x - iy$ has no derivative. This occurs because Δz is a function of two variables, x and y, and thus can approach zero in many different ways. But a complex function can have a derivative only if Eq. (8-15) yields the same value for $f'(z)$ regardless of the way in which Δz approaches zero. Consequently, to insure existence of a derivative, restrictions are imposed on $f(z)$.

Functions that meet these restrictions are called **analytic, regular,** or **holomorphic.** A complex function $w = f(z) = u + iv$ is analytic at a point z_0 if dw/dz exists at z_0 and every point in the neighborhood of z_0. The function is analytic in a domain of the z plane if it is analytic at every point in the domain.

Note that $w = z^2$ is analytic at $z = 0$ because its derivative $dw/dz = 2z$ exists not only at $z = 0$ but also at every point in the neighborhood of $z = 0$. In contrast, $w = |z|^2$ is not analytic at $z = 0$ because its derivative exists only at $z = 0$.

The following are both necessary and sufficient conditions for w to be analytic at a point z_0:

1. u and v must be real, single-valued, continuous functions of x and y.

2. The partial derivatives of u and v at z_0 and in its neighborhood must be continuous and must satisfy the **Cauchy-Riemann equations:**

$$\frac{\partial u}{\partial x} = \frac{\partial v}{\partial y} \qquad \text{or} \qquad \frac{\partial u}{\partial r} = \frac{1}{r}\frac{\partial v}{\partial \theta} \tag{8-16}$$

$$\frac{\partial u}{\partial y} = -\frac{\partial v}{\partial x} \qquad \text{or} \qquad \frac{\partial v}{\partial r} = -\frac{1}{r}\frac{\partial u}{\partial \theta} \tag{8-17}$$

(The second equation in each case is the polar form.)

As an example, let us use Eqs. (8-16) and (8-17) to show that $w = z^2$

is analytic at $z = 0$.   For this function, $u = x^2 - y^2$ and $v = 2xy$.   Thus, u and v are real, single-valued, continuous functions of x and y.   Furthermore, the partial derivatives of u and v at $z = 0$ and its neighborhood (and at all other points of the z plane) exist, are continuous, and satisfy the Cauchy-Riemann equations:

$$\frac{\partial u}{\partial x} = 2x = \frac{\partial v}{\partial y} \quad \text{and} \quad \frac{\partial u}{\partial y} = -2y = -\frac{\partial v}{\partial x}$$

Therefore, $w = z^2$ is analytic not only at $z = 0$ but throughout the z plane and has a derivative for all x and y.

If $w = f(z) = u + iv$ is analytic at z_0, then the derivative at z_0 is

$$f'(z) = \frac{\partial u}{\partial x} + i\frac{\partial v}{\partial x} = \frac{\partial v}{\partial y} - i\frac{\partial u}{\partial y} \qquad (8\text{-}18)$$

In polar form, with u and v as functions of the modulus r and the amplitude θ, the derivative is

$$f'(z) = (\cos \theta - i \sin \theta)\left(\frac{\partial u}{\partial r} + i\frac{\partial v}{\partial r}\right) \qquad (8\text{-}19)$$

It is not always necessary to apply the Cauchy-Riemann equations and other conditions to determine if a function is analytic.   You can often use the following instead:

•Where two functions are analytic, their sum, product, and quotient (if the denominator is not zero) also are analytic.

•An analytic function of an analytic function is analytic, too.

•Derivatives and integrals of analytic functions are analytic.

Thus, once you have become familiar with the elementary analytic functions, you often can recognize at a glance the more complicated ones.

Analytic functions are important because ordinary rules of differentiation and integration apply to them.   They can be represented by power series.   They also possess other properties that make them useful in solving engineering problems.

If $w = u + iv$ is analytic, for instance, both u and v satisfy the **Laplace equations**:

$$\frac{\partial^2 u}{\partial x^2} + \frac{\partial^2 u}{\partial y^2} = 0 \qquad \frac{\partial^2 v}{\partial x^2} + \frac{\partial^2 v}{\partial y^2} = 0 \qquad (8\text{-}20)$$

Hence, u and v are **harmonic functions**.   They are called **harmonic conjugates**.   Later, we shall see how these relations can be used to solve problems.

Another significant property of an analytic function w is that the curves $u =$ constant and $v =$ constant are orthogonal. We shall also make use of this property later.

8-5   *Elementary Analytic Functions.* In Art. 8-4, we noted that if you can recognize the elementary analytic functions, you can, in general, determine easily whether more complicated functions are analytic. At this stage, therefore, you will be introduced to some elementary analytic functions.

First, let us note that $z^k = (x + iy)^k$ is analytic throughout the z plane when k is a positive real constant. If k is negative; that is, $f(z) = 1/z^k$, the function is analytic except where $z = 0$. Similarly, real powers of $z - z_0$, where z_0 is a complex number, are analytic, except at z_0 for negative powers. We shall consider the case of complex powers a little later.

Since z^k is analytic and the sum of analytic functions is analytic, it follows that sums of powers of z are analytic. Hence, every polynomial

$$P(z) = a_0 + a_1z + a_2z^2 + \cdots + a_nz^n \tag{8-21}$$

where a_i is a complex number (constant) and $z = x + iy$, is analytic at every point of the z plane. (Such functions are called **entire.**)

The exponential function is defined by

$$e^z = \exp z = e^x(\cos y + i \sin y) \tag{8-22}$$

It is analytic everywhere in the z plane. Its absolute value is e^x and its amplitude is y. If x is set equal to 0, then

$$e^{iy} = \exp iy = \cos y + i \sin y \tag{8-23}$$

Also,

$$e^{-iy} = \exp(-iy) = \cos y - i \sin y \tag{8-24}$$

These relations indicate that the exponential function is periodic with period $2\pi i$:

$$\exp(z + 2\pi i) = \exp z \tag{8-25}$$

Simultaneous solution of Eqs. (8-23) and (8-24) gives for every real number y

$$\cos y = \frac{1}{2}(e^{iy} + e^{-iy}) \tag{8-26}$$

$$\sin y = \frac{1}{2i}(e^{iy} - e^{-iy}) \tag{8-27}$$

It is logical to extend these results to define trigonometric functions of complex variables.

We therefore define the cosine and sine functions by

$$\cos z = \frac{1}{2}(e^{iz} + e^{-iz}) \tag{8-28}$$

$$\sin z = \frac{1}{2i}(e^{iz} - e^{-iz}) \tag{8-29}$$

They are analytic everywhere in the z plane.   Other trigonometric functions and formulas can be obtained from sines and cosines in the same way as for real variables.

The sine and cosine functions can also be written as

$$\sin z = \sin (x + iy) = \sin x \cosh y + i \cos x \sinh y \tag{8-30}$$
$$\cos z = \cos (x + iy) = \cos x \cosh y - i \sin x \sinh y \tag{8-31}$$

When $x = \pi/2$, $\sin z = \cosh y$, and when $x = 0$, $\cos z = \cosh y$.   But $\cosh y = \frac{1}{2}(e^y + e^{-y})$.   Consequently, when y increases without bound, so does $\cosh y$.   Therefore, unlike the sine and cosine functions of real variables, $\sin z$ and $\cos z$ are not bounded in absolute value.   Furthermore, Eqs. (8-30) and (8-31) indicate that $\sin z$ and $\cos z$ can be zero only when $y = 0$. Thus, $\sin z = 0$ implies $z = \pm n\pi$, where $n = 0, 1, 2, \ldots$; and $\cos z = 0$ implies $z = \pm \frac{1}{2}(2n - 1)\pi$.

The preceding relations also make it logical to define the hyperbolic sine and cosine by

$$\sinh z = \frac{e^z - e^{-z}}{2} \tag{8-32}$$

$$\cosh z = \frac{e^z + e^{-z}}{2} \tag{8-33}$$

The other hyperbolic functions and formulas can be obtained from these in the same way as for real variables.   Sinh z and cosh z are entire functions; that is, they are analytic throughout the z plane.   Tanh z is analytic in every domain where $\cosh z \neq 0$.

The logarithmic function is defined by

$$\log_e z = \log_e re^{i\theta} = \log_e r + i\theta \qquad r > 0 \tag{8-34}$$

where r is the modulus and θ the argument of z.   Log$_e$ z is multivalued, for θ can be set equal to $\phi \pm 2n\pi$, with $-\pi < \phi < \pi$ and $n = 0, 1, 2, \ldots$.

The principal value of $\log_e z$ is obtained by setting $n = 0$.   It is not defined for $z = 0$ or some points on the negative real axis.   But it is analytic where defined.

Any particular value of n determines a branch of the complex logarithmic function. Each specific branch is analytic.

For our purposes, we need define only one more elementary function z^c, where c is a complex number. Extending once again the concepts of functions of real variables, we specify

$$z^c = e^{c \log_e z} = \exp (c \log_e z) \tag{8-35}$$

It is multivalued. Its principal branch is computed for the principal value of $\log_e z$ and is single-valued and analytic in the domain $r > 0$, $-\pi < \phi < \pi$.

8-6 Mapping by Complex Functions. The complex function $w = f(z) = u(x,y) + iv(x,y)$ relates two pairs of variables u, v and x, y. Consequently, two planes are needed for graphical representation of w. One plane is used for the variable $z = x + iy$, the other for the variable $w = u + iv$. Points in one plane are said to be mapped into or transformed into corresponding points (images) in the other. Let us now investigate some simple transformations.

The linear function $w = z + c$, where c is a complex constant, maps a region of the z plane into one in the w plane of the same size, shape, and orientation. But each point is translated along the vector c (Fig. 8-4).

Mapping with the linear function $w = kz$, where k is a complex constant, consists of a rotation through the angle amp k and a magnification by a factor equal to the modulus of k, or $|k|$ (Fig. 8-5).

Mapping with the general linear function $w = kz + c$ is a combination of the transformations illustrated in Figs. 8-4 and 8-5. The result is a translation along the vector c, a rotation amp k, and a magnification by $|k|$.

The transformation $w = z^n$, where n is a positive integer, maps the wedge-shaped region in Fig. 8-6a onto the upper half of the w plane (Fig. 8-6b). It transforms a circular arc, $r = r_0$, $\theta_0 < \theta < \theta_0 + 2\pi/n$ into the circle $\rho = r_0^n$. Thus, the function $w = z^2$ maps the first quadrant of the z plane onto the upper half of the w plane. Also, $w = z^2$ maps a semicircle in the z plane onto a circle in the w plane.

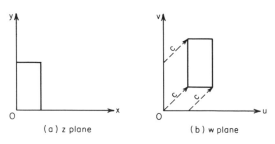

(a) z plane   (b) w plane

Fig. 8-4

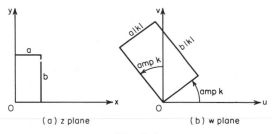

Fig. 8-5

Let us next examine the transformation $w = 1/z$. Note that $1/z = \bar{z}/z\bar{z} = \bar{z}/|z|^2$. The transformation may be resolved into two steps. The first shifts each point z_0 along its radius vector to a point z_1 whose distance from the origin is $1/|z_0|$. This is called an inversion with respect to the unit circle $r = 1$. The second step transforms z_1 into its conjugate $\bar{z}_1$. This is equivalent to a reflection in the real axis. Thus, points outside the unit circle in the z plane are mapped inside the unit circle $|w| = 1$ in the w plane. Points inside the unit circle in the z plane are mapped outside $|w| = 1$. Points on the unit circle are reflected in the real axis. Circles not passing through $z = 0$ transform into circles not passing through $w = 0$. Every circle through $z = 0$ transforms into a straight line in the w plane. Straight lines in the z plane, except lines through $z = 0$, transform into circles through $w = 0$. Images of lines through $z = 0$ are lines through $w = 0$.

Finally, let us look at the transformation $w = e^z$. By Eq. (8-22), $e^z = e^x(\cos y + i \sin y)$. By setting x equal to a constant c, you will observe that the transformation maps lines parallel to the y axis, $x = c$, into circles in the w plane ($\rho = e^c$). When you set y equal to a constant k, you will find that lines parallel to the x axis, $y = k$, map into rays ($\phi = k$). Thus, a rectangle with sides parallel to the x and y axes maps onto the portion of a circular sector cut off by two circles (Fig. 8-7). Also, the infinite strip $0 \leq y \leq \pi$ maps onto the upper half of the w plane.

8-7 *Conformal Mapping.* A mapping or transformation that preserves angles is called conformal.

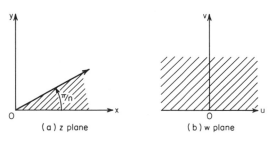

Fig. 8-6

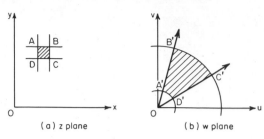

Fig. 8-7

Mapping done with an analytic function f(z) is conformal for all points z for which f'(z) ≠ 0.

In general, under a transformation with an analytic function, shapes in the z plane are rotated and expanded or contracted in the w plane. (Some examples are given in Art. 8-6.) But except where $f'(z) = 0$, the angles between intersecting straight lines or between tangents to intersecting curves at the point of intersection are invariant under the transformation.

Thus, every conformal transformation maps orthogonal curves into orthogonal curves. In particular, the lines x = constant, y = constant in the z plane transform into mutually orthogonal trajectories in the w plane. Similarly, the lines u = constant, v = constant in the w plane map into orthogonal trajectories in the z plane. Furthermore, a conformal mapping retains the sense, or sign, of angles.

But while angles are preserved, you should bear in mind that lines are rotated:

An analytic mapping function f(z) turns the directed tangent to a curve at a point z_0 through the angle amp $f'(z_0)$.

As an example, let us return to the analytic mapping function $w = e^z$ discussed in Art. 8-6 and illustrated in Fig. 8-7. The lines AB and DA in the z plane are perpendicular (Fig. 8-7a). AB maps onto a ray $A'B'$ and DA onto an arc $D'A'$ with its center on that ray (Fig. 8-7b). Hence AB and DA remain normal to each other.

8-8 Solving Laplace's Equation by Mapping. In Art. 8-4, your attention was called to a very important property of analytic functions. They are harmonic functions; they satisfy Laplace's equation [Eq. (8-20)]. You should now note still another important characteristic, which makes conformal mapping useful in solving engineering problems:

An harmonic function remains harmonic in conformal mapping.

Suppose, for example, that $\phi(x,y)$ is a solution of

$$\frac{\partial^2 \phi}{\partial x^2} + \frac{\partial^2 \phi}{\partial y^2} = 0 \qquad (8-36)$$

Then, when an analytic function $z = x + iy = f(u + iv)$ transforms ϕ into a function of u and v, $\phi(u,v)$ will satisfy

$$\frac{\partial^2 \phi}{\partial u^2} + \frac{\partial^2 \phi}{\partial v^2} = 0 \qquad (8\text{-}37)$$

wherever the mapping is conformal.

This characteristic often makes conformal mapping useful in solving problems involving Laplace's equation. In Art. 7-6, we noted that this equation is the elliptic type. Usually, boundary conditions specify the function ϕ in Eq. (8-36), or its normal derivative, or a linear combination of these at every point of the closed boundary of the region R within which you are seeking a solution. For the Laplace and other elliptic equations, these boundary conditions completely determine the solution within the boundary. But unless R has a simple shape, you may find it difficult to get a solution directly. You may, however, be able to find the solution more easily in a simpler region, such as a circle or a half plane. If so, look for an analytic mapping function that will transform the given region R into the simpler region. When you get the solution for the simpler region, the inverse transformation will provide the solution for R.

As an example of the use of this technique, let us determine the paths of the particles of an ideal fluid approaching with uniform flow a long circular cylinder perpendicular to the direction of flow. We will assume the fluid to be incompressible and free of viscosity. The flow is irrotational in the region we are considering, and there are no sources or sinks there.

We have observed that complex numbers and two-dimensional vectors are isomorphic with respect to addition. Hence, we can represent the velocity V of the fluid at any point in the z plane by

$$V = V_1 + iV_2 \qquad (8\text{-}38)$$

where V_1 is the velocity component parallel to the x axis and V_2 the component parallel to the y axis. Because the fluid is incompressible and the flow uniform, a function $\phi(x,y)$ exists such that

$$V_1 = \frac{\partial \phi}{\partial x} \qquad V_2 = \frac{\partial \phi}{\partial y} \qquad (8\text{-}39a)$$

ϕ is called the velocity potential, and the trajectories $\phi = $ constant are called **equipotentials.** V at any point is normal to the equipotential there.

With no sinks and sources and no changes in the paths of the particles of the fluid with time, we have a condition of steady-state flow. In a given time, the amount of fluid entering the region we are considering equals the amount leaving during that period. So ϕ must satisfy Laplace's equation [Eq. (8-36)].

As a solution of Laplace's equation, ϕ is harmonic. Consequently, it has an harmonic conjugate $\psi(x,y)$ to which it is related by the Cauchy-Riemann equations [Eqs. (8-16) and (8-17)]. The function ψ is called the **stream function,** and the trajectories $\psi = $ constant are called the **streamlines** of the flow. Since ϕ and ψ are harmonic conjugates, the equipotentials and streamlines are orthogonal. Hence, V at any point is tangent to the streamline there. Also,

$$V_1 = \frac{\partial \psi}{\partial y} \qquad V_2 = -\frac{\partial \psi}{\partial x} \qquad\qquad (8\text{-}39b)$$

Let us take the flow of the fluid parallel to the x axis. Let the cylinder be represented in the z plane by the circle $|z| = 1$, with center at the origin of coordinates (Fig. 8-8). Our objective is to determine $\psi(x,y)$ to satisfy Laplace's equation with the boundary condition that a streamline $\psi = k$ lies along the x axis for $x < -1$ and $x > 1$ and along the circle for $-1 \le x \le 1$. Also, all other streamlines must lie outside the circle. And at large distances from the cylinder, the streamlines should be parallel or nearly parallel to the x axis.

As a first step, let us take advantage of the fact that harmonic conjugates, by definition, are components of an analytic function

$$F(z) = \phi(x,y) + i\psi(x,y) \qquad\qquad (8\text{-}40)$$

This function is called the **complex potential** of the flow. Let us also take advantage of symmetry and consider only the half of the z plane

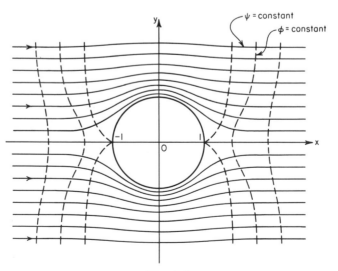

Fig. 8-8

above the x axis. The required solution then is the imaginary part of $F(z)$ in the upper half of the z plane.

Because of the shape of the boundary, we would find it difficult to obtain the solution directly. We do know, however, the solution for unobstructed uniform flow. Let us assume that it takes place in the upper half of the w plane, parallel to the u axis. Then, $V_1 = c$ and $V_2 = 0$, and by Eqs. (8-39b) the stream function is $\psi = cv$, where c is a real constant. The equipotentials must be $\phi = cu$ to satisfy the Cauchy-Riemann equations. Hence, the complex potential $F(w) = \phi + i\psi = cu + icv = cw$.

Now, we need an analytic function to map the upper half of the w plane onto the upper half of the z plane. If you had a catalog of conformal maps available, you would find that

$$w = z + \frac{1}{z} \tag{8-41}$$

serves the purpose. By substituting $x + iy$ for z, you can verify that $v = 0$ when $y = 0$ or when $x^2 + y^2 = 1$. Consequently, the u axis maps onto the x axis for $x < -1$ and $x > 1$ and onto the upper half of the circle $x^2 + y^2 = 1$ when $-1 \le x \le 1$. Also, for very large values of z, w is nearly equal to z. So streamlines at large distances from the origin in the z plane will be practically unaltered from their condition in the w plane.

Using Eq. (8-41) for the transformation of $F = cw$, we get, with $z = re^{i\theta}$,

$$F = c\left(z + \frac{1}{z}\right) = c\left[\left(r + \frac{1}{r}\right)\cos\theta + i\left(r - \frac{1}{r}\right)\sin\theta\right]$$

The imaginary part gives us the streamlines, or paths of the fluid particles:

$$\psi = c\left(r - \frac{1}{r}\right)\sin\theta = \text{constant} \tag{8-42}$$

The streamline $\psi = 0$ consists of the circle $r = 1$ and the x axis. For very large values of r, the streamline is given approximately by $r\sin\theta = y = \text{constant}$ (see Fig. 8-8). The velocity components, if desired, can be obtained by applying Eqs. (8-39) to the real or imaginary part of F.

8-9 Integrals of a Complex Variable. Let f be a complex function of a real variable t such that

$$f(t) = u(t) + iv(t) \tag{8-43}$$

Here, u and v are real functions that are continuous over an interval (a,b), except possibly for a finite number of finite discontinuities. The definite integral of f is defined by

$$\int_a^b f(t)\, dt = \int_a^b u(t)\, dt + i \int_a^b v(t)\, dt \tag{8-44}$$

Thus, we can obtain the integral of a complex function by integrating real functions.

A line integral of a function f of a complex variable z from a point $z = a$ to a point $z = b$ is defined in terms of the values of $f(z)$ along a contour C extending from a to b.   (A **contour** is a continuous chain of a finite number of smooth arcs.)

$$\int_C f(z)\, dz = \int_a^b f[\phi(t) + i\psi(t)] \left(\frac{d\phi}{dt} + i\frac{d\psi}{dt} \right) dt \tag{8-45}$$

where $z = x + iy$

$$x = \phi(t)$$
$$y = \psi(t)$$

If $f = u + iv$, the line integral can be written in terms of real line integrals.

$$\int_C f(z)\, dz = \int_C (u\, dx - v\, dy) + i \int_C (u\, dy + v\, dx) \tag{8-46}$$

Conventionally, the positive direction of integration around a closed contour is counterclockwise. Reversal of the direction of integration reverses the sign of the integral.

The length of a contour C is given by

$$L = \int_C |dz| \tag{8-47}$$

The polar form is often useful in evaluating line integrals around closed contours. For example, let us evaluate on a circle C with radius r and center z_0

$$F = \int_C \frac{dz}{(z - z_0)^{n+1}} \tag{8-48}$$

To put the integrand in polar form, set $z - z_0 = re^{i\theta}$, where θ goes from 0 to 2π as z ranges along the circle. $dz = rie^{i\theta}\, d\theta$. Substitution in Eq. (8-48) gives

$$F = \int_0^{2\pi} \frac{rie^{i\theta}}{r^{n+1}e^{i(n+1)\theta}}\, d\theta = \frac{i}{r^n} \int_0^{2\pi} e^{-in\theta}\, d\theta \tag{8-49}$$

If $n = 0$, Eq. (8-49) yields

$$\int_C \frac{dz}{z - z_0} = i \int_0^{2\pi} d\theta = 2\pi i \tag{8-50}$$

When $n \neq 0$, Eq. (8-49) becomes

$$F = \frac{i}{r^n} \int_0^{2\pi} (\cos n\theta - i \sin n\theta) \, d\theta = 0 \tag{8-51}$$

If a function $f(z)$ is analytic in a region R, it satisfies the Cauchy-Riemann equations there. But, as pointed out in Art. 2-6, when these equations are satisfied, the line integral of the function between any two points in R is independent of the path. And the line integral of the function around a closed contour is zero. This result is summarized by the **Cauchy-Goursat** theorem:

If $f(z)$ is analytic at all points inside and on a closed contour C,

$$\int_C f(z) \, dz = 0 \tag{8-52}$$

The converse is also true (**Morera's theorem**):

If $f(z)$ is continuous throughout a simply connected domain R and if for every closed contour C inside R, Eq. (8-52) holds, then $f(z)$ is analytic throughout R.

The Cauchy-Goursat theorem also holds if inside C there are a finite number of nonintersecting closed contours C_1, C_2, . . . , and the integral is taken over C and all the C_j. Since the positive direction for integrating C is counterclockwise, the positive direction for the interior contours C_j is clockwise.

As an example, let us evaluate $\int_C dz/[z^2(z^2 + 9)]$ on the circles $|z| = 2$ described counterclockwise and $|z| = 1$ described clockwise (Fig. 8-9). The integrand is analytic except at $z = 0$ and $z = \pm 3i$. These points lie outside the annular region between the two circles. Hence, by the Cauchy-Goursat theorem, the integral is zero.

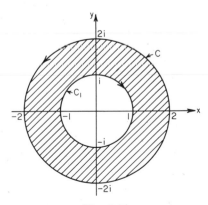

Fig. 8-9

From the Cauchy-Goursat theorem for multiply-connected regions, another important conclusion follows:

The line integral of an analytic function $f(z)$ around any closed contour C_1 equals the line integral of $f(z)$ around any other closed curve C_2 into which C_1 can be continuously deformed without passing through a point where $f(z)$ is not analytic.

Thus, in the preceding example, the line integral around $|z| = 2$ counterclockwise equals the line integral around $|z| = 1$ counterclockwise. If the exterior contour encloses more than one interior closed contour, then the integral over the exterior curve equals the sum of the integrals around the interior curves, with the integration executed in the same direction for all curves.

Still another important result, called Cauchy's integral formula, can be deduced from the Cauchy-Goursat theorem.

$$f(z_0) = \frac{1}{2\pi i} \int_C \frac{f(z)}{z - z_0} \, dz \tag{8-53}$$

where $f(z)$ is analytic everywhere inside and on a closed contour C, z_0 is any point inside C, and integration is counterclockwise.

This formula indicates that every analytic function is completely determined in the interior of a closed region when the values of the function are given on the boundary.

Equation (8-53) can be used to evaluate line integrals. As an example, let us evaluate $F = \int_C e^{-z} \, dz/(z + 1)$ over the circle $|z| = 2$. Let $f(z) = e^{-z}$ and $z_0 = -1$. Then, apply Cauchy's formula. Notice that e^{-z} is analytic and that -1 lies within the given circle. Then,

$$F = 2\pi i f(z_0) = 2\pi i e$$

The values of an analytic function $f(z)$ on a boundary also determines the values of derivatives of $f(z)$ at interior points. For repeated differentiation of Eq. (8-53) yields

$$f'(z_0) = \frac{1}{2\pi i} \int_C \frac{f(z)}{(z - z_0)^2} \, dz \tag{8-54}$$

$$f''(z_0) = \frac{2!}{2\pi i} \int_C \frac{f(z)}{(z - z_0)^3} \, dz \tag{8-55}$$

and in general,

$$f^{(n)}(z_0) = \frac{n!}{2\pi i} \int_C \frac{f(z)}{(z - z_0)^{n+1}} \, dz \tag{8-56}$$

8-10   Power Series for Complex Numbers. If $f(z)$ is analytic at all points inside a circle C with center at z_0 and radius r, then at each point z

inside C, $f(z)$ can be expanded in a **Taylor series**:

$$f(z) = f(z_0) + f'(z_0)(z - z_0)$$
$$+ \frac{f''(z_0)}{2!}(z - z_0)^2 + \cdots + \frac{f^{(n)}(z_0)}{n!}(z - z_0)^n + \cdots \quad (8\text{-}57)$$

When $z_0 = 0$, this reduces to **Maclaurin's series**.

The maximum radius of the circle of convergence is the distance from z_0 to the nearest point at which $f(z)$ is not analytic (singular point).

Both Taylor's and Maclaurin's series for analytic functions are identical in form with the corresponding series for real functions. For example, compare

$$e^x = 1 + \sum_{n=1}^{\infty} \frac{x^n}{n!} \qquad e^z = 1 + \sum_{n=1}^{\infty} \frac{z^n}{n!} \quad (8\text{-}58)$$

A function cannot be expanded in a Taylor series around points or neighborhoods of points where the function is not analytic. An alternative, however, is available. Laurent's series is valid for the ring formed by two concentric circles if the function is analytic on and between the the two circles. The series converges even if there are singular points within the inner circle.

Consider an annular region R between two circles C_1, with radius r_1, and C_2, with radius r_2, $r_2 < r_1$. Let z_0 be the common center of the circles. If $f(z)$ is analytic throughout R, then at each point in R **Laurent's expansion** gives

$$f(z) = a_0 + a_1(z - z_0) + a_2(z - z_0)^2 + \cdots + a_n(z - z_0)^n + \cdots$$
$$+ \frac{b_1}{z - z_0} + \frac{b_2}{(z - z_0)^2} + \cdots + \frac{b_n}{(z - z_0)^n} + \cdots \quad (8\text{-}59)$$

where

$$a_n = \frac{1}{2\pi i} \int_{C_1} \frac{f(z)\, dz}{(z - z_0)^{n+1}} \qquad n = 0, 1, 2, \ldots \quad (8\text{-}60)$$

$$b_n = \frac{1}{2\pi i} \int_{C_2} \frac{f(z)\, dz}{(z - z_0)^{-n+1}} \qquad n = 1, 2, \ldots \quad (8\text{-}61)$$

Integration around each circle should be counterclockwise.

You may discover that Eq. (8-60) has the same form as Eq. (8-56). But this does not imply that you can use the derivatives of $f(z)$ for the coefficients a_n. For Eq. (8-56), like Eq. (8-53), holds only if $f(z)$ is analytic everywhere inside the closed contour. Yet, there may be many points within C_2, and therefore within C_1, where $f(z)$ is not analytic.

The coefficients a_n and b_n are often found by some means other than the integrals in Eqs. (8-60) and (8-61). The Laurent expansion over a given ring is unique. Therefore, an expansion of the Laurent type found by any process is the only Laurent series.

For example, Laurent's series for e^z/z^2 can be obtained by dividing each term of Maclaurin's series for e^z [Eq. (8-58)] by z^2. Laurent's series for $1/z(1-z)^2$, expanded about $z=0$, can be found in two steps. First, expand $(1-z)^{-2}$ by the binomial theorem or by dividing 1 by $1-2z+z^2$. Second, divide the resulting series by z to get $1/z + 2 + 3z + 4z^2 + \cdots$. Note also that the Laurent series for $(z-1)^2$ is $(z-1)^2$ and for $1/(z-1)^2$ is $1/(z-1)^2$. Thus, many of the coefficients in a Laurent series may be zero.

8-11   *Residues and Poles.* In Art. 8-10, we observed that the coefficients in Laurent's series are defined as integrals. Yet, it is practical to determine the coefficients by some other means. When we have obtained a coefficient, we also have the value of the defining integral. Consequently, we have another technique for evaluating integrals of complex functions, which can be adapted to integration of real functions. For the purpose, we use residues and poles.

If $f(z)$ has an **isolated singularity** (point where it is not analytic) at $z = z_0$, the residue of $f(z)$ at z_0 is the coefficient b_1 of $1/(z - z_0)$ in Laurent's series for $f(z)$. But from Eq. (8-61) with $n = 1$, we find that

$$b_1 = \frac{1}{2\pi i} \int_C f(z)\, dz \tag{8-62}$$

Given the residue, we can evaluate the line integral over any closed contour around an isolated singularity.

This result can be generalized to include several singular points. Let $R_1, R_2, \ldots, R_n$ be the residues at singular points inside a closed contour C within and on which $f(z)$ is analytic, except at those points. If C encloses a finite number of singular points, then

$$\int_C f(z)\, dz = 2\pi i(R_1 + R_2 + \cdots + R_n) \tag{8-63}$$

This is known as the **residue theorem.**

Let $f(z)$ be expanded in a Laurent series. Also, let $b_1, b_2, \ldots, b_m,$ $b_{m+1}, \ldots, b_n$ be the coefficients of the negative powers of $z - z_0$. If coefficients $b_{m+1}, b_{m+2}, \ldots, b_n$ are all zero, but $b_m \neq 0$, the isolated singular point z_0 is called a pole of order m of the function $f(z)$. If

$m = 1$, z_0 is called a **simple pole.** If m is unbounded; that is, Laurent's series contains an infinite number of terms with powers of $z - z_0$, then z_0 is called an **essential singularity** of $f(z)$. The following theorem may be an aid in determining the order of a pole:

If $\phi(z) = (z - z_0)^m f(z)$ *is analytic at* z_0 *and* $\phi(z_0) \neq 0$, $f(z)$ *has a pole of order* m *at* z_0.

In practice, poles usually are found by expressing $f(z)$ as a fraction in lowest terms. The presence of a factor $(z - z_0)^m$ in the denominator indicates that $f(z)$ has a pole of the mth order at z_0. For example, $1/z(z - 1)(z - 2)$ has first-order poles at $z = 0$, $z = 1$, and $z = 2$. And $1/z^2(z^2 + 1)$ has a second-order pole at $z = 0$ and first-order poles at $z = \pm i$.

If $f(z)$ has a pole of order m at z_0, the residue of $f(z)$ at z_0 is

$$b_1 = \frac{1}{(m-1)!}\left[\frac{d^{m-1}\phi}{dz^{m-1}}\right]_{z=z_0} \qquad m > 1 \qquad (8\text{-}64a)$$

$$b_1 = \lim_{z \to z_0} (z - z_0)f(z) \qquad m = 1 \qquad (8\text{-}64b)$$

To illustrate the use of Eqs. (8-64), let us consider the function $f(z) = 1/z(z - 1)^2$. It has a simple pole at $z = 0$, and $\phi(z) = 1/(z - 1)^2$. Hence, the residue at $z = 0$ is

$$\lim_{z \to 0} [1/(z - 1)^2] = 1$$

by Eq. (8-64b). At $z = 1$, $f(z)$ has a second-order pole, and $\phi(z) = 1/z$. By Eq. (8-64a), the residue at $z = 1$ is

$$b_1 - \frac{1}{1!}\frac{d}{dz}\left(\frac{1}{z}\right)_{z=1} = -\frac{1}{z^2}\Big]_{z=1} = -1$$

As an example of the use of residues in integrating real functions, let us evaluate $F = \int_0^{2\pi} d\theta/(1 + \alpha \sin \theta)$ for $0 < \alpha < 1$. Let $z = e^{i\theta}$. Then, $d\theta = dz/iz$, and, by Eq. (8-29), $\sin \theta = (z - z^{-1})/2i$. Substitution for θ in F yields for integration over the circle $|z| = 1$:

$$F = \int_C \frac{dz}{iz[1 + \alpha(z - z^{-1})/2i]} = \frac{2}{\alpha}\int_C \frac{dz}{z^2 + 2iz/\alpha - 1}$$

Solving $z^2 + 2iz/\alpha - 1 = 0$, we find that for $0 < \alpha < 1$ the root satisfying $|z_0| \leq 1$ is

$$z_0 = -\frac{i}{\alpha}(1 - \sqrt{1 - \alpha^2})$$

This is a simple pole.   Hence, the residue at z_0 is, by Eq. (8-64b),

$$b_1 = \lim_{z \to z_0} \frac{z - z_0}{z^2 + 2iz/\alpha - 1} = \lim_{z \to z_0} \frac{1}{z + \dfrac{i}{\alpha}(1 + \sqrt{1 - \alpha^2})} = \frac{\alpha}{2i\sqrt{1 - \alpha^2}}$$

Substitution of F and b_1 in Eq. (8-62) gives

$$F = \left(\frac{2}{\alpha}\right) 2\pi i b_1 = \frac{2}{\alpha}(2\pi i)\frac{\alpha}{2i\sqrt{1 - \alpha^2}} = \frac{2\pi}{\sqrt{1 - \alpha^2}}$$

8-12   Bibliography

L. V. AHLFORS, "Complex Analysis," 2d ed., McGraw-Hill Book Company, New York.

E. F. BECKENBACH, "Construction and Application of Conformal Maps," *Natl. Bur. Std. Appl. Math. Ser.*, vol. 18.

G. F. CARRIER, M. KROOK, and C. E. PEARSON, "Functions of a Complex Variable," McGraw-Hill Book Company, New York.

R. V. CHURCHILL, "Complex Variables and Applications," McGraw-Hill Book Company, New York.

J. W. DETTMAN, "Applied Complex Variables," The Macmillan Company, New York.

Z. NEHARI, "Conformal Mapping," McGraw-Hill Book Company, New York.

PROBLEMS

1. Reduce to lowest terms:
   (a) $(1 + 2i)(1 - 3i)/(2 - i) - 2i$.
   (b) $(1 + i)/(1 - i) - (1 - i)/(1 + i)$.
2. Solve
   (a) $w = u + iv = 0$ for x and y when $u = x + y + 2$ and $v = x^2 + y$.
   (b) $z^2 - z + 1 = 0$ for z.
   (c) $z^4 + 1 = 0$ for z.
3. Determine $(\cos\theta + i\sin\theta)^3$ by De Moivre's theorem and by the binomial expansion, and use the results to express $\cos 3\theta$ and $\sin 3\theta$ in powers of $\sin\theta$ and $\cos\theta$.
4. What region of the z plane is determined by $|z - z_0| \le a$, where a is a real number and z_0 a complex number?
5. Are the following analytic?   If not, explain why not.   If any one is analytic, give the derivative of the function.
   (a) $z - \bar{z}$.
   (b) $x^2 + iy^2$.
   (c) $xy + iy$.
   (d) $x^3 - 3xy^2 + i(3x^2y - y^3)$.
6. Given $u = x^3 - 3xy^2$, find its harmonic conjugate v.
7. Give numerical values for the following to four significant figures:
   (a) e^{1+i}.
   (b) $\sin(1 + i)$.

(c) $\log_e (1 - i)$.

(d) i^i.

8. Prove

(a) $de^z/dz = e^z$.

(b) $d \log_e z/dz = 1/z$ if $z \neq 0$.

(c) $d \sin z/dz = \cos z$.

9. Find the velocities and paths of particles of an ideal liquid moving with uniform flow around a 90° bend in a channel. The left bank forms a sharp 90° angle. The right bank is rounded. (It may be assumed hyperbolic in the region of interest.) Assume that the flow satisfies Laplace's equation.

10. (a) A thin, very long, metal plate has insulated faces. The plate is a uniform π ft wide. One long edge is held at 0°C, the other long edge at 100°C. The steady-state temperature T satisfies Laplace's equation. Find a simple solution for the temperature distribution in the plate away from its ends.

(b) A thin, very long, very wide, rectangular plate has insulated faces. At the center of one long edge, for a distance of 2 ft, a very small fraction of the length, the edge is maintained at 100°C. Elsewhere, the edge is held at 0°C. The steady-state temperature T satisfies Laplace's equation. Find the temperature distribution in the plate. [*Hint:* Use the solution to Prob. 10a and the conformal transformation $w = \log_e (z - 1)/(z + 1)$.]

(c) A thin, very long, rectangular plate has insulated faces. The plate is a uniform π ft wide. The long edges are maintained at 0°C. One end is held at 100°C. The steady-state temperature T satisfies Laplace's equation. Find the temperature distribution throughout the plate. (*Hint:* Use the solution to Prob. 10b and the conformal transformation $w = \sin z$.)

11. Evaluate the following line integrals:

(a) $\int_C z^2 \, dz$, where C is the straight line from $z = 0$ to $z = 2 + i$.

(b) $\int_C z^2 \, dz$, where C is the x axis from $x = 0$ to $x = 2$ and the line from $(0,2)$ to $z = 2 + i$.

(c) $\int_C \bar{z} \, dz$, where C is the semicircle $|z| = 1$ from $(1,0)$ to $(-1,0)$ passing through $(0,1)$.

(d) $\int_C \bar{z} \, dz$, where C is the semicircle $|z| = 1$ from $(1,0)$ to $(-1,0)$ passing through $(0,-1)$.

12. Evaluate the following line integrals:

(a)

$\int_C z e^{-z} \, dz$, where C is the unit circle

(b)

$\int_C \frac{1}{z} \sin z \, dz$, where C is the ellipse $9x^2 + 4y^2 = 1$

(c)

$\int_C \frac{1}{z} \cos z \, dz$, where C is the ellipse $9x^2 + 4y^2 = 1$

(d)
$$\int_C \frac{\tan z\, dz}{z^2}, \text{ where } C \text{ is the unit circle}$$

(e)
$$\int_C \frac{2z^2 + z}{z^2 - 1}\, dz, \text{ where } C \text{ is the circle } |z| = 2$$

(f)
$$\int_C \frac{2z^2 + z}{(z - 1)^2}\, dz, \text{ where } C \text{ is the circle } |z| = 2$$

13. Expand in a Taylor series:
   (a) $1/z$ about $z = 1$.
   (b) $(z - 1)/(z + 1)$ about $z = 0$.
   (c) $(z - 1)/(z + 1)$ about $z = 1$.

14. Find Laurent's expansion for
   (a) $z/(z - 1)(z - 3)$ for $0 < |z - 1| < 2$.
   (b) $(1 + 2z)/(z^2 + z^3)$ for $0 < |z| < 1$.
   (c) Cosecant z for $0 < |z| < \pi$.

15. Find the residues of the following:
   (a)
   $$\frac{z^2 - 2z + 6}{z - 2} \text{ at } z = 2$$

   (b)
   $$\frac{\sinh z}{z^4} \text{ at } z = 0$$

   (c)
   $$\frac{1 + z}{z(2 - z)} \text{ at } z = 0 \text{ and } z = 2$$

   (d)
   $$\frac{e^z}{(z - 1)^2} \text{ at } z = 1$$

16. Evaluate the following integrals by using residues:
   (a)
   $$\int_C \frac{1 + z}{z(2 - z)}\, dz, \text{ where } C \text{ is the circle } |z| = 1$$

   (b)
   $$\int_C \frac{1 \dot{+} z}{z(2 - z)}\, dz, \text{ where } C \text{ is the circle } |z| = 3$$

   (c)
   $$\int_{-\pi}^{\pi} \frac{d\theta}{1 + \sin^2 \theta}$$

   (d)
   $$\int_0^{2\pi} \frac{\cos 2\theta\, d\theta}{1 - 2\alpha \cos \theta + \alpha^2}, \text{ where } -1 < \alpha < 1$$

ANSWERS

1. (a)

$$\frac{(1+2i)(1-3i)}{2-i} - 2i = \frac{7-i}{2-i} - 2i = \frac{7-i}{2-i}\frac{2+i}{2+i} - 2i$$

$$= \frac{15+5i}{5} - 2i = 3 - i$$

See Art. 8-3.

   (b)

$$\frac{1+i}{1-i} - \frac{1-i}{1+i} = \frac{(1+i)^2 - (1-i)^2}{(1-i)(1+i)} = 2i$$

See Art. 8-3.

2. (a) Since $w = 0$, $u = 0$, and $v = 0$. Simultaneous solution yields $x = 2$, $y = -4$; $x = -1$, $y = -1$. See Art. 8-1.

   (b) If $az^2 + bz + c = 0$, $z = -b/2a + (1/2a)\sqrt{b^2 - 4ac}$. Hence, the solution of $z^2 - z + 1 = 0$ is $z = \frac{1}{2} \pm \frac{i}{2}\sqrt{3}$.

   (c) By Eq. (8-13), $z = (-1)^{\frac{1}{4}} = \cos[(\pi + 2\pi k)/4] + i\sin[(\pi + 2\pi k)/4]$ $= (\sqrt{2}/2)(1+i)$; $(\sqrt{2}/2)(-1+i)$; $(\sqrt{2}/2)(-1-i)$; $(\sqrt{2}/2)(1-i)$.

3. By Eq. (8-13), $(\cos\theta + i\sin\theta)^3 = \cos 3\theta + i\sin 3\theta$. Also, by the binomial expansion,

$$(\cos\theta + i\sin\theta)^3 = \cos^3\theta + 3i\cos^2\theta\sin\theta - 3\cos\theta\sin^2\theta - i\sin^3\theta$$

Equating real and imaginary parts of the equivalent expressions, we get

$$\cos 3\theta = \cos^3\theta - 3\cos\theta\sin^2\theta$$
$$\sin 3\theta = 3\cos^2\theta\sin\theta - \sin^3\theta$$

4. Take $\mathbf{z}$, $\mathbf{z}_0$, and $\mathbf{z} - \mathbf{z}_0$ as vectors such that $\mathbf{z}_0 + (\mathbf{z} - \mathbf{z}_0) = \mathbf{z}$. $\mathbf{z}_0$ is a fixed vector, and the vector $\mathbf{z} - \mathbf{z}_0$ may extend in any direction. If the magnitude of $\mathbf{z} - \mathbf{z}_0$ equals a, a constant, $|\mathbf{z} - \mathbf{z}_0| = a$ represents a circle with the point z_0 as center and a as radius. Therefore, $|\mathbf{z} - \mathbf{z}_0| \leq a$ represents the region in and on the circle. See Art. 8-2.

5. (a) No. $z - \bar{z} = x + iy - x + iy = 0 + 2iy$, and $\partial(0)/\partial x \neq \partial(2y)/\partial y$. See Eq. (8-16).

   (b) No. Let $u = x^2$ and $v = y^2$. Then,

$$\left(\frac{\partial u}{\partial x} = 2x\right) \neq \left(\frac{\partial v}{\partial y} = 2y\right)$$

See Art. 8-4.

   (c) No. $\partial xy/\partial x \neq \partial y/\partial y$. See Art. 8-4.

   (d) Yes. Let $u = x^3 - 3xy^2$ and $v = 3x^2y - y^3$. Then

$$\frac{\partial u}{\partial x} = 3x^2 - 3y^2 = \frac{\partial v}{\partial y} \quad \text{and} \quad \frac{\partial u}{\partial y} = -6xy = -\frac{\partial v}{\partial x}$$

Thus, Eqs. (8-16) and (8-17) are satisfied. By Eq. (8-18),

$$f'(z) = 3x^2 - 3y^2 + 6ixy = 3(x^2 - y^2 + 2ixy) = 3(x + iy)^2 = 3z^2$$

6. u and v must satisfy the Cauchy-Riemann equations. By Eq. (8-16),

$$\frac{\partial u}{\partial x} = 3x^2 - 3y^2 = \frac{\partial v}{\partial y}$$

Integration with respect to y gives $v = 3x^2y - y^3 + f(x)$. Also, by Eq. (8-17),

$$\frac{\partial v}{\partial x} = 6xy + f'(x) = -\frac{\partial u}{\partial y} = 6xy$$

Hence, $f'(x) = 0$ and $f(x) = c$, a constant. So, $v = 3x^2y - y^3 + c$.

7. (a) By Eq. (8-22), $e^{1+i} = e(\cos 1 + i \sin 1) - 2.718(0.5403 + 0.8415i) = 1.469 + 0.2287i$.

(b) By Eq. (8-29),

$$\sin (1 + i) = \frac{1}{2i} (e^{-1+i} - e^{1-i})$$

$$= \frac{1}{2i} [e^{-1}(\cos 1 + i \sin 1) - e(\cos -1 + i \sin -1)]$$

$$= -\frac{i}{2} [(e^{-1} - e) \cos 1 + i(e^{-1} + e) \sin 1]$$

$$= \tfrac{1}{2}[2.350(0.5403)i + 3.086(0.8415)] = 1.299 + 0.6348i$$

(c) By Eq. (8-34),

$$\log_e (1 - i) = \log_e \sqrt{2} - i \left(\frac{\pi}{4} + 2k\pi\right) = 0.3466 - i \left(\frac{\pi}{4} + 2k\pi\right)$$

$$k = 0, 1, 2, 3, \ldots$$

(d) By Eq. (8-35), $i^i = e^{i \log i}$. Now, $\log_e i = \log_e 1 + i(\pi/2 + 2k\pi) = i(\pi/2 + 2k\pi)$. Therefore, $i^i = \exp [-(\pi/2 + 2k\pi)]$, $k = 0, 1, 2, 3, \ldots$.

8. (a) Since e^z is an entire function (Art. 8-5), we can get its derivative from Eq. (8-18). First, write $e^z = e^x(\cos y + i \sin y)$, from Eq. (8-22). Next, set $u = e^x \cos y$ and $v = e^x \sin y$. By Eq. (8-18),

$$\frac{de^z}{dz} = \frac{\partial u}{\partial x} + i\frac{\partial v}{\partial x} = e^x \cos y + ie^x \sin y = e^z$$

(b) Each branch of $\log_e z$ is analytic when $z \neq 0$ (Art. 8-5). By Eq. (8-35), $z = \exp \log_e z$. For each k then, $dz = (\exp \log_e z)d \log_e z$. Dividing through by $z\, dz$, we get

$$\frac{d \log_e z}{dz} = \frac{1}{z}$$

(c) Since $\sin z$ is an entire function (Art. 8-5), we can get its derivative from Eq. (8-18). First, write $\sin z = \sin x \cosh y + i \cos x \sinh y$, from Eq. (8-30). By Eqs. (8-18) and (8-31),

$$\frac{d \sin z}{dz} = \cos x \cosh y - i \sin x \sinh y = \cos z$$

9. Let us assume the flow makes the turn near the origin in the first quadrant of the z plane. Now, we know the solution for unobstructed straight uniform flow. Let us assume that this flow occurs in the upper half of the w plane, parallel to the u axis. Then, the velocity components parallel and perpendicular to the u axis are $V_1 = c$, a constant, and $V_2 = 0$. As in Art. 8-8, the stream function $\psi = cv$, the equipotential $\phi = cu$, and the complex potential $F = cw$.

We saw in Art. 8-6 that $w = z^2$ maps the first quadrant of the z plane onto the upper half of the w plane. Using this transformation, we obtain $F = cz^2 = c(x^2 - y^2) + 2icxy$. Hence, $\psi = 2cxy$. The streamlines $\psi = $ constant are rectangular hyperbolas. The left bank is the streamline $xy = 0$, and the right bank, the streamline $xy = k^2$, a constant such that the maximum width of channel occurs at $x = y = k$. The velocity component parallel to the x axis, by Eqs. (8-39b), is $V_1 = 2cx$. The velocity component parallel to the y axis is $V_2 = -2cy$. Thus, the speed of flow is

$$|V| = 2c \sqrt{x^2 + y^2} = 2cr$$

where r is the distance from the origin. The speed is proportional to distance from the origin.

10. (a) Assume the edge of the plate at 0°C to lie along the x axis of the z plane. The edge at 100°C is at $y = \pi$. Notice that the solution is independent of x. One solution then is the harmonic function $T = 100y/\pi$. (See Art. 8-8.)

(b) Assume the plate to occupy the upper half of the z plane, with the portion of the edge at 100°C lying on the x axis between $x = -1$ and $x = 1$.

Let us examine the transformation $w = u + iv = \log_e (z - 1)/(z + 1)$. By Eq. (8-34), $w = \log_e |z - 1| - \log_e |z + 1| + i[\text{amp}\,(z - 1) - \text{amp}\,(z + 1)]$. Now,

$$\frac{z - 1}{z + 1} = \frac{x - 1 + iy}{x + 1 + iy} \frac{x + 1 - iy}{x + 1 - iy} = \frac{x^2 + y^2 - 1 + 2iy}{(x + 1)^2 + y^2}$$

Hence,

$$v = \text{amp}\,(z - 1) - \text{amp}\,(z + 1) = \arctan \frac{2y}{x^2 + y^2 - 1} \qquad 0 \leq v \leq \pi$$

[See Eq. (8-6).] Thus, when $y = 0$, $v = 0$ or π. If $y > 0$, then $v > 0$, with a range from 0 to π, while u can have any real value. Hence, w transforms the upper half of the z plane into the infinite strip in the w plane between $v = 0$ and $v = \pi$. When $y = 0$ and $-1 < x < 1$, amp $(z - 1) = \pi$

and amp $(z + 1) = 0$; so $v = \pi$. Thus, the part of the x axis where $T = 100°C$ maps onto $v = \pi$. The rest of the x axis maps onto $v = 0$.

Problem 10a has a solution $T = 100v/\pi$ for the infinite strip with a width π, when the edge $v = 0$ is at $0°C$ and the edge $v = \pi$ is at $100°C$. Then, w transforms this temperature distribution into the required solution

$$T = \frac{100}{\pi} \arctan \frac{2y}{x^2 + y^2 - 1}$$

[Notice that the isotherms, $T = $ constant, are circles with centers on the y axis that pass through $(\pm 1, 0)$.]

(c) Assume the plate to lie in the z plane, with the end at $100°C$ along the x axis between $x = -\pi/2$ and $x = \pi/2$. The y axis lies midway between the long edges.

Let us examine the transformation $w = u + iv = \sin z$. By Eq. (8-30), $w = \sin x \cosh y + i \cos x \sinh y$. So $u = \sin x \cosh y$ and $v = \cos x \sinh y$. When $x = \pm\pi/2$, $v = 0$, and $u = \pm \cosh y$. Thus, $|u| > 1$. When $y = 0$, again $v = 0$, and $u = \sin x$. As x varies between $-\pi/2$ and $\pi/2$, u ranges from -1 to 1. Furthermore, because $y \geq 0$, $v \geq 0$. Hence, w transforms the plate area into the upper half of the w plane. The plate end at $100°C$ maps onto the u axis between $u = -1$ and $u = 1$. The edges at $0°C$ map onto the u axis where $u < -1$ and $u > 1$.

From Prob. 10b the solution for the half plane is

$$T = \frac{100}{\pi} \arctan \frac{2v}{u^2 + v^2 - 1}$$

Then, w transforms this into the required solution

$$T = \frac{100}{\pi} \arctan \frac{2 \cos x \sinh y}{\sin^2 x \cosh^2 y + \cos^2 x \sinh^2 y - 1}$$

This, however, can be simplified. Substitution of $\sin^2 x + \cos^2 x$ for 1 in the denominator reduces it to $\sinh^2 y - \cos^2 x$. Now, dividing numerator and denominator by $\sinh^2 y$ yields

$$T = \frac{100}{\pi} \arctan \frac{2 \cos x/\sinh y}{1 - (\cos x/\sinh y)^2} = \frac{200}{\pi} \arctan \frac{\cos x}{\sinh y}$$

$$0 < T < 100$$

on using the relation between the tangent of twice an angle and the tangent of the angle. The isotherms, $T = c$, a constant, are given by $\cos x = \sinh y \tan \pi c/2$. These are curves that pass through $(\pm\pi/2, 0)$.

11. (a) $z^2 = u + iv = x^2 - y^2 + 2ixy$. By Eq. (8-46),

$$\int_C z^2 \, dz = \int_C [(x^2 - y^2) \, dx - 2xy \, dy] + i \int_C [(x^2 - y^2) \, dy + 2xy \, dx]$$

On the line from $z = 0$ to $z = 2 + i$, we have $x = 2y$ and $dx = 2dy$. Hence,

$$\int_C z^2 \, dz = \int_0^1 (6y^2 - 4y^2) \, dy + i \int_0^1 (3y^2 + 8y^2) \, dy$$

$$= \left[\frac{2}{3} y^3\right]_0^1 + i \left[\frac{11}{3} y^3\right]_0^1 = \frac{2 + 11i}{3}$$

(b) For the lines $y = 0$ and $x = 2$, we have, from Eq. (8-46),

$$\int_C z^2 \, dz = \int_0^2 x^2 \, dx + \int_0^1 - 4y \, dy + i \int_0^1 (4 - y^2) \, dy$$

$$= \left[\frac{1}{3} x^3\right]_0^2 - [2y^2]_0^1 + i \left[4y - \frac{1}{3} y^3\right]_0^1 = \frac{2 + 11i}{3}$$

(c) By Eq. (8-4c) and the requirement $|z| = r = 1$, $\bar{z} = e^{-i\theta}$, for $0 < \theta < \pi$. $dz = ie^{i\theta} \, d\theta$. Hence,

$$\int_C \bar{z} \, dz = \int_0^\pi (e^{-i\theta})(ie^{i\theta} \, d\theta) = i[\theta]_0^\pi = \pi i$$

(d) From Prob. 11c,

$$\int_C \bar{z} \, dz = \int_\pi^{2\pi} i \, d\theta = i\pi$$

12. (a) ze^{-z} is analytic on and in the unit circle. By the Cauchy-Goursat theorem, $\int_C ze^{-z} \, dz = 0$. See Art. 8-9.

(b) $(\sin z)/z$ is analytic on and in the ellipse. By the Cauchy-Goursat theorem, therefore, $\int_C [(\sin z)/z] \, dz = 0$. See Art. 8-9.

(c) $\cos z$ is analytic on and in the ellipse. By Eq. (8-53), $\int_C (\cos z)/z \, dz = 2\pi i \cos 0 = 2\pi i$.

(d) $\tan z$ is analytic on and in the unit circle. By Eq. (8-54),

$$\int_C \frac{\tan z}{z^2} \, dz = 2\pi i \left[\frac{d \tan z}{dz}\right]_{z=0} = 2\pi i \sec^2 0 = 2\pi i$$

(e) $$\int_C \frac{2z^2 + z}{z^2 - 1} \, dz = \int_C \left(2 + \frac{1.5}{z - 1} - \frac{0.5}{z + 1}\right) dz$$

$$= 0 + 1.5(2\pi i) - 0.5(2\pi i) = 2\pi i$$

by Eqs. (8-50) and (8-52).

(f) $2z^2 + z$ is analytic on and in the circle. By Eq. (8-54),

$$\int_C \frac{2z^2 + z}{(z - 1)^2} \, dz = 2\pi i \left[\frac{d(2z^2 + z)}{dz}\right]_{z=1} = 2\pi i(4 + 1) = 10\pi i$$

13. (a) $d^n(1/z)/dz^n = (-1)^n n! z^{-n-1}$, where $n = 1, 2, 3, \ldots$, and $z \neq 0$. By Eq. (8-57), for expansion about $z = 1$,

$$\frac{1}{z} = 1 - (z - 1) + (z - 1)^2 - (z - 1)^3 + \cdots + (-1)^n(z - 1)^n + \cdots$$

$$|z - 1| < 1$$

(b) $f(z) = (z - 1)/(z + 1) = 1 - 2/(z + 1)$. From the solution to Prob. 13a,

$$\frac{d^n}{dz^n}\frac{1}{z + 1} = (-1)^n n!(z + 1)^{-n-1}$$

Hence, by Eq. (8-57), for expansion about $z = 0$,

$$f(z) = 1 - 2[1 - z + z^2 - \cdots + (-1)^n z^n + \cdots]$$
$$= -1 + 2z - 2z^2 + \cdots - 2(-1)^n z^n + \cdots \qquad |z| < 1$$

(c) From the solution to Prob. 13b and Eq. (8-57), for expansion about $z = 1$,

$$f(z) = 1 - 2\left[\frac{1}{2} - \frac{z - 1}{4} + \frac{(z - 1)^2}{8} - \cdots + (-1)^n \frac{(z - 1)^n}{2^{n+1}} + \cdots\right]$$
$$= \frac{z - 1}{2} - \frac{(z - 1)^2}{4} + \cdots - (-1)^n \frac{(z - 1)^n}{2^n} + \cdots$$

$$|z - 1| < 2$$

14. (a)

$$f(z) = \frac{z}{(z - 1)(z - 3)} = -\frac{1}{2(z - 1)} + \frac{3}{2(z - 3)}$$

From the solution to Prob. 13a:

$$\frac{d^n}{dz^n}\frac{1}{z - 3} = (-1)^n n!(z - 3)^{-n-1}$$

Hence, by Eq. (8-57), for expansion about $z = 1$,

$$f(z) = -\frac{1}{2(z - 1)} + \frac{3}{2}\left[-\frac{1}{2} - \frac{z - 1}{4} - \frac{(z - 1)^2}{8} - \cdots\right.$$
$$\left. - \frac{(z - 1)^n}{2^{n+1}} - \cdots\right]$$

(b) By Eq. (8-57), for expansion about $z = 0$,

$$\frac{1 + 2z}{z^2 + z^3} = \frac{1}{z^2}\left(2 - \frac{1}{z + 1}\right) = \frac{1}{z^2}(2 - 1 + z - z^2 + z^3 - \cdots)$$
$$= \frac{1}{z^2} + \frac{1}{z} - 1 + z^2 - z + \cdots$$

(c) $\csc z = 1/(\sin z)$. $\sin z = z - z^3/3! + z^5/5! - \cdots$. Hence, by division of 1 by $\sin z$,

$$\csc z = \frac{1}{z} + \frac{1}{3!}z - \left(\frac{1}{5!} - \frac{1}{3!3!}\right)z^3 + \cdots$$

15. (a)

$$\frac{z^2 - 2z + 6}{z - 2} = z + \frac{6}{z - 2} = (z - 2) + 2 + \frac{6}{z - 2}$$

Hence, the residue at $z = 2$ is 6.   See Art. 8-11.

(b) Division of the Maclaurin expansion of sinh z by z^4 gives

$$\frac{\sinh z}{z^4} = \frac{1}{z^3} + \frac{1}{3!}\frac{1}{z} + \frac{1}{5!}z + \frac{1}{7!}z^3 + \cdots$$

So the residue at $z = 0 = 1/3! = \frac{1}{6}$.   See Art. 8-11.

(c) Both $z = 0$ and $z = 2$ are simple poles. $(1 + z)/(2 - z)$ is analytic and not zero at $z = 0$.   By Eq. (8-64b), the residue at $z = 0$ is

$$\lim_{z\to 0}\frac{1 + z}{2 - z} = \frac{1}{2}$$

Similarly, $-(1 + z)/z$ is analytic and not zero at $z = 2$.   Hence, the residue at $z = 2$ is

$$\lim_{z\to 2} -\frac{1 + z}{z} = \frac{-3}{2}$$

(d) $z = 1$ is a pole of order 2. e^z is analytic and not zero at $z = 1$.   By Eq. (8-64a), the residue at $z = 1$ is

$$\frac{d}{dz}e^z\Big]_{z=1} = e^z\Big]_{z=1} = e$$

16. (a) The circle encloses the isolated singularity $z = 0$.   From the solution to Prob. 15c, the residue at $z = 0$ is $\frac{1}{2}$.   Hence, by Eq. (8-62),

$$\int_C \frac{1 + z}{z(2 - z)}\,dz = 2\pi i\left(\frac{1}{2}\right) = \pi i$$

(b) The circle encloses the two isolated singularities at $z = 0$ and $z = 2$. From the solution to Prob. 15c, the residues at those points are $\frac{1}{2}$ and $-\frac{3}{2}$, respectively.   By Eq. (8-63),

$$\int_C \frac{1 + z}{z(2 - z)}\,dz = 2\pi i\left(\frac{1}{2} - \frac{3}{2}\right) = -2\pi i$$

(c) Substitute $(1 - \cos 2\theta)/2$ for $\sin^2 \theta$.   Then, let $2\theta = \phi$, $d\theta = d\phi/2$.   The limits of integration change from $\theta = -\pi$ and π to $\phi = -2\pi$ and 2π.   Thus,

$$\int_{-\pi}^{\pi} \frac{d\theta}{1 + \sin^2 \theta} = \int_{-2\pi}^{2\pi} \frac{d\phi}{3 - \cos \phi}$$

Let $z = e^{i\phi}$.   Then, $d\phi = dz/iz$ and, by Eq. (8-26), $\cos \phi = (z + z^{-1})/2$. We have to integrate twice around the unit circle, since ϕ ranges from -2π to 2π.   With these substitutions, the integral becomes

$$2\int_C \frac{1}{3 - (z + z^{-1})/2}\frac{dz}{iz} = -\frac{4}{i}\int_C \frac{dz}{z^2 - 6z + 1}$$

There are simple poles at $z_0 = 3 \pm 2\sqrt{2}$, but only $3 - 2\sqrt{2}$ lies within the unit circle. The residue at that pole, by Eq. (8-64b), is

$$b_1 = \lim_{z \to z_0} \frac{1}{z - 3 - 2\sqrt{2}} = \frac{1}{3 - 2\sqrt{2} - 3 - 2\sqrt{2}} = -\frac{1}{4\sqrt{2}}$$

By Eq. (8-62), then, the integral equals $2\pi i b_1(-4/i) = \pi\sqrt{2}$.

(d) For the given value of α, the denominator of the integrand can never be zero. Let $z = e^{i\theta}$. Then, $d\theta = dz/iz$, and by (8-26), $\cos\theta = (z + z^{-1})/2 = (z^2 + 1)/2z$, and $\cos 2\theta = (z^2 + z^{-2})/2 = (z^4 + 1)/2z^2$. Hence, for integration around the unit circle, the integral becomes

$$\int_C \frac{z^4 + 1}{2z^2} \frac{1}{1 - 2\alpha(z^2 + 1)/2z + \alpha^2} \frac{dz}{iz} = -\frac{1}{2i} \int_C \frac{(z^4 + 1)\,dz}{z^2(\alpha z - 1)(z - \alpha)}$$

The first-order pole at $z = \alpha$ and the second-order pole at $z = 0$ lie within the unit circle. By Eq. (8-64b), the residue at $z = \alpha$ is

$$\lim_{z \to \alpha} -\frac{1}{2i} \frac{z^4 + 1}{z^2(\alpha z - 1)} = -\frac{\alpha^4 + 1}{2i\alpha^2(\alpha^2 - 1)}$$

By Eq. (8-64a), the residue at $z = 0$ is

$$\frac{d}{dz}\left[-\frac{1}{2i} \frac{z^4 + 1}{(\alpha z - 1)(z - \alpha)} \right]_{z=0} = -\frac{\alpha^2 + 1}{2i\alpha^2}$$

By Eq. (8-63), the integral equals

$$2\pi i\left[-\frac{\alpha^4 + 1}{2i\alpha^2(\alpha^2 - 1)} - \frac{\alpha^2 + 1}{2i\alpha^2} \right] = \frac{2\pi\alpha^2}{1 - \alpha^2}$$

NINE

Probability

Preceding chapters of this book presented methods for predicting from given initial conditions an outcome that is certain to occur. This and the following chapter discuss methods that can be used when incomplete information is available for results to be determined with certainty. The fundamental mathematical models that we shall use for this purpose are based on the theory of probability.

This theory provides a means of reaching conclusions and making decisions despite uncertainties. It has wide application. For one thing, it is fundamental in statistical theory, which we shall examine in Chap. 10. In engineering, both probability and statistical theory are widely used for such purposes as traffic control, quality control, interpretation of data, adjustments of observations to minimize effects of accidental errors, predicting characteristics or behavior of a population from those of a sample, and scheduling maintenance and production. When you become familiar with probability theory, you will find that it has applications in almost everything you do. For example, you can use it in athletics, games, gambling, investments, and evaluating prospects for advancement in employment.

9-1   *Simple Probability.* In many ways, the concept of probability is like that of temperature. To measure temperature quantitatively, we select two physical conditions, freezing and boiling of water, and assign numerical values to them. Then, we establish a numerical scale for intermediate conditions. We treat probability similarly.

We seek to define probability as a measure of the likelihood that a specific event will occur. One definition of simple probability, for example, is the following:

Among several events equally likely to occur, the probability that a given event will occur is the ratio of the number of favorable cases to the total number of possible cases.

Let E be the occurrence of an event and E' the failure of the event to happen. Also, let $n - n_E + n_{E'}$ be the total number of possible events, where n_E is the number of ways that E may occur and $n_{E'}$ the number of ways that E' may occur. Then, by definition, the probability of E is

$$p(E) = \frac{n_E}{n} = \frac{n_E}{n_E + n_{E'}} \tag{9-1}$$

Subtraction of both sides of the equation from 1 yields the probability $q(E)$ that E will not happen

$$q(E) = p(E') = \frac{n_{E'}}{n} = 1 - p(E) \tag{9-2}$$

If E cannot happen, $n_E = 0$; hence, $p(E) = 0$. If E must always occur, $n_E = n$; therefore, $p(E) = 1$. Furthermore, Eq. (9-1) requires that $0 \leq p(E) \leq 1$.

Here is a typical application: What is the probability of rolling a 10 on one throw with a pair of dice? Each face of either die is equally likely to turn up (unless the dice are loaded). Each die, therefore, may show any one of six numbers. Thus, the total number of ways the pair of dice may fall is $6 \times 6 = 36$. But a 10 can be rolled in only three ways— two 5s, or a 6 on one die and a 4 on the other, or vice versa. So by Eq. (9-1), the probability of rolling a 10 is $p(10) = \frac{3}{36} = \frac{1}{12}$.

Notice the assumption that the dice are not loaded. It points out the trouble with this definition. What is meant by equally likely events? We cannot define them as events with equal probability of occurrence, for then we would be defining probability in terms of itself, and this is unacceptable. We cannot omit the modifier, equally likely, because it conveys the idea of the kind of events we are considering. Yet finding an acceptable definition is very difficult, if not impossible.

For example, consider a race run under rules that apply equally to each contestant. If you knew nothing about the capabilities of each contestant, is it correct to assume that each contestant is equally likely to win? Is it wise to assume before you have seen a pair of dice rolled several times that each face is equally likely to turn up? No; these assumptions may lead to false conclusions.

We believe we know what we mean by equally likely events, but we cannot define them without using other undefined terms. So we may as well accept equally likely events as undefined terms. But we must be careful not to use the preceding definition of probability under conditions where our intuitive notion of equally likely events may lead to false conclusions.

The definition also breaks down when we cannot determine the number of ways in which events can occur. For example, how can we use Eq. (9-1) to obtain the probability that a 40-year-old man will live to age 50? We cannot count ways of occurrence in this case. We need a definition of probability that applies to this type of problem.

Such a definition can be based on observations. Let us consider the occurrence of events as the consequence of trials in an experiment. We call the number of times n_E that an event E occurs in n trials the **frequency** of E. Also, we call n_E/n the **relative frequency** of E. Let us assume that n_E/n approaches a limit as n becomes infinitely large. Then, we can offer the definition:

The probability of an event E in one trial of a given experiment is the limit of the relative frequency of E as the number of trials become infinitely large.

This definition enables us to compute probability when we cannot count ways of occurrence. But it has the disadvantage that it is impracticable to perform an infinite number of trials. In practice, we have to be satisfied with the relative frequency determined from a large number of trials, or from observations made on a very large sample. For example, to determine the probability that a 40-year-old man will live to age 50, we would take a sample of the population containing a large number of 40-year-old men. Next, we would count the number of these men still alive 10 years later. Then we would assume that the ratio of the number alive to the number in the original group is a close approximation to the probability that 40-year-old men in the whole population will live to age 50.

Despite its shortcomings, the definition of probability as the limit of relative frequency has wide application in statistics. The definition also provides useful information consistent with that given by the preceding definition: $0 \leq p(E) \leq 1$. For a certain event, $p(E) = 1$; for an impossible event, $p(E) = 0$. For other events, probabilities are given by

$$p(E) = \lim_{n \to \infty} \frac{n_E}{n} \tag{9-3}$$

where n_E is the number of times E occurs in n trials or in a population with n elements. Furthermore, the probability that E', or not E, will occur in a single trial is $q(E) = 1 - p(E)$.

Since the preceding definitions contain undefined or imprecise terms, we might as well accept probability as an undefined term at the start. Then, we can treat probabilities as undefined objects of mathematical models. These models can be defined from information obtained from the preceding definitions. Each model is to describe a specific idealized experiment or observation with a set S of theoretically possible results, or events. Let these events be represented by E_i, where i ranges from 1 to n, and n may be ∞.

First, we shall require that the set S comprise an algebra of events, isomorphic with a Boolean algebra. Next, we shall require that each event E_i be the union, or sum, of a set of mutually exclusive events e_i. Hence, an event $E_i = \Sigma e_i$ occurs when at least one of the e_i happens.

Let us denote by I the certain event that at least one of the events E_i will occur. Then, I is the set formed by the union of all events, ΣE_i, and thus the union of all the mutually exclusive events e_j comprising ΣE_i. I is called a **sample space**. All the events e_j in I comprise the **points** of the space.

Then, just as we associate a vector with points in physical space, we associate a probability with each point of a sample space. Finally, we offer the definition:

The probability $p(E)$ of an event E is the sum of the probabilities $\Sigma p(e_i)$ of the points e_i included in E.

Now, we require that the probabilities $p(E)$ obey the following axioms:

1. $p(E) \geq 0$ for every event E in I.
2. $p(E) = 1$ when the event E is certain.
3. For every countable set of mutually exclusive events E_1, E_2, . . . , E_n, the probability that at least one of these events (E_1, E_2, E_n, any combination of them, or all) will occur is the sum of the probabilities of each of the events:

$$p(E_1 + E_2 + \cdots + E_n) = p(E_1) + p(E_2) + \cdots + p(E_n)$$
$$(9\text{-}4)$$

4. For any two events E_1 and E_2, the probability that both will occur (joint event E_1E_2) given that E_1 occurs is the product of the probability $p(E_1)$ and the probability of E_2 conditioned on the occurrence of E_1 [written $p(E_2|E_1)$]:

$$p(E_1E_2) = p(E_1)p(E_2|E_1)$$
$$(9\text{-}5)$$

This is called the **law of compound probabilities**.

Axioms 1 to 3 imply that $p(E) = 0$ if E is an impossible event.   Hence, $0 \leq p(E) \leq 1$.

If a sample space contains a finite number of points, it is called a finite sample space.   Suppose in such a space a probability p_i is associated with each of the n points e_i in I under the condition $\sum_{1}^{n} p_i = 1$.   Then, by definition the probability of an event E composed of the union of the first m points is $\sum_{1}^{m} p_i$.

If all the points of a sample space are assigned the same probability, the space is called uniform.   If a uniform space contains n points, the probability of each point is $1/n$.   Then, the probability of an event with m points is

$$p(E) = \sum_{1}^{m} p_i = \frac{m}{n} \tag{9-6}$$

This result corresponds to Eq. (9-1).

Rolling of an idealized cubic die is an example of a uniform space containing six points.   We assign a probability of $\frac{1}{6}$ to each face.

If two such dice are rolled, the sample space I comprises $6 \times 6 = 36$ events, since each of the six numbers on one die may occur with any of six numbers on the other die.   We assign a probability of $\frac{1}{36}$ to each point of this space.

What is the probability of rolling a 6 and a 4?   The event E in this case consists of the two points, 6 on one die and 4 on the other, and vice versa.   By Eq. (9-6), the required probability equals the sum of the probabilities of the points in E, or $\frac{1}{36} + \frac{1}{36} = \frac{1}{18}$.

Now, what is the probability that the sum of the numbers upturned equals 10?   We now have a different sample space with 36 points.   Each point represents a possible sum of any two numbers from 1 to 6.   Again, we assign a probability of $\frac{1}{36}$ to each point.   The event E in this case comprises three points, two points corresponding to 6 on one die and 4 on the other, and one point corresponding to 5 on each die.   By Eq. (9-6), the probability of the sum 10 is then $\frac{2}{36} + \frac{1}{36} = \frac{3}{36} = \frac{1}{12}$.

As an example of a nonuniform space, let us assume that one die is loaded so that the probability of a 6 is 0.90.   If this die is rolled with an idealized die, what is the probability that the sum will be 7?   To obtain 7, the idealized die must show 1, for which the probability is $\frac{1}{6}$.   As will be shown later, the probability of a 7 in this way is the product of the probabilities, or $0.90(\frac{1}{6}) = 0.15$.   If the second die was also loaded but to show 1, the probability of a 7 with the loaded 6 showing would increase to $(0.90)(0.90) = 0.81$.   The total probability of 7 is larger because of the off-chance of rolling it with some other combination, 5 and 2, or 4 and 3.

9-2 *Permutations and Combinations.* For finite, uniform sample spaces, you can use Eqs. (9-1) and (9-2) for computing simple probabilities. For this purpose, you will find it worth while to be familiar with combinatorial analysis, since Eq. (9-1) requires determination of the number of ways in which events may occur. As a refresher, this article presents some of the principles and formulas of combinatorial analysis.

Suppose that an event E_1 can happen in three ways and then an event E_2 can happen in two ways. The joint event, both E_1 and E_2, can happen in six ways. This results because for each way in which E_1 occurs, there are two ways in which E_2 may occur. Since E_1 occurs in three ways, E_1E_2 can occur in $3 \times 2 = 6$ ways.

In general, if an event E_1 can happen in m ways and an event E_2 in n ways, E_1E_2 can occur in mn ways.

For example, in how many ways can you draw first a red ball and then a black ball, without peeking, from a box containing five red balls and four black ones? There are five ways of picking a red ball on the first draw. There are four ways of picking a black ball on the second draw. Thus, each way of picking a red ball may be followed by four ways of picking a black one. Hence, there are $5 \times 4 = 20$ ways of picking two balls from the box in the order red, black. Similarly, there are $4 \times 5 = 20$ ways of picking two balls from the box in the order black, red.

Now, in how many ways can we pick one red ball and one black one, regardless of the order of drawing? We have found that we can pick red, black in 20 ways and black, red in 20 ways. Hence, if we ignore the order of drawing, we can pick a red ball and a black one in $20 + 20 = 40$ ways; that is, we can pick red, black or black, red in 40 ways.

In general, if E_1 and E_2 are mutually exclusive events and E_1 can occur in m ways and E_2 in n ways, either E_1 or E_2 can happen in m + n ways.

If E_1 and E_2 are not mutually exclusive, E_2 may occur in some cases when E_1 happens. Suppose that these cases arise in c ways, while E_1 occurs in m ways and E_2 in n ways. Both m and n include c; so $m + n$ contains c twice. Hence, E_1 or E_2 can happen in $m + n - c$ ways.

Permutations are arrangements. For example, AB and BA are two permutations of the letters A and B. Similarly, AB, AC, BA, BC, CA, and CB are six permutations of the three letters A, B, C taken two at a time.

The last example represents the solution to the problem of determining the number of permutations when three items are taken two at a time. Without writing down and counting each permutation, we could have obtained the solution as follows: The first item can be chosen in three different ways. After that is done, the second position can be filled in only two different ways. Then, both positions can be filled in 3×2 different ways.

If we are dealing with n objects to be taken r at a time, we can denote the number of permutations by $P(n,r)$ (sometimes written $_nP_r$ or P_r^n). In this case, the first position can be filled in n ways, the second in $n - 1$ ways, the third in $n - 2$ ways, and the last in $n - r + 1$ ways. Hence, the number of permutations of n different items taken r at a time is

$$P(n,r) = n(n - 1)(n - 2) \cdots (n - r + 1) \tag{9-7}$$

Notice that the right-hand side of Eq. (9-7) contains r terms.

For example, taking the first eight letters of the alphabet three at a time, we can arrange them in $8 \times 7(8 - 3 + 1) = 336$ ways.

Permutations of n different things taken n at a time total

$$P(n,n) = n! \tag{9-8}$$

But if r items are alike, the number of permutations reduce to $n!/r!$ If, in addition, s are alike, $P = n!/r!s!$ And if t also are alike,

$$P = \frac{n!}{r!s!t!} \tag{9-9}$$

For example, the number of permutations of the eight letters in *engineer* taken all at a time is $8!/3!2! = 3,360$, when account is taken of the fact that e occurs three times and n twice.

Any collection of items composes a combination. For example, AB and BA are the same combination of the letters A and B. And only the three combinations AB, BC, and AC can be formed from the three letters A, B, C taken two at a time.

The last example represents the solution to the problem of determining the number of ways in which two items can be selected from a set of three items in any order. Without writing down and counting each combination, we could have obtained the solution as follows: The number of permutations of three things taken two at a time is 3×2. The two objects in each combination can be arranged in two ways. So if there are C possible combinations, the number of permutations also equals $2C$. Hence, $C = 3 \times 2/2 = 3$.

We will represent the number of combinations of n different items taken r at a time by $C(n,r)$ $\left[\text{sometimes written } _nC_r \text{ or } \binom{n}{r} \right]$. The r objects in any one combination of n things taken r at a time can be arranged in $r!$ ways. The $C(n,r)$ combinations, therefore, can be arranged in a total of $r!C(n,r)$ ways, equal to $P(n,r)$. Consequently,

$$C(n,r) = \frac{P(n,r)}{r!} = \frac{n(n - 1) \cdots (n - r + 1)}{r!} = \frac{n!}{r!(n - r)!} \tag{9-10}$$

As an example, let us compute the number of ways we can form a committee of three from a group of twenty-one.   With $n = 21, r = 3$, and $n - r + 1 = 19$, we obtain

$$C(21,3) = \frac{21 \times 20 \times 19}{3!} = 1{,}330$$

Note that the last term on the right in Eq. (9-10) is symmetrical in r and $n - r$.   Hence,

$$C(n,r) = C(n, n - r) \tag{9-11}$$

This relationship can often be used to reduce the amount of calculations. For example, since $C(21,3) = C(21,18)$, we can use the preceding computation to determine that a committee of 18 can be formed from a group of 21 in 1,330 ways.

Note also that

$$C(n,n) = C(n,0) = 1 \tag{9-12}$$

Some useful probability formulas are derived from the fact that the expansion of a binomial raised to a power can be expressed conveniently in terms of combinations:

$$(p + q)^n = p^n + C(n,1)p^{n-1}q + C(n,2)p^{n-2}q^2 + \cdots \\ + C(n, n - r)p^r q^{n-r} + \cdots + q^n \tag{9-13}$$

For example, we can get one formula at once by taking $p = q = 1$:

$$C(n,1) + C(n,2) + C(n,3) + \cdots + C(n,n) = 2^n - 1 \tag{9-14}$$

Suppose, for example, that a shop has 10 different articles for sale. How many choices has a purchaser?   The purchaser can buy one, or any two, or any three, etc., or all ten.   So the number of choices is $2^{10} - 1 = 1{,}023$.

9-3   *Bayes' Theorem.* From Axiom 4 in Art. 9-1, we can derive Bayes' theorem, a formula for the probability that an event that has already happened did so in a particular way.

Consider a sample space composed of mutually exclusive events e_i, so that $I = e_1 + e_2 + \cdots + e_n$.   Each point e_i is associated with a probability $p(e_i)$.   Suppose that occurrence of an event E is conditioned on occurrence of one of the events e_i.   By Axiom 4, the probability that both E and this e_i will occur is

$$p(Ee_i) = p(e_i)p(E|e_i) \tag{9-15}$$

Now, $E = Ee_1 + Ee_2 + \cdots + Ee_n = \Sigma Ee_i$, since E occurs when at least one of the joint events Ee_i does. By Axiom 3 then,

$$p(E) = p(\Sigma Ee_i) = \Sigma p(Ee_i) = \Sigma p(e_i)p(E|e_i) \qquad (9\text{-}16)$$

By Axiom 4 also,

$$p(Ee_i) = p(E)p(e_i|E) = p(e_i)p(E|e_i) \qquad (9\text{-}17)$$

From this and Eq. (9-16), we get the desired relation between the probability of an event e_i given that E occurs and the probability of E given that e_i happens (Bayes' Theorem):

$$p(e_i|E) = \frac{p(e_i)p(E|e_i)}{p(E)} = \frac{p(e_i)p(E|e_i)}{\Sigma p(e_i)p(E|e_i)} \qquad (9\text{-}18)$$

To illustrate how Eq. (9-18) can be used to determine the probable causes of an event, let us solve the following problem: On a large construction project, records are kept of tests of concrete strength and the results of investigations into causes of any low-strength concrete produced. The table below gives the findings in the first three columns, with relative frequencies treated as probabilities. Possible conditions of the concrete mix are given in column 1 as mutually exclusive events e_i. The probability of these events occurring, as deduced from project records, is listed in the second column. From the results of investigations, the third column gives $p(E|e_i)$, probability of low-strength concrete occurring given each e_i, where E represents the event low-strength concrete. A test report is submitted showing that the latest mix of concrete has undesirably low strength. What is the probability of each item in column 1 being the cause?

Probabilities of Factors Causing Low-strength Concrete E

(1) Factors affecting concrete strength	(2) $p(e_i)$	(3) $p(E|e_i)$	(4) $p(e_i)p(E|e_i)$	(5) $p(e_i|E)$
Poor quality cement e_1	0.01	0.90	0.009	0.05
Insufficient cement e_2	0.05	0.80	0.040	0.24
Excessive water e_3	0.10	0.80	0.080	0.47
Excessive admixture e_4	0.04	0.10	0.004	0.02
Inaccurate batching of sand and stone e_5	0.10	0.30	0.030	0.18
Acceptable batching and materials e_6	0.70	0.01	0.007	0.04
	1.00		$p(E) = 0.170$	1.00

The solution can be obtained from Eq. (9-18). Column 4, the products of columns 2 and 3, gives the numerator of the right-hand side of Eq. (9-18). The sum of the values in column 4 equals $p(E)$, the probability of low-strength concrete on this project, roughly $\frac{1}{6}$. Column 5, obtained by dividing each value in column 4 by $p(E)$, tabulates the required probability that each factor in column 1 caused the low-strength concrete reported. The results indicate that excessive water, with the highest probability, 0.47, is the most likely cause. Insufficient cement, with the next highest probability, 0.24, is the second most probable cause.

9-4   Compound Probabilities. Events we dealt with in Art. 9-3 and other events to which Axiom 4 of Art. 9-1 applies are dependent on each other. The occurrence of one event is contingent on the occurrence of another. Often, however, you may have to determine the probability of joint events that are independent of each other.

Two events E_1 and E_2 are statistically, or stochastically, independent if and only if

$$p(E_1 E_2) = p(E_1)p(E_2) \tag{9-19}$$

So $p(E_1|E_2) = p(E_1)$ if $p(E_2) \neq 0$ and $p(E_2|E_1) = p(E_2)$ if $p(E_1) \neq 0$.
    The outcomes of successive rolls of dice are independent events. Each roll has no affect on the following one. Hence, Eq. (9-19) can be used to solve such problems as: What is the probability of rolling two 10s in succession with two throws of a pair of dice? The probability of rolling a 10 on the first roll is $\frac{1}{12}$ (Art. 9-1). The probability of rolling a 10 on the second roll is also $\frac{1}{12}$, since this roll is independent of the first. By Eq. (9-19), the probability of two 10s in succession is

$$p = \frac{1}{12}\frac{1}{12} = \frac{1}{144}$$

As another example, a bag contains six black balls and three white balls. What is the probability of picking two white balls in two tries, if one ball is taken out at a time and not replaced? On the first try, the total number of ways of drawing a ball is $6 + 3$, and a white ball can be drawn in three ways. By Eq. (9-1), the probability of a white ball on the first try is $\frac{3}{9} = \frac{1}{3}$. Now, eight balls remain in the bag, and only two are white. Hence, the probability of a white ball on the second try is $\frac{2}{8} = \frac{1}{4}$. By Eq. (9-19), therefore, the probability of picking two white balls in succession is $(\frac{1}{3})(\frac{1}{4}) = \frac{1}{12}$.

9-5   Total Probabilities. Axiom 3 of Art. 9-1 provides the means of determining the probability that any one of a set of mutually exclusive events will occur. For example, by Eq. (9-4), the mathematical representation of Axiom 3, the probability that either of two mutually exclusive events E_1 and E_2 will occur is

$$p(E_1 + E_2) = p(E_1) + p(E_2) \qquad (9\text{-}20)$$

Equations (9-4) and (9-20) can be used to solve such problems as: What is the probability of drawing either a black or a white ball on one try from a box containing three black, four white, and five red balls? The total number of ways of drawing a ball is $3 + 4 + 5 = 12$. A black ball can be drawn in three ways. By Eq. (9-1), the probability of drawing a black ball is $\frac{3}{12} = \frac{1}{4}$. The probability of drawing a white ball, by Eq. (9-1), is $\frac{4}{12} = \frac{1}{3}$. And by Eq. (9-20), the probability of picking either a black ball or a white one is $\frac{1}{4} + \frac{1}{3} = \frac{7}{12}$.

Equation (9-4) can be used for problems with more than two events, such as: What is the probability of rolling a 3 or less with a die? The problem may sometimes be phrased: What is the probability of at least a 1, 2, or 3 turning up in one roll of a die? The probability of rolling any number is $\frac{1}{6}$. Hence, by Eq. (9-4), the probability of 1, 2, or 3 = $\frac{1}{6} + \frac{1}{6} + \frac{1}{6} = \frac{1}{2}$.

Suppose that events E_1 and E_2 are not mutually exclusive. E_2 may occur in some cases when E_1 happens. As pointed out in Art. 9-2, if E_1 can occur in m ways, E_2 in n ways, and E_1E_2 in c ways, $E_1 + E_2$ can happen in $m + n - c$ ways. If N is the total number of cases, then, by Eq. (9-1),

$$p(E_1 + E_2) = \frac{m + n - c}{N} = \frac{m}{N} + \frac{n}{N} - \frac{c}{N} \qquad (9\text{-}21)$$

But $m/N = p(E_1)$, $n/N = p(E_2)$, and $c/N = p(E_1E_2)$. Hence, Eq. (9-21) can be written

$$p(E_1 + E_2) = p(E_1) + p(E_2) - p(E_1E_2) \qquad (9\text{-}22)$$

This is known as the **theorem of total probability.**

As an example, what is the probability that at least one 4 will appear when two dice are rolled? The probability of a 4 on either die is $\frac{1}{6}$. The probability of a 4 on both dies is $(\frac{1}{6})(\frac{1}{6}) = \frac{1}{36}$. By Eq. (9-22), the probability of at least one 4 is $\frac{1}{6} + \frac{1}{6} - \frac{1}{36} = \frac{11}{36}$.

For more than two events, the probability of at least one occurring can be determined with the aid of a Venn diagram for ascertaining intersecting events. For example, for three events,

$$p(E_1 + E_2 + E_3) = p(E_1) + p(E_2) + p(E_3) - p(E_1E_2)$$
$$- p(E_1E_3) - p(E_2E_3) + p(E_1E_2E_3) \qquad (9\text{-}23)$$

9-6 *Probability of Repetitions.* From now on, we will often encounter the concept of probability as a function of a random variable, or variate, which may or may not have a numerical value. A random, or stochastic, variable corresponds on a reciprocal one-to-one basis with the points of a sample space.   Specification of one state or value of the variate determines a point of the sample space.   Selection of a point of the space determines a value of the stochastic variable.

Since a probability is associated with each point of the space, each value of the variate determines the probability of a point.   Hence, if x is the value of the variate, we can express the probability of a corresponding point as $p = f(x)$. $f(x)$ is called a **frequency function** because its value is proportional to the expected frequency of a specific value of x in a given number of observations.   The set of probabilities p corresponding to the set x is called the **probability distribution** of x.

One example is the **binomial frequency function,** or **binomial distribution:**

$$p_{rn} = f(r) = C(n,r)p^r q^{n-r} \tag{9-24}$$

where p is the probability that an event happens in a single trial and q the probability that it will fail. $f(r)$ determines the probability p_{rn} that the event will occur exactly r times in n trials.

To illustrate let us assume that a coin is tossed four times.   What is the probability of exactly r heads?   Instead of using Eq. (9-24) directly at first, let us apply fundamentals, so that you can obtain a better understanding of the significance of it.

First, let $r = 0$.   All four tosses must yield tails.   If $q = \frac{1}{2}$ is the probability of a tail on one toss, the probability of no heads is $f(0) = q^4 = \frac{1}{16}$.   Next, let $r = 1$.   If the head occurs on the jth toss, the coin must show a tail on the three other tosses.   The probability of this joint event is pq^3, where $p = \frac{1}{2}$ is the probability of a head on one toss. Since there are four tosses, or four values for j, the probability of at least one such joint event is, by Eq. (9-4), $4pq^3 = C(4,1)pq^3 = \frac{4}{16}$.   Then, let $r = 2$.   If the heads occur on the jth and kth tosses, the coin must show tails on the other two tosses.   The probability of this joint event is $p^2 q^2$.   Since j can be any one of four tosses and k any one of three other tosses, in any order, the joint event can occur in $C(4,2)$ ways.   Therefore, the probability that at least one of the joint events will happen is $C(4,2)p^2 q^2 = \frac{6}{16}$.   We can continue in this manner to calculate $f(r)$ or use Eq. (9-24) directly: For exactly three heads, $f(r) = C(4,3)p^3 q = \frac{4}{16}$. These computations yield the following binomial distribution for number of heads in four tosses of a coin:

$r =$	0	1	2	3	4
$f(r) =$	$\frac{1}{16}$	$\frac{4}{16}$	$\frac{6}{16}$	$\frac{4}{16}$	$\frac{1}{16}$.

Notice that the function is symmetrical about $r = 2$.

Another type of distribution can be obtained from Eq. (9-13). If p is the probability that an event will occur in one trial and q the probability that it will fail, $p + q = 1$, by Axiom 2, Art. 9-1. Then, Eq. (9-13) becomes, on using Eq. (9-11) to set $C(n,r) = C(n, n - r)$,

$$1 = p^n + C(n, n - 1)p^{n-1}q + C(n, n - 2)p^{n-2}q^2$$
$$+ \cdots + C(n,r)p^r q^{n-r} + \cdots + q^n \quad (9\text{-}25)$$

Now, p^n is the probability that the event will occur exactly n times in n trials; q^n is the probability that the event will not occur at all; and by Eq. (9-24), the typical term $C(n,r)p^r q^{n-r}$ is the probability that the event will happen exactly r times in n trials. Hence, the sum on the right-hand side of Eq. (9-25) gives the probability that the event will happen either n times, or $n - 1$ times, or $n - 2$ times, or zero times, which is a certainty.

The **cumulative binomial distribution function** formed by the first r terms of the sum in Eq. (9-25)

$$p_r = p^n + C(n, n - 1)p^{n-1}q + C(n, n - 2)p^{n-2}q^2$$
$$+ \cdots + C(n,r)p^r q^{n-r} \quad (9\text{-}26)$$

gives the probability that an event will happen at least r times in n trials.

For example, what is the probability of throwing at least three 6s in five throws with a single die? The probability of a 6 on one throw is $p = \frac{1}{6}$, and of not succeeding is $q = \frac{5}{6}$. With $n = 5$ and $r = 3$, Eq. (9-26) gives the probability as $(\frac{1}{6})^5 + C(5,4)(\frac{1}{6})^4(\frac{5}{6}) + C(5,3)(\frac{1}{6})^3(\frac{5}{6})^2 = 276/7,776$.

By extending the right-hand side of Eq. (9-26) to but not including q^n and substituting in Eq. (9-25), we find:

The probability that an event will happen at least once in n trials is

$$p_1 = 1 - q^n \quad (9\text{-}27)$$

This equation can be used to correct a mistaken concept of the meaning of probability. Some gamblers believe that if the probability of an event is $1/n$, the event is certain to occur at least once in n trials. They are wrong. The probability, by Eq. (9-27), is less than 1 by the nth power of the probability that the event will not occur in one trial. For example, the probability of rolling at least one 3 in six rolls of a die is $1 - (\frac{5}{6})^6 = 31,031/46,656$, or about $\frac{2}{3}$.

Suppose now that we wish to obtain the probability that an event will occur at most once in two trials. This is equivalent to requiring that we desire the probability that the event will fail to happen at least once in the two trials. Hence, we can use Eq. (9-26) for the calculation with

p and q interchanged; that is, $p_1' = q^2 + C(2,1)pq$.   In general, we can conclude:

The probability that an event will happen at most r times in n trials is the cumulative binomial distribution function

$$p_r' = q^n + C(n,1)pq^{n-1} + C(n,2)p^2q^{n-2} + \cdots + C(n,r)p^rq^{n-r}$$

$$(9\text{-}28)$$

For example, what is the probability of rolling no more than two 6s in five throws with a single die?   Here, $p = \frac{1}{6}$, $q = \frac{5}{6}$, $n = 5$, and $r = 2$.   Hence, the required probability, by Eq. (9-28), is $p_2' = (\frac{5}{6})^5 + C(5,1)(\frac{1}{6})(\frac{5}{6})^4 + C(5,2)(\frac{1}{6})^2(\frac{5}{6})^3 = 7,500/7,776$.

Probability distributions or functions when continuous may be represented graphically by curves and when discrete (for a variate that assumes only isolated values), by histograms.   A probability histogram is a form of bar chart.   The bars are centered over each variate, and the bar height gives the probability corresponding to the value of the variate.   Typical histograms are shown in Figs. 9-1 and 9-2.

Figure 9-1*a* shows the binomial distribution for the probability of an event occurring exactly x times in 20 trials, when the probability p of it happening in one try is 0.2.   Similarly, Fig. 9-1*b* shows the binomial distribution when $p = 0.4$ and $n = 20$.   Notice that this histogram is more nearly symmetrical than the one in Fig. 9-1*a*.   The highest probability in Fig. 9-1*b* is at $x = 8$, whereas the peak in Fig. 9-1*a* is at $x = 4$. For $p = 0.5$ and $n = 20$, the binomial distribution is symmetrical about a peak at $x = 10$.   Also, notice that the highest probability, 0.2182, for

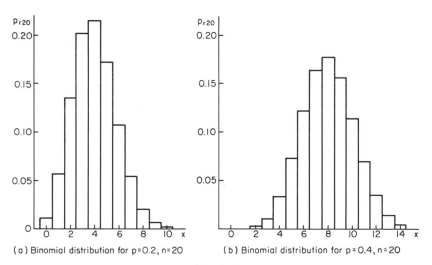

(a) Binomial distribution for p=0.2, n=20   (b) Binomial distribution for p=0.4, n=20

Fig. 9-1

$p = 0.2$ is larger than the peak, 0.1797, for $p = 0.4$, which, in turn, is larger than the peak, 0.1762, for $p = 0.5$.

Histograms of the types shown in Fig. 9-2 are often used in statistical studies. Figure 9-2a shows the probability that each value of the variate will be equaled or exceeded in 20 trials, when the probability of x in one trial is 0.4. Similarly, Fig. 9-2b shows the probability of each value of the variate being equaled or not exceeded when $p = 0.4$ and $n = 20$.

Books containing statistical tables usually include tables of values for binomial distributions. (See Bibliography, Art. 9-10.)

9-7 *Expectation.* The peaks in Fig. 9-1 occur at the most probable values of the variate. These values provide a good estimate of the number of times an event will occur on the average if a large number of observations are made. For example, for an event with $p = 0.2$, you might estimate that it will happen on the average four times in every twenty trials, because the peak is at $x = 4$. For an event with $p = 0.4$, you might estimate that it will happen eight times, because the peak is at $x = 8$.

Another estimate is provided by the expectation, or expected value, $E(x)$ of a variate x. Graphically, $E(x)$ is the value of x at the center of gravity of the area of the diagram for the frequency function for x. For a discrete variable x_i with probabilities p_i, since $\Sigma p_i = 1$,

$$E(x) = \frac{\Sigma x_i p_i}{\Sigma p_i} = x_1 p_1 + x_2 p_2 + \cdots + x_n p_n \qquad (9\text{-}29)$$

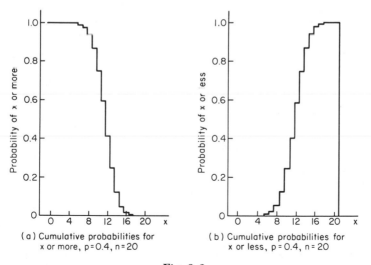

(a) Cumulative probabilities for
x or more, p=0.4, n=20

(b) Cumulative probabilities for
x or less, p=0.4, n=20

Fig. 9-2

For a continuous variable x with frequency function $f(x)$, since $\int_{-\infty}^{\infty} f(x)\, dx = 1$,

$$E(x) = \frac{\int_{-\infty}^{\infty} xf(x)\, dx}{\int_{-\infty}^{\infty} f(x)\, dx} = \int_{-\infty}^{\infty} xf(x)\, dx \tag{9-30}$$

As an example, suppose that a contractor studying the long-time records of his company found that the probability of making a profit exceeding 20 percent or losing more than 10 percent was zero. The records also showed the following probabilities: 10 percent loss, 0.10; no profit, 0.20; 10 percent profit, 0.60; and 20 percent profit, 0.20. What is his expectation of profit? By Eq. (9-29),

$$E(x) = -10\% \times 0.10 + 0 \times 0.10 + 10\% \times 0.60 + 20\% \times 0.20 = 9\%$$

Thus, he can expect about 9 percent profit over a large number of jobs.

If all values of the variate x_i are equally likely to occur, $p_i = 1/n$ and $E(x)$ equals the **arithmetic mean** $(1/n)\Sigma x_i$.

If p and q are the probability of an event occurring and not occurring, respectively, in one trial, the expected value for one trial is, by Eq. (9-29), $1p + 0q = p$. Hence, the expected value, or expected number of occurrences, in n trials, is

$$E(x) = np \tag{9-31}$$

where x = number of occurrences.

As an example, suppose that maintenance records show that the probability of a certain type of lamp burning out within 2 years is 0.20. What is the expectation for a group of 1,000 such lamps in a 2-year period? By Eq. (9-31), $E(x) = 1{,}000 \times 0.20$, or 200 lamps will burn out.

For a binomial distribution, the value of the variate having the highest probability, or the most probable value of the variate, equals the expected value when np is an integer. Otherwise, the most probable value is within 1 of np. For large values of n, the probability of the most probable value of the variate is given approximately by

$$p_{\max} = \frac{1}{\sqrt{2\pi npq}} \tag{9-32}$$

where $q = 1 - p$. For example, for the binomial distribution in Fig. 9-1a, Eq. (9-32) gives 0.22 for the peak probability, compared with the exact value of 0.2182.

When a process is random, an increase in the number of trials or observations tends to increase also the absolute deviations of the variate x from its expected value $|x - np| = |x - E(x)|$. Division by n gives the relative deviation $|x/n - p| = |x/n - E(x/n)|$. When n is large, this tends to become small. Thus, the more observations made, the closer the relative frequency comes to the probability of the event occurring in one trial. This relation is often used in statistical investigations.

9-8   *Poisson Distribution.* Conditions often arise where the possible number of observations n are so large that it is impracticable to count them or where it is impossible to determine the number—for example, the number of times lightning strikes per year or the number of atoms in a given mass that may decay radioactively. In such cases, the binomial distribution cannot be used, since it requires that n be known.

Under other conditions, experience may indicate that a different probability distribution may be more applicable than the binomial. The Poisson distribution is one alternative. It is often applicable when the number of observations is large and the probability of occurrence of an event is small. It offers the advantage over a binomial distribution that n need not be known. In the Poisson distribution, probability is a function of np, the expected value of the variate. The hourly flow of highway traffic and hourly volume of telephone calls often have a Poisson distribution. The number of radioactive atoms decaying per second in a given mass also occurs in a Poisson distribution when the expectation of decaying atoms is small. And the number of defective items produced under strict quality control in a large-volume process often corresponds to a Poisson distribution.

The Poisson distribution gives the probability $P(x)$ of an event occurring exactly x times in n trials, or in a specific interval of time or space, or in a given size of population as

$$P(x) = e^{-m} \frac{m^x}{x!} \tag{9-33}$$

where $m -$ the expected value of $x = np$ and $e = 2.71828 \ldots$.

As an example, let us compute the probability of five atoms decaying per millisecond in a gram of material when, on the average, two decay per millisecond. The expected number of decays $np = 2$. By Eq. (9-33), the probability is $e^{-2}2^5/5! = 0.036$.

Books containing statistical tables usually have tables of Poisson distribution functions. (See Bibliography, Art. 9-10.) Figure 9-3 shows a histogram of the Poisson distribution for $m = 4$. A comparison with Fig. 9-1a indicates that there is a close resemblance between the two

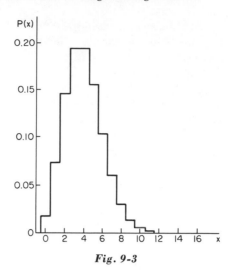

Fig. 9-3

histograms.   Actually, Poisson's distribution approximates the binomial for small p, if n is sufficiently large that np is moderate.

9-9   *Normal Distribution.* In Arts. 9-7 and 9-8, we examined two types of distributions of discrete stochastic variables.   Let us now consider continuous variates.   These may be actually continuous, or discrete variables considered continuous, or discrete variables approximated by continuous variables to simplify computations.   An example of a continuous variate is the probability of a specific magnitude of error in a measurement.   An example of a discrete variate considered continuous is the number of lamps per 100 lamps burning out per year in a large factory.   Actually, the number must be an integer.   But if the total number of lamps being recorded is not exactly 100 or if records are averaged over several years, the number of burned-out lamps per 100 may be any number up to 100, not necessarily an integer.   For example, the chief engineer of an industrial company with many buildings might find the records showing 5.7 lamps per 100 in one plant, 3.4 in another, 4.6 in still another, etc.   Hence, the variate may be considered continuous.   Furthermore, the shape of the histograms when discrete variates are plotted, such as those in Figs. 9-1 and 9-3, may suggest the possibility of representing each distribution by a smooth curve given by an equation that will simplify computations.

The function $f(x)$ for such a curve, or for a similar curve for an actually continuous variable, is called a **frequency function,** or **probability density function.**

The function $F(x)$ for the cumulative probability that the random variable x will not exceed specific values X is called the **distribution function.** For example, the function for a smooth curve approximating the histogram in Fig. 9-2b is a distribution function.

Strictly, a random variable is continuous if and only if its distribution function $F(x)$ is continuous and has a piecewise continuous derivative $f(x)$, the probability density of x:

$$f(x) = \frac{dF(x)}{dx} \tag{9-34}$$

A probability density $f(x)$ has the following properties: It is a single-valued, positive, real number for all real values of x. The total area under the curve $y = f(x)$ equals 1.

$$\int_{-\infty}^{\infty} f(x)\, dx = 1 \tag{9-35}$$

The probability that $x \leq X$ is given by the area under the curve $y = f(x)$ to the left of the ordinate at $x = X$.

$$P[x \leq X] = \int_{-\infty}^{X} f(x)\, dx = F(x) \tag{9-36}$$

If a and b are any two values of x, with $a < b$, the probability that x will be between a and b is given by the area of the curve $y = f(x)$ between ordinates at $x = a$ and $x = b$.

$$P[a < x \leq b] = \int_{a}^{b} f(x)\, dx = F(b) - F(a) \tag{9-37}$$

Suppose that we pick $a < X$ and $b > X$ and let a and b approach X. Since the integral in Eq. (9-37) will become zero when $a = b = X$, the probability that x will be exactly X is zero. Hence, for a continuous variable, we do not accept the value of the density function $f(X)$ as the probability of X, whereas for a discrete variable, the value of the frequency function $f(x_i)$ for $x_i = X$ gives the probability of X.

This characteristic of a continuous variable is not a serious handicap. In engineering problems, you generally will be concerned with $F(x)$ rather than $f(x)$. For example, you usually will want to know the probability that there will not be more than X defects per 1,000 units or the probability that the flow in a stream will not exceed X cfs. In those cases when you do wish the probability of X, you can use Eq. (9-37) by integrating over a range about X depending on the accuracy you desire. For example, you might decide on $X \pm 0.01$, or $X \pm 0.1$, or $X \pm 0.5$, or some similar range.

The distribution function $F(x)$ has the following properties: It is a nondecreasing function of x. Also, $0 < F(x) \leq 1$. And $F(b) - F(a)$ equals the probability that x will be between a and b, where $a < b$.

The mean, or expected value, $\bar{X}$ of a continuous variable x with density function $f(x)$ is defined by

$$\bar{X} = \int_{-\infty}^{\infty} x f(x)\ dx \tag{9-38}$$

The **variance** of x is defined by

$$\sigma^2 = \int_{-\infty}^{\infty} (x - \bar{X})^2 f(x)\ dx \tag{9-39}$$

σ, the square root of the variance, is called the **standard deviation** of x. Variance and mean are related by

$$\sigma^2 + \bar{X}^2 = \int_{-\infty}^{\infty} x^2 f(x)\ dx \tag{9-40}$$

One distribution of a continuous random variable widely used for statistical purposes is called normal distribution. The **normal density function** is

$$f(x) = \frac{1}{\sigma \sqrt{2\pi}}\, e^{-t^2/2} \tag{9-41}$$

where σ is the standard deviation of x and $t = (x - \bar{X})/\sigma$, with $\bar{X}$ the expected value of X.

This function is also known as the Gaussian, or error, function. It is plotted in Fig. 9-4 for $\bar{X} = 8$ and $\sigma = 2.19$. Notice that the curve is bell-shaped, symmetric about $x = 8$, and asymptotic to the x axis.

If we hold σ constant but change $\bar{X}$, the curve shifts along the x axis, to remain symmetric about $\bar{X}$, where the curve has a maximum. In contrast, if we hold $\bar{X}$ constant and change σ, the peaked portion becomes

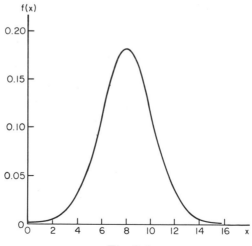

Fig. 9-4

wider or narrower. Thus, σ is a measure of the deviation of observed values of x from the mean. The larger σ is, the greater the spread of the curve, the smaller the probability of $\bar{X}$, and the further from $\bar{X}$ that relatively large probabilities occur. For the area under the curve must remain equal to unity.

The **normal distribution function** is

$$F(x) = \frac{1}{\sigma \sqrt{2\pi}} \int_{-\infty}^{x} e^{-t^2/2} \, dx \tag{9-42}$$

It gives the probability that x will not be exceeded.

If we replace x and dx by t and dt, the resulting function, called the **standard normal distribution,** is independent of σ and in a very convenient form for tabulating numerical values of the function.

$$F(t) = \frac{1}{\sqrt{2\pi}} \int_{-\infty}^{t} e^{-t^2/2} \, dt \tag{9-43}$$

Books containing statistical tables usually contain tables giving values of the standard normal distribution function and the **standard normal density function** $e^{-t^2/2}/\sqrt{2\pi}$. Alternatively, some tables give values when the integration limits are 0 and t or $-t$ and t. (See Bibliography, Art. 9-10.) The standard normal distribution function is plotted in Fig. 9-5.

To learn the significance of normal distributions, let us examine the situation where surveyors make a large number of measurements of the distance between two points. After they correct for all systematic errors,

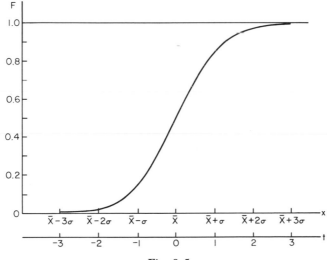

Fig. 9-5

such as wind, temperature, slope, curvature of the earth, they find that few of their measurements are identical. What is the true distance between the two points?

Let $\bar{X}$ be this distance, d_i the ith measurement of it, and x_i the error in the ith measurement. Then, $\bar{X} = d_i + x_i$. Summing from $i = 1$ to $i = n$, the number of observations, we get

$$n\bar{X} = \sum d_i + \sum x_i \quad \text{or} \quad \bar{X} = \frac{1}{n}\sum d_i + \frac{1}{n}\sum x_i \tag{9-44}$$

Let us now assume that the arithmetic mean of the measurements $\Sigma d_i/n$ is the best estimate we can make of $\bar{X}$. In that case, Eq. (9-44) yields

$$\Sigma x_i = 0 \tag{9-45}$$

Thus, our assumption requires that the errors be random and their algebraic sum should be zero.

To derive the normal density function, Gauss postulated in addition that the maximum probability, at $\bar{X}$, remains constant with change in $\bar{X}$. These assumptions lead to a differential equation for which the solution is the density function

$$f(x) = \frac{h}{\sqrt{\pi}} e^{-h^2 x^2} \tag{9-46a}$$

where h is a measure of precision of the observations. For practical purposes, Eq. (9-46a) can be modified to give the approximate probability of an error x. If Δx is the range of error, within which x may lie without being assigned a different value, the probability of x is

$$p(x) = \frac{h}{\sqrt{\pi}} e^{-h^2 x^2} \Delta x \tag{9-46b}$$

The probability that an error lies between $-X$ and X is given by the error function

$$\text{erf } x = \frac{2}{\sqrt{\pi}} \int_0^x e^{-z^2} dz = 2F(x\sqrt{2}) - 1 \tag{9-47}$$

where $x = hX$, and $F(x\sqrt{2})$ is the standard normal distribution function with $t = x\sqrt{2}$. Note that $h = 1/\sigma\sqrt{2}$. The probability that an error lies between X_1 and X_2 is given by

$$p = \frac{1}{\sqrt{\pi}} \int_{x_1}^{x_2} e^{-x^2} dx = F(x_2\sqrt{2}) - F(x_1\sqrt{2}) \tag{9-48}$$

Suppose, for example, that after all systematic errors have been eliminated, the arithmetic mean of a set of observations is 90.3. If the standard deviation is 2.2, what is the probability that a corrected observation will lie in the range 90.3 ± 3.1? Let us take $t = 3.1/2.2 = 1.4$. With this, we can use Fig. 9-5 in two ways to obtain the solution. One way is to use the property of a distribution function $F(t)$ that the probability of a deviation from the mean lying between a and b is $F(b) - F(a)$. Using Fig. 9-5 in this manner, we get $F(1.4) - F(-1.4) = 0.92 - 0.08 = 0.84$. The second method of solution is to use Eqs. (9-47) and (9-48), with $X = 3.1$, $x = 3.1h = 3.1/2.2 \sqrt{2} = 1.4/\sqrt{2}$, and $x\sqrt{2} = 1.4$. From Fig. 9-5, $F(1.4) = 0.92$. Then, Eq. (9-47) gives the probability $2 \times 0.92 - 1 = 0.84$.

A comparison of Fig. 9-4 with Fig. 9-1b indicates a close resemblance between the two diagrams. The similarity suggests that the normal density function can be substituted for the binomial distribution. If so, the substitution would be advantageous when both n, the number of observations, and np, the expected value, are large. We observed in Art. 9-8 that we can substitute the Poisson distribution for the binomial when n is large and np is not. Calculations are lengthy for the binomial distribution when n is large.

Laplace showed that for large n and np the binomial distribution can be approximated by a normal distribution. Let the deviation from the expected value be $r - np$ and let $t = (r - np)/\sqrt{npq}$, where r is the number of times an event occurs in n trials, p the probability that the event will occur in one trial, and q the probability that the event will not occur in one trial. Since the smallest change possible in r is 1, the smallest change possible in t is $\Delta t = 1/\sqrt{npq}$. Then, the **Laplace,** or **normal, approximation** to the probability of r occurrences in n trials is

$$p_{rn} = \frac{1}{\sqrt{2\pi}} e^{-t^2/2} \, \Delta t \tag{9-49}$$

Cumulative probability is obtained by summing the right-hand side of Eq. (9-49). As Δt approaches zero, the sum approaches an integral. It becomes the standard normal distribution function when the integration limits are $-\infty$ and t.

Chapter 10 discusses applications of normal distributions in statistics.

9-10 *Bibliography*

M. ABRAMOVITZ and I. A. STEGUN, "Handbook of Mathematical Functions with Formulas, Graphs, and Mathematical Tables," Superintendent of Documents, Government Printing Office, Washington, D.C.

R. S. BURINGTON and D. C. MAY, "Handbook of Probability and Statistics with Tables," McGraw-Hill Book Company, New York.

R. C. DUBES, "Theory of Applied Probability," Prentice-Hall, Inc., Englewood Cliffs, N.J.

W. FELLER, "An Introduction to Probability Theory and Its Applications," vol. 1, John Wiley & Sons, Inc., New York.

B. W. LINDGREN and G. W. McELRATH, "Introduction to Probability and Statistics," The Macmillan Company, New York.

E. PARZEN, "Modern Probability Theory and Its Applications," John Wiley & Sons, Inc., New York.

E. O. THORP, "Elementary Probability," John Wiley & Sons, Inc., New York.

G. A. WADSWORTH and J. G. BRYAN, "Introduction to Probability and Random Variables," McGraw-Hill Book Company, New York.

PROBLEMS

1. In one toss of a pair of idealized dice, what is the probability of rolling the sum (a) 7 and (b) 8?

2. Records kept by management of a factory showed that over the years 1,500,000 items were produced and 5,000 were defective. What is the probability of this factory producing a defective item if production conditions have not changed?

3. What is the probability that a number x chosen at random in the interval between 0 and 1 will fall between $\frac{3}{8}$ and $\frac{7}{8}$?

4. A coin is tossed twice. The probability of a head is $\frac{1}{2}$ on each toss.

    (a) What is the probability of two heads?

    (b) What is the probability of exactly one head?

    (c) What is the probability of at least one head?

    (d) What is the probability of at most one head?

    (e) What is the probability of two heads if either the first or second toss is a head?

5. A pair of dice are loaded so as to have the following probabilities: For die 1, $p(6) = 0.90$; for all other numbers, $p(x) = 0.02$. For die 2, $p(1) = 0.90$; for all other numbers, $p(x) = 0.02$. What is the probability of 7 on the first roll?

6. (a) How many four-digit numbers contain two 3s and two 5s?

    (b) How many three-digit numbers have no zero and no digit repeated?

    (c) How many three-digit numbers have no zero?

7. A 52-card poker deck is shuffled. What is the probability of dealing

    (a) 10, J, Q, K, and A of spades in succession? (Do not carry out the multiplication.)

    (b) A royal flush in spades, 10, J, Q, K, and A? (Do not carry out the multiplication.)

8. (a) A paint company sent out 17 containers, 10 with white, 4 with black, and 3 with red paint. During transfer, all labels were torn off. Deliverymen reattached them at random. What is the probability that a customer who ordered four white, three black, and two red received the colors and quantities he ordered?

    (b) Before the order in Prob. 8a was delivered, one customer was offered but rejected one container. What is the probability that it held white or black paint?

9. A sewage-treatment process breaks down when certain pollutants, dis-

charged by factories into the sewage system, are present. Test show relative frequency of breakdown with each pollutant as follows: oils, 0.90; acids, 0.80; alkalies, 0.60; other causes, 0.001. Records show that the relative frequency of occurrence of each pollutant is oils, 0.015; acids, 0.020; alkalies, 0.025; and normal pollutants, 0.94. Assuming relative frequencies may be taken as probabilities, what is the probability of process breakdown? The plant superintendent receives a report that the process is not working. What chemical is the most probable cause?

10. With 52-card poker decks, what is the probability of:

(a) Picking an ace in one try from each of two decks.

(b) Picking at least one ace in one draw from each of two decks.

(c) Drawing two aces in two tries from one deck, without replacing the first draw.

(d) Selecting an ace or king in one try from one deck.

11. The probability that engineer A can design a satisfactory machine for a new process is 0.80 and that engineer B can do it is 0.60.

(a) What is the probability of success if both engineers work independently?

(b) If engineer C with a probability of design success of 0.50 is added to the project, what is the probability of obtaining a satisfactory design?

(c) Solve Prob. 11*b* by computing the probability that all three engineers will fail.

12. Maintenance records show that a lamp has a probability of burning out of 0.10 in the first year.

(a) If two new lamps are installed simultaneously, what is the probability that at least one will still be in service one year later?

(b) What is the probability of at least one out of three new lamps installed at the same time still being in service at year end?

13. The quantity of water that must be supplied to a building for plumbing fixtures depends on the maximum number of fixtures that will be in use at the same time. Investigations show that the probability of one type of fixture being used is 0.1. In a building with five such fixtures:

(a) What is the probability of exactly two fixtures being used simultaneously?

(b) What is the probability of at least three fixtures being in use at the same time?

(c) What is the probability of at most three fixtures being in use at the same time?

(d) What is the probability of at least one fixture being in use?

14. Records kept for a factory indicate that the relative frequency of a defective item is 0.0001. Among how many items manufactured by this factory would there be more than an even chance ($p \geq \frac{1}{2}$) of finding a defective item?

15. All switches in a certain electric circuit operate together. For the circuit to be closed, two gaps in it must be closed by switches, each of which have a probability of failing of 0.1.

(a) If the two switches are in series and open, what is the probability that the circuit will remain open despite a close signal?

(b) If another set of two switches in series are connected in parallel with the switches in Prob. 15*a*, what is the probability that the circuit will remain open?

16. In a game of chance, two coins are tossed. What is your expectation if you lose $3 when two heads appear and win $1 for a head and tail and $2 for two tails?

17. A coin is to be tossed repeatedly until it shows a head. What is the expected number of tosses?

18. Factory records show a relative frequency of 25 defective items for every 10,000 manufactured.

(a) What is the expected number of defects in 100,000 items?

(b) What is the probability, as computed by the Poisson approximation, that a shipment of 1,000 items will contain at most three defective items?

(c) How many items should be included in a shipment so there is a probability of at least 0.95 that 1,000 items are good. (Assume that the expected number of defective items per 1,000 is 2.5.)

19. On one lane of a highway, cars travel at 30 mph and are spaced about 0.1 mile apart. A signal light is installed, visible for 1 mile. It has a 1-min-long red light and a 3-min-long green light.

(a) How many cars in that lane may be expected to stop for the red light if deceleration effects are ignored?

(b) What is the probability that exactly three cars will stop?

(c) What is the probability that at most five cars will stop?

20. The precision constant h of a series of measurements of an angle is 0.6. The instrument can be read to 0.1 sec.

(a) What is the probability of a 2-sec error?

(b) What is the probability of an error not exceeding 2 sec?

21. A series of measurements with a standard deviation of 8 has a mean of 850.

(a) What is the probability of a reading as low as 840 (840 or less)?

(b) What is the probability of a reading as high as 865 (865 or more)?

(c) What is the probability of a reading between 840 and 865?

22. Factory records show a relative frequency of defective items of 0.01. One day when 10,000 items were made, 120 were defective. Using a normal approximation to the binomial distribution, compute the probability of producing at least this many defective items.

23. In a shipment of wood studs for a housing project, 80 percent had a length of 8 ft or more. A sample of 100 studs is taken at random. What is the probability that the sample contains at least 30 studs less than 8 ft long? (Use the normal approximation to the binomial distribution.)

ANSWERS

1. (a) There are 36 possible events. A 7 can occur in six ways—two ways with 1 and 6, two ways with 2 and 5, and two ways with 3 and 4. By Eq. (9-1), $p(7) = \frac{6}{36} = \frac{1}{6}$.

(b) An 8 can occur in five ways—two ways with 2 and 6, two ways with 3 and 5, and one way with 4 and 4. By Eq. (9-1), $p(8) = \frac{5}{36}$.

2. Probability can be given only approximately, since n is large but not infinite. By definition, p is about $5,000/1,500,000 = \frac{1}{300}$. See Art. 9-1.

3. Consider the unit interval divided into eight equal segments. The sample space then comprises eight points, selection of any of the eight segments, and is uniform. Hence, assign a probability of $\frac{1}{8}$ to each point. The successful event $\frac{3}{8} < x < \frac{7}{8}$ contains four points, because it occurs when x falls on any of the four segments between $\frac{3}{8}$ and $\frac{7}{8}$. By Eq. (9-6), the probability is $4(\frac{1}{8}) = \frac{1}{2}$. See Art. 9-1.

4. (a) The sample space contains four points, HH, HT, TH, and TT, with equal probability of occurrence. Hence, the probability of HH is $\frac{1}{4}$. See Art. 9-1.

(b) Two of the points in the sample space contain exactly one head. Hence, the probability of exactly one head is $2(\frac{1}{4}) = \frac{1}{2}$. See Art. 9-1.

(c) There will be at least one head so long as TT does not occur. The probability of TT is $\frac{1}{4}$. Hence, the probability of at least one head is $1 - \frac{1}{4} = \frac{3}{4}$. See Art. 9-1.

(d) There will be at most one head when only TT, TH, and HT occur. Hence, the probability is $\frac{3}{4}$. See Art. 9-1.

(e) The sample space now consists of HH, HT, and TH, each with equal probability of occurrence. Hence, $p(HH) = \frac{1}{3}$. See Art. 9-1.

5. The event $E = $ rolling of 7 contains the following points and probabilities: 6 on die 1 and 1 on die 2, $p = (0.90)(0.90) = 0.81$; and 1 on die 1 and 6 on die 2, two points with 5 on either die and 2 on the other, and two points with 4 on either die and 3 on the other, all with $p = (0.02)(0.02) = 0.0004$. By Eq. (9-4), the probability of 7 is $0.81 + 0.0004 + 2 \times 0.0004 + 2 \times 0.0004 = 0.812$.

6. (a) By Eq. (9-9), there are $4!/2!2! = 6$ such numbers.

(b) By Eq. (9-7), there are $P(9,3) = 9 \times 8 \times 7 = 504$ such numbers.

(c) There are $9^3 = 729$ such numbers.

7. (a) There are $P(52,5)$ ways of dealing five cards in order. Only one way meets the requirement. Hence, the probability of 10, J, Q, K, and A of spades in succession is $1/(52 \cdot 51 \cdot 50 \cdot 49 \cdot 48)$. See Arts. 9-1 and 9-2.

(b) There are $C(52,5)$ ways of dealing a five-card hand. A royal flush in spades can be obtained in 5! ways. Hence, the probability is $5!/52 \cdot 51 \cdot 50 \cdot 49 \cdot 48$. See Arts. 9-1 and 9-2.

8. (a) There are $C(17,9)$ ways of selecting nine containers from the seventeen. White-paint containers for the order can be picked in $C(10,4)$ ways. Black paint containers can be chosen in $C(4,3)$ ways. And red-paint containers can be selected in $C(3,2)$ ways. So the order can be filled in $C(10,4)C(4,3)C(3,2)$ ways. (See Art. 9-2.) Hence, the probability of the customer getting the containers ordered is, by Eq. (9-1),

$$C(10,4)C(4,3)C(3,2)/C(17,9) = 252/2,431$$

(b) The probability of white paint is $\frac{10}{17}$. The probability of black paint is $\frac{4}{17}$. By Eq. (9-20), the probability of white or black paint is

$^{10}\!/_{17} + {}^{4}\!/_{17} = {}^{14}\!/_{17}$. Alternatively, the probability of red paint is $^{3}\!/_{17}$ and of not getting red paint is $1 - {}^{3}\!/_{17} = {}^{14}\!/_{17}$.

9. Arrange the computations for solution with Eqs. (9-16) and (9-18).

Probability of Pollutants Causing Process Breakdown E

Pollutants affecting process	$p(e_i)$	$p(E|e_i)$	$p(e_i)p(E|e_i)$	$p(e_i|E)$
Oils e_1	0.015	0.90	0.0135	0.30
Acids e_2	0.020	0.80	0.0160	0.35
Alkalies e_3	0.025	0.60	0.0150	0.33
Other e_4	0.940	0.001	0.0009	0.02
	1.000		$P(E) = 0.0454$	1.00

The probability of process breakdown is 0.0454. Acids are the most probable cause.

10. (a) For either deck, $p(A) = {}^{4}\!/_{52} = {}^{1}\!/_{13}$. By Eq. (9-19), $p(AA) = ({}^{1}\!/_{13})({}^{1}\!/_{13}) = {}^{1}\!/_{169}$.

(b) By Eq. (9-22), probability of at least one ace is ${}^{1}\!/_{13} + {}^{1}\!/_{13} - {}^{1}\!/_{169} = {}^{25}\!/_{169}$.

(c) On the first draw, $p(A) = {}^{4}\!/_{52}$. On the second draw, p (second A given first A) $= {}^{3}\!/_{51}$, since the 51 cards remaining contain only three aces after the first ace is drawn. By Eq. (9-5), the probability of two aces is $({}^{4}\!/_{52})({}^{3}\!/_{51}) = {}^{1}\!/_{221}$.

(d) $p(A) = p(K) = {}^{4}\!/_{52} = {}^{1}\!/_{13}$. By Eq. (9-20), $p(A + K) = {}^{1}\!/_{13} + {}^{1}\!/_{13} = {}^{2}\!/_{13}$.

11. (a) By Eq. (9-22), $p(A + B) = 0.80 + 0.60 - (0.80)(0.60) = 0.92$.

(b) By Eq. (9-23), $p(A + B + C) = 0.80 + 0.60 + 0.50 - (0.80)(0.60) - (0.80)(0.50) - (0.60)(0.50) + (0.80)(0.60)(0.50) = 0.96$.

(c) By Eq. (9-2), $q(A) = 1 - 0.80 = 0.20$, $q(B) = 0.40$, and $q(C) = 0.50$. By Eq. (9-19), $q(ABC) = (0.20)(0.40)(0.50) = 0.04$. Hence, $p(A + B + C) = 1 - 0.04 = 0.96$.

12. (a) The probability of both lamps failing is, by Eq. (9-19), $(0.10)(0.10) = 0.01$. Hence, the probability of at least one being in service is, by Eq. (9-2), $1 - 0.01 = 0.99$.

(b) The probability of all three lamps failing is, by Eq. (9-19), $(0.10)^3 = 0.001$. The probability of at least one remaining in service is, by Eq. (9-2), $1 - 0.001 = 0.999$.

13. (a) Let $p = 0.1$, $q = 0.9$. By Eq. (9-24),

$$p_{5,2} = C(5,2)(0.1)^2(0.9)^3 = 10(0.00729) = 0.0729$$

(b) $p_3 = (0.1)^5 + C(5,4)(0.1)^4(0.9) + C(5,3)(0.1)^3(0.9)^2 = 0.00856$

(c) By Eq. (9-28),

$$p_3' = (0.9)^5 + C(5,1)(0.1)(0.9)^4 + C(5,2)(0.1)^2(0.9)^3$$
$$+ C(5,3)(0.1)^3(0.9)^2$$
$$= 0.99954$$

(d) By Eq. (9-27),

$$p_1 = 1 - (0.9)^5 = 0.40951$$

14. Let x = number of items to be examined. Then, $(1 - 0.0001)^x$ is the approximate probability that the x items do not include a defective item. So, $(0.9999)^x \leq \frac{1}{2}$, or $(1.0001)^x \geq 2$. Taking logarithms and solving for x yields $x \geq (\log 2)/\log 1.0001 = 6,932$.

15. (a) The circuit will remain open if at least one switch fails to close. By Eq. (9-27), the probability of this is $1 - (0.9)^2 = 0.19$.

(b) The circuit will remain open if at least one switch in both sets fails to close. By Eq. (9-19), the probability of this is $(0.19)^2 = 0.0361$.

16. The probabilities are $p(HH) = p(TT) = \frac{1}{4}$ and $p(HT) = \frac{1}{2}$. By Eq. (9-29), the expectation is $-\$3(\frac{1}{4}) + \$1(\frac{1}{2}) + \$2(\frac{1}{4}) = \$\frac{1}{4}$.

17. The expectation of heads in n tosses is to be 1. By Eq. (9-31), $E(H) = 1 = np = n/2$. Hence, $n = 2$.

18. (a) By Eq. (9-31), the expectation is about $0.0025 \times 100,000 = 250$.

(b) By Eq. (9-31), the expectation is $0.0025 \times 1,000 = 2.5$ defective items. By Eq. (9-33), the probability of at most 3 is

$$p(0) + p(1) + p(2) + p(3) = e^{-2.5}\left(\frac{2.5^0}{0!} + \frac{2.5^1}{1!} + \frac{2.5^2}{2!} + \frac{2.5^3}{3!}\right) = 0.758$$

(c) The factory should ship $1,000 + x$ items, with a probability of 0.95 or more that there will be at most x defective items. By Eq. (9-33),

$$e^{-2.5}\left(\frac{2.5^0}{0!} + \frac{2.5^1}{1!} + \frac{2.5^2}{2!} + \cdots + \frac{2.5^x}{x!}\right) \geq 0.95$$

By trial, $x = 5$. So, 1,005 items should be shipped.

19. (a) At 30 mph, a car goes 0.5 mile in 1 min. While the red signal shows, therefore, every car within 0.5 mile of it will have to stop. With a spacing of 0.1 mile, the expectation is that $0.5/0.1 - 5$ cars will stop.

(b) By Eq. (9-33), the probability is $e^{-5}5^3/3! = 0.14$.

(c) By Eq. (9-33), the probability of at most 5 cars stopping is

$$\sum_0^5 p(i) = e^{-5}\left(\frac{5^0}{0!} + \frac{5^1}{1!} + \frac{5^2}{2!} + \frac{5^3}{3!} + \frac{5^4}{4!} + \frac{5^5}{5!}\right) = 0.62$$

20. (a) By Eq. (9-46b), the probability of an error X such that $1.95 < X < 2.05$ is, with $hx = 0.6 \times 2 = 1.2$,

$$p = \frac{0.6}{\sqrt{\pi}} e^{-1.44}(0.1) = 0.008$$

(b) The probability is given by Eq. (9-47) with $x = 0.6 \times 2 = 1.2$ and $x\sqrt{2} = 1.7$. From Fig. 9-5, $F(1.7) = 0.96$. Hence, by Eq. (9-47), $p = \text{erf } 1.2 = 2 \times 0.96 - 1 = 0.92$.

21. (a) Let $t = (840 - 850)/8 = -1.25$. From Fig. 9-5, $p(x \leq 840) = 0.11$.

(b) Let $t = (865 - 850)/8 = 1.88$. From Fig. 9-5, $p(x \leq 865) = 0.97$. Hence, $p(x \geq 865) = 1 - 0.97 = 0.03$.

(c) From Fig. 9-5, $p(840 < x < 865) = 0.97 - 0.11 = 0.86$.

22. $\sqrt{npq} = \sqrt{10,000 \times 0.01 \times 0.99} = 9.95$. Let $t = (r - np)/\sqrt{npq} = (120 - 100)/9.95 = 2.01$. From Fig. 9-5, $p(x \leq 120) = 0.98$. Then, $p(x \geq 120) = 1 - 0.98 = 0.02$.

23. $\sqrt{npq} = \sqrt{100 \times 0.20 \times 0.80} = 4$. Let $t = (r - np)/\sqrt{npq}) = (30 - 20)/4 = 2.5$. From Fig. 9-5, $p(x \leq 30) = 0.99$. Hence, $p(x \geq 30) = 1 - 0.99 = 0.01$.

TEN

Statistics

In this chapter, we continue our discussion of probability theory. We will examine its applications in statistics.

Statistical mathematics is that branch of mathematics dealing with numerical measurements and analysis of numerical data and enabling development and application of methods in which probability theory is used to evaluate uncertainties.

We will give special attention to the application of statistics to analysis of numerical data. This is intended to help you in making decisions in the face of uncertainties.

Methods used for the purpose are similar to those you would use in carrying out an experiment. You formulate an hypothesis, then test it with experimentally obtained data. In statistics, however, a measure of uncertainty is attached to conclusions reached.

Many statistical models are available for testing hypotheses. You need a good knowledge of statistical theory to be able to select the best model, one that will yield the correct conclusions in the most efficient way.

In this chapter, we consider numerical data as specific values of random variables. Statisticians call the set of all possible values of a random variable a **population.** For example, a population might comprise the number of men 40 years or older in a given year, or the diameters of all tubes produced by a factory, or the weights of all ingots shipped by a mill. Also, statisticians call any portion of a population a **sample.**

Generally, the numbers that are to be analyzed comprise one or more samples. Your problem may be to determine whether or not the samples came from the same population. Or the problem may be to compute from the data of a sample certain numerical characteristics of the population.

Suppose, for example, that a chemical manufacturer contracts to deliver to your company a product 99.0 percent pure. Three tests of a batch received show purities of 98.8, 98.9, and 99.1 percent. Should you accept or reject the batch? Here, the population consists of 99.0 percent pure batches produced by this manufacturer. The batch received is a sample. The problem is: Did it come from this population?

Again, chemical analysis of several samples of a water supply shows an average content of chlorides of 100 ppm. The problem is: What is the average content of chlorides in the reservoir from which the water comes?

Methods used to reach conclusions about a population from a sample taken from the population are called **statistical inference.** This chapter presents such methods. But before we can tackle them, we need the preliminary information in the first few articles.

In the following, we often will find it desirable to describe a population by giving **parameters,** numerical values of certain characteristics. Corresponding numerical values obtained from samples are called **statistics.** To avoid confusion, statisticians use different symbols for parameters and corresponding statistics, usually Greek letters for population characteristics and corresponding Roman letters for sample characteristics. In general, this chapter follows this practice.

If a sample is to provide reliable information about a population, the sample must be representative. Thus, it must be properly chosen from the population. Space limitations preclude our discussing sampling methods in this chapter; but you will find it worth while to investigate these methods in future study. Suffice it to say that a sample should be randomly selected if inferences drawn from it about the population are to be valid.

10-1 *Measures of Central Tendency.* Chapter 9 pointed out that the relative frequency of an event obtained from a large number of observations approximates the probability of the event in a single trial. For practical reasons, we usually treat relative frequencies as probabilities, even though the number of observations are not very large.

For analysis purposes, we may collect data and plot the observations with respect to their frequency, or relative frequency, of occurrence. The resulting histograms or curves may or may not resemble the theoretical frequency or distribution diagrams in Chap. 9. In any event, to be able to generalize our observations, we must first be able to describe them briefly in numerical terms.

Two numbers are usually most useful for this purpose. One number is the observation or measurement, about which the other observations tend to cluster. The second number is a measure of the spread of the data. This article describes some of the statistics used for the first number, to measure cluster, or central tendency, of data. These statistics are alternatives to such imprecise terms as average, typical, normal, usual, and representative. Let us start with one of great importance:

The **mean** is the center of gravity of the observations. For a continuous variable x, the mean is given by

$$\bar{x} = \frac{\int_{-\infty}^{\infty} xf(x)\ dx}{\int_{-\infty}^{\infty} f(x)\ dx} \tag{10-1}$$

where $f(x)$ is the density function giving the relation between observations and relative frequency. The denominator in Eq. (10-1) should equal 1, since the area under a probability density function equals 1. For a discrete variable x_i, the mean is the **arithmetic mean**

$$\bar{x} = \frac{\Sigma x_i}{n} \tag{10-2}$$

where n is the number of observations and the summation is from $i = 1$ to $i = n$.

The mean is important for several reasons. First, it is a useful base for measuring the spread of the data. When we adopt for this purpose the squares of deviations from the measure of central tendency, choice of the mean minimizes the sum of these squares. Also, the mean of a series of measures in which only random errors are present is the most probable value or best estimate of the characteristic being measured.

In general, the mean of a sample is the best estimate of the mean of the population from which the sample is taken.

Furthermore, regardless of the distribution of a population, the means of samples drawn from it will be approximately normally distributed. The larger the sample sizes, the closer the distribution of the sample means will be to the Gaussian. This characteristic of the mean enables us to use a normal distribution for probability computations concerning a population mean, though the population may not be normally distributed.

The **median** is the middle measure in a set in which all measures have been arranged according to magnitude. Thus, equal numbers of items are above and below it. (For an even number of measures, the median is the mean of the two center measures.) One of the most important characteristics of the median is that the sum of the absolute values of the deviations from it is a minimum.

In some cases, the median may be a more useful measure of central tendency than the mean. For example, a few extremely high or low observations may cause an unduly large shift in the mean, yet the median will remain unchanged and be more representative of the population. Such a characteristic may be desirable in studies of income, for instance.

If a series of measures are arranged according to size, the p **percentile** is the measure below which p percent of the measures lie. Thus, the 90th percentile is the item larger than 90 percent of all the measures in the set.   The 25th percentile is called the first quartile, and the 75th percentile is called the third quartile. The 50th percentile is the median. The multiple-of-ten percentiles are known as deciles; thus the 30th percentile is the third decile.

The **mode** of a set of measures is the one that occurs most often. It corresponds to the observation at the highest point of a frequency curve.

To illustrate the preceding measures of central tendency, let us examine the following data: In a factory where 1 man earns \$50,000, 9 men earn \$10,000 each, 40 men earn \$6,000 each, and 10 men earn \$4,000 each, what are the mean wage, median wage, first quartile, and mode?

By Eq. (10-2), the mean is

$$\bar{x} = \frac{50{,}000 + 9 \times 10{,}000 + 40 \times 6{,}000 + 10 \times 4{,}000}{1 + 9 + 40 + 10} = 7{,}000$$

The median is \$6,000, since as many men in this factory earn \$6,000 or more as earn \$6,000 or less.   The first quartile also is \$6,000, since 25 percent or 15, of the men earn that amount or less.   The mode also is \$6,000, since most men earn that amount.

You also are likely to encounter one other measure of central tendency, the **geometric mean** GM. For a set with n measures, the geometric mean is the nth root of the product of the measures. For example, the geometric mean of 5, 10, and 20 is

$$GM = (5 \times 10 \times 20)^{1/3} = 10$$

We take the cube root because there are three numbers in the set.   Geometric means are used principally in studies involving average rates of change.

10-2   *Measures of Variability*. Compactness of a frequency distribution of a set of measures about a point of central tendency may be described by such terms as dispersion, spread, scatter, deviation, and variability.   This article describes some of the statistics used to measure compactness.

The **range** of a set of measures is the difference between the largest and smallest values in the set.   Its chief advantage is ease of computation.   But since range is determined from only two items in a set, it is not sensitive to variations in the rest of the data and is thus usually not a good measure of compactness.

The **semi-interquartile range**, or **quartile deviation**, is half the difference between the first and third quartiles.   It has disadvantages similar to those of the range.

The **mean deviation** is the arithmetic mean of the absolute values of the deviations of the observations from the median.   (In the past, these deviations were taken from the mean.   The median is preferred because of ease of computation.)   Use of mean deviations is limited, principally because computation of absolute values is difficult to handle in development of theory.   (Signed deviations are not a good measure of scatter of data because they offset each other when added.)

Standard deviation of a population is defined by

$$\sigma = \sqrt{\frac{\Sigma(x_i - \bar{x})^2}{n}} = \sqrt{\frac{\Sigma e_i^2}{n}} \tag{10-3}$$

where x_i denotes a set of observations, $e_i = x_i - \bar{x}$ represents the deviations of the observations from their mean $\bar{x}$, n is the number of observations in the set, and the summations are from $i = 1$ to $i = n$.   The positive sign is always taken with the square root.

For a continuous variable x, standard deviation of a population is given by

$$\sigma = \left[\frac{\int_{-\infty}^{\infty} (x - \bar{x})^2 f(x)\, dx}{\int_{-\infty}^{\infty} f(x)\, dx}\right]^{\frac{1}{2}} \tag{10-4}$$

where $f(x)$ is the density function relating observations and relative frequency.   Notice that the numerator in Eq. (10-4) is the moment of inertia of the density function about the mean.   The denominator is the area of the density function diagram.   Hence, standard deviation is equivalent to the radius of gyration of the diagram, a geometric characteristic often used in stress analysis.

As a measure of variability, standard deviation overcomes the disadvantage of mean deviation by eliminating the need for absolute values.

By requiring squares of deviations, the standard deviation sums only positive numbers, regardless of the sign of any deviation. Thus, for a given number of observations, the larger the deviations, the larger the measure of variability—a desirable characteristic for such a measure.

We encountered standard deviation previously in the discussion of the normal density function in Art. 9-9. We observed there that the larger the standard deviation, the broader the base of the bell-shaped curve becomes. Hence, the further away from the mean relatively large values of the variable occur.

Fewer computations are required and there is less chance of error if either of the following is used in Eq. (10-3) instead of e_i:

$$\sum e_i{}^2 = \sum x_i{}^2 - \frac{1}{n}\left(\sum x_i\right)^2 = \sum x_i{}^2 - \bar{x}\sum x_i \tag{10-5}$$

For example, let us determine the standard deviation of the set 3, 8, 9, 10, 12, 15. The squares of these six numbers, $x_i{}^2$, are 9, 64, 81, 100, 144, 225. The sum of the numbers $\Sigma x_i = 57$ and the sum of their squares $\Sigma x_i{}^2 = 623$. Using in Eq. (10-3) the first expression on the right of Eq. (10-5), we get

$$\sigma = \sqrt{\frac{623}{6} - \left(\frac{57}{6}\right)^2} = 3.68$$

In Art. 10-1, we observed that the mean of a sample is the best estimate of the mean of the population from which the sample is taken. The standard deviation of a sample, however, will usually be smaller than the standard deviation of the population if we use Eq. (10-3) for both the statistic and the parameter. The reason is that the sum of the squares of the sample deviations from the sample mean is a minimum. Hence, the sum of the squares of the sample deviations from the population mean will be larger, since the population mean is usually different from the sample mean. Consequently, the standard deviation of a sample is not computed from Eq. (10-3) but from a formula that provides a better estimate of the standard deviation of the population.

Standard deviation of a sample is defined by

$$s = \sqrt{\frac{\Sigma(x_i - \bar{x})^2}{n-1}} = \sqrt{\frac{\Sigma e_i{}^2}{n-1}} \tag{10-6}$$

As for Eq. (10-3), computations for Eq. (10-6) can be reduced by use of Eq. (10-5).

You may sometimes encounter another measure of variability derived from the standard deviation, the **coefficient of variation.** It is the ratio of standard deviation to mean. It is useful in comparing distributions with different means. Note that the standard deviation has the same dimension as the variate, whereas the coefficient of variation is dimensionless.

Another important measure of variability is variance, closely related to standard deviation. Variance is preferred to standard deviation because it is easier to use in analyses of causes of variability.

Variance is the square of the standard deviation, for both a population and a sample taken from it. Thus, for a population, variance is the mean of the squares of the deviations from the population mean. It is denoted by σ^2. Variance of a sample is defined by

$$s^2 = \frac{\Sigma(x_i - \bar{x})^2}{n - 1} = \frac{\Sigma e_i^2}{n - 1} \tag{10-7}$$

Notice that the dimension of variance is the square of the dimension of x_i, whereas standard deviation has the same dimension as x_i. For this reason, standard deviation often is preferred for reporting spread of measurements.

If a variable X is composed of several statistically independent variables x_i, the variance of X equals the sum of the variances of x_i, if the population is very large. Thus, the contribution of each variable to the scatter of X can be determined. Analysis can then indicate what steps should be taken to improve the compactness of X. This additive property of variance makes it a very useful measure of the variability of observations. Analysis of variance is a very important branch of statistics.

If X can be expressed as a function of the statistically independent variables x_i, $X = f(x_1, x_2, \ldots, x_n)$, then the variance of X is related to the variances of its components by

$$\sigma^2(X) = \left(\frac{\partial X}{\partial x_1}\right)^2 \sigma^2(x_1) + \left(\frac{\partial X}{\partial x_2}\right)^2 \sigma^2(x_2) + \cdots + \left(\frac{\partial X}{\partial x_n}\right)^2 \sigma^2(x_n)$$

$$\tag{10-8}$$

where the partial derivatives are to be evaluated at $\bar{x}_i$.

Suppose in a hydraulic laboratory measurements of the flow of water over a weir indicate a mean head $\bar{H}$ of 5 ft and mean velocity $\bar{V}$ of 10 fps. The variance of head is 0.04 and of velocity, 0.25. What variance may be expected in the flow Q, cfs, if $Q = 10(H + V^2/64.4)^{3/2}$, where H is head, ft, and V velocity, fps?

The variance of Q may be obtained from Eq. (10-8), since H and V are statistically independent. For use in this equation, we first calculate

$$\frac{\partial Q}{\partial H} = 15\left(H + \frac{V^2}{64.4}\right)^{1/2} = 15\left(5 + \frac{100}{64.4}\right)^{1/2} = 38.3$$

$$\frac{\partial Q}{\partial V} = \frac{30V}{64.4}\left(H + \frac{V^2}{64.4}\right)^{1/2} = \frac{300}{64.4}\left(5 + \frac{100}{64.4}\right)^{1/2} = 11.9$$

Substitution in Eq. (10-8) then gives for the variance of Q

$$\sigma^2(Q) = (38.3)^2(0.04) + (11.9)^2(0.25) = 58.8 + 35.7 = 94.5$$

The contribution to the variance of Q from measurements of head is far larger than that from velocity. Thus, to decrease the variation in Q, it would be worth while to improve the precision of head measurements or to decrease the head.

10-3   *Estimation of Population Means from Samples.* Two statistical theorems enable a good estimate to be made of population means from a random sample. Before the theorems are presented, however, we need a preliminary explanation.

Assume an infinitely large population. Suppose that we select at random from it many samples of size n. Let us compute the mean $\bar{x}_i$ of each sample. Then, $\bar{x}_i$ is a random variable, and like any other random variable, it will have a frequency function and distribution. One theorem states:

For any population, with any distribution, the mean of the means of samples of size n drawn from the population equals the population mean. If the population has a finite variance σ^2, the variance $\bar{\sigma}^2$ of the sample means equals the population variance divided by n.

$$\bar{\sigma}^2 = \frac{\sigma^2}{n} \tag{10-9}$$

A second theorem, called the **central-limit theorem,** states:

If a population, with any distribution, has a finite variance σ^2 and a mean μ, the distribution of means of samples of size n drawn from the population approaches the normal distribution with mean μ and variance σ^2/n as n increases.

The closer the distribution of the population is to the normal distribution, the smaller n need be for the distribution of sample means to be very nearly normal. In general, n may have to be 30 or more to obtain a normal distribution for sample means.

Now, Eq. (10-6) gives an estimate s, based on measurements comprising a sample, of the standard deviation σ of the population from which the sample is taken. Hence, it follows from Eq. (10-9) that we can estimate the standard deviation $\bar{s}$ of the means from

$$\bar{s} = \frac{s}{\sqrt{n}} \tag{10-10}$$

(Standard deviation of the means is sometimes called **standard error of the mean.**)

As an example of the use of these theorems, suppose that 100 measurements of the internal diameters of randomly selected tubing indicate a mean of 1.880 in. with a variance of 0.050. For all this size tubing, within what range can we say internal diameters lie, with a probability of 95 percent of being correct?

Let us assume that the sample mean approximates the population mean, and the sample variance 0.050 can be used as the population variance. By Eq. (10-10) then, the standard deviation of the means $\bar{s} = \sqrt{0.05/100} = 0.0224$. Now, from Table 10-1, we find that there is a 95 percent probability that $\pm 1.96\sigma$ will not be exceeded in a normal distribution, where σ is the standard deviation of the distribution. Hence, by the central-limit theorem, we can say that there is a 95 percent probability that the internal diameters of the tubing will lie in the range $1.880 \pm 1.96 \times 0.0224 = 1.880 \pm 0.044$.

TABLE 10-1 Probability That $(x - \bar{x})/\sigma$ or $(|x - \bar{x}|)/\sigma$ Will Not Be Exceeded in a Normal Distribution

$\dfrac{x - \bar{x}}{\sigma}$	Probability, percent	$\dfrac{|x - \bar{x}|}{\sigma}$	Probability, percent
1.00	34.13	±1.00	68.26
1.28	39.97	±1.28	80
1.64	44.95	±1.64	90
1.96	47.50	±1.96	95
2.33	49.01	±2.33	98
2.58	49.51	±2.58	99
3.29	49.95	±3.29	99.9

10-4 *Confidence Intervals and Levels.* The example in Art. 10-3 illustrates how in statistics a measure of uncertainty is attached to a conclusion. In that case, the range established for tubing diameter is called the confidence interval and the 95 percent probability that it will not be exceeded is called the confidence level, or confidence coefficient.

In general, the confidence level is the probability, expressed in percent, that a statement is true.

Suppose that we are trying to determine a parameter θ, such as the mean, of a population. We draw a large random sample from the population and compute a statistic t corresponding to θ from the sample. If we are concerned with the population mean, we compute the sample mean. If we draw many samples, we could get a frequency function $f(t)$. From it, we can learn the probability that t will fall between any two values t_1 and t_2.

The confidence interval of a parameter θ is the set of values between t_1 and t_2, inclusive, such that

$$p(t_1 \leq \theta \leq t_2) = \gamma \qquad (10\text{-}11)$$

where γ is a specified probability. t_1 and t_2 are called the confidence limits.

The confidence level is the probability γ, percent, associated with a confidence interval.

Table 10-1 lists confidence levels for normal distributions. For example, we can say at the 95 percent confidence level that if a single measure is selected at random from a normal distribution, it will lie within 1.96 standard deviations σ of the mean. At the 98 percent level, we can say it will lie within 2.33σ of the mean, and at the 99 percent level, we can say it will lie within 2.58σ of the mean. Thus, the 99 percent confidence interval for the true mean of a population is $\bar{x} \pm 2.58\bar{s}$, where $\bar{x}$ is the mean of a random sample and $\bar{s}$ is the standard deviations of the means of samples. The term $(x - \bar{x})/\sigma$ in the table headings are called **standard deviates.**

Note that for a normal distribution, the larger the confidence level, the larger the confidence interval. Given a confidence level, the confidence interval is the product of a constant and the standard deviation. So to control the confidence-interval size, we must control the standard deviation. For example, for a mean, the standard deviation is inversely proportional to the number of measurements in a sample. Hence, we must increase the number of observations or improve the precision of the measurements.

As an example of the use of confidence level, let us examine the situation where a manufacturer is packaging 1-lb units by machine. Records indicate that actual weights are normally distributed with a standard deviation of 0.04 lb. The packaging machine must be adjusted regularly so that packages are neither too heavy nor too light. Does the machine need adjustment if nine packages selected at random have a mean weight of 1.05 lb? The decision that the machine is performing satisfactorily should be at a confidence level of 95 percent. Thus, the machine will not have to be adjusted if the mean package weight w lies between some acceptable minimum mean weight w_1 and some acceptable maximum mean weight w_2. The problem is to determine w_1 and w_2.

Since the standard deviation of the population (unit weights) is 0.04, the standard deviation of the means is, by Eq. (22-9), $0.04/\sqrt{9} = 0.0133$. From Table 10-1, the confidence limits for a confidence level of 95 percent are $\pm 1.96\sigma = \pm 1.96 \times 0.0133 = \pm 0.026$. Hence, the confidence interval is 1.0 ± 0.026, and $w_1 = 0.974$ and $w_2 = 1.026$. Since the mean weight of the samples is $1.05 > 1.026$, we can conclude, with less than a

5 percent probability of being wrong, that the machine is producing over-weight packages and should be adjusted.

10-5   *Significance Tests.* The technique used in solving the example in Art. 10-4 is known as statistical inference.   It is based on statistical theorems stating that if a sample is very large, a statistic, such as the mean, determined from the sample may be expected to nearly equal the corresponding parameter, for example, the mean, of the population.   The larger and more representative the sample, the closer the approximation, and the smaller the probability that the estimated parameter will differ from its true value by more than any specified amount.

To solve the preceding example, we set up a mathematical model, a normal distribution of packaged unit weights.   In this case, we were informed that records based on observations indicated that such a mathe-matical model would be in agreement with reality.   Often however, in practical problems, you may not have such information and may have to make assumptions to obtain a solution.   Statisticians have devised many tests, called significance tests, for comparing theory and observa-tions.   We shall discuss some of these in following articles.

In a typical problem, you may be concerned with a population, a set of specified measures, such as the weight of a package, as in the preceding example, diameter of an object, or the number of defective products in a factory output.   You will be given some description of this population in terms of one or more parameters, such as the mean and variance. Also, you will be given a sample comprising n observations of a variable x.   You will be required to determine whether this sample may be con-sidered a random sample drawn from the given population.   For example, is the mean of the sample acceptably close to the population mean?

In a significance test, you start with an hypothesis: The sample has been drawn from the population.   Such an assumption is called the **null hypothesis,** because it assumes that there is no difference between the sample statistic and population parameter.   It may be denoted by

$$H_o: O = E \tag{10-12}$$

where O indicates observed or computed value and E, expected or given parameter.

A test procedure divides all possible values of O into two sets.   One set, the **critical region,** is the region of rejection of the hypothesis.   The second set is a region of indecision.   If the sample statistic lies in the critical region, you reject the hypothesis; otherwise, you do not reject it.   When you reject the hypothesis, you will generally accept an alter-native value of the parameter.

The critical region is usually determined by the specification, *at the start of a significance test,* that the hypothesis is to be rejected only if, in so doing, you have a probability α, percent, or less, of being wrong. This probability α that the statistic lies in the critical region is called the **significance level** of the test.

Numerically, significance level, percent, equals $100 -$ confidence level, percent. For example, from Table 10-1, the significance level for $\pm 1.96\sigma$ is $100 - 95 = 5$ percent. Note, however, that the significance level for either 1.96σ or -1.96σ is, from Table 10-1, $50 - 47.50 = 2.50$ percent. In this case, the significance level is the probability that 1.96σ will be exceeded, or the probability that a smaller value than -1.96σ will occur.

With the hypothesis stated and significance level established, the general test procedure is as follows:

Assume that the hypothesis is true. Then, the sample statistic O should approximate the population parameter E when the sample size n is large. Define a *positive* measure δ of the deviation of O from E. (This is equivalent to selecting a specific significance test, such as chi-squared or Student's t.) δ should have a known distribution $f(\delta)$. From $f(\delta)$, determine a deviation δ' such that α is the probability that δ' will be exceeded. (Or you can determine δ' from the probability $100 - \alpha$ that δ' will not be exceeded.)

If $\delta > \delta'$, the deviation of O from E is significant at the α level. Hence, $H_o: O = E$ can be rejected with a probability of α of being wrong.

If $\delta \leq \delta'$, the deviation of O from E may be due to random fluctuations. The sample may be consistent with the population. More information is needed to justify a decision.

In making such a significance test, you may make either of two types of errors in reaching a conclusion. False rejection of an hypothesis is a type I error. False acceptance is a type II error.

If the significance level is set so low that there is a small probability of a type I error, false rejection, there is a high probability of a type II error, false acceptance. Usually, significance levels of 1 to 5 percent give close to the minimum probabilities of both types of errors.

Statisticians use a measure, called **power of a test,** for comparing tests as to ability to evaluate hypotheses. The power of a test is the probability of rejecting an hypothesis when it is false. Thus, power of a test also equals 1 minus the probability of a type II error.

Examples of the use of the preceding test procedure are given in the following articles.

10-6 *Chi-squared Test.* Written χ^2, the chi-squared test often is useful with hypotheses concerning sample variance or stating that a population has a specified distribution.

The test procedure given in Art. 10-5 starts with a null hypothesis $H_o: O = E$, the sample statistic O is not different from the population parameter E. Then, the procedure requires selection of a positive measure δ of the deviation of O from E, and δ is to have a known distribution $f(\delta)$. For the χ^2 test, $\chi^2 = \delta$.

In this case, the density function contains two variables, χ^2 and degrees of freedom m, which will be explained later. Thus,

$$f(\chi^2,m) = \frac{e^{-\chi^2/2}\chi^{m-2}}{2^{m/2}\Gamma\left(\dfrac{m}{2}\right)} \qquad \chi^2 > 0 \tag{10-13}$$

Because of the complexity of this formula, probabilities of χ^2 are obtained from tables. Most books of statistics with tables contain χ^2 distribution tables. Table 10-2 is a small example of such tables. It gives the probability that each given value of χ^2 will be exceeded, for degrees of freedom m up to 29. For larger values of m, the distribution of χ^2 approaches the normal distribution with mean m and standard deviation $\sqrt{2m}$. Hence, χ^2 can be approximated by

$$\chi^2 = \frac{(\sqrt{2m - 1} + t)^2}{2} \tag{10-14}$$

where t is the value of $(x - \bar{x})/\sigma$ in Table 10-1, corresponding to a probability equal to 50 percent minus the given level of significance.

TABLE 10-2 Minimum Value of χ^2 for Significance at Various Levels

Degrees of freedom	Levels of significance			
	5%	2%	1%	0.1%
1	3.84	5.41	6.64	10.83
2	5.99	7.82	9.21	13.82
3	7.82	9.84	11.34	16.27
4	9.49	11.67	13.28	18.47
5	11.07	13.39	15.09	20.52
6	12.59	15.03	16.81	22.46
7	14.07	16.62	18.48	24.32
8	15.51	18.17	20.09	26.13
9	16.92	19.68	21.67	27.88
10	18.31	21.16	23.21	29.59
11	19.68	22.62	24.73	31.26
12	21.03	24.05	26.22	32.91
13	22.36	25.47	27.69	34.53
14	23.69	26.87	29.14	36.12
19	30.14	33.69	36.19	43.82
24	36.42	40.27	42.98	51.18
29	42.56	46.69	49.59	58.30

One variate that has the distribution given by Eq. (10-13) is

$$\chi^2 = \sum \frac{(O - E)^2}{E} \qquad (10\text{-}15)$$

where O denotes an observed value, E an expected value, and $O - E$ the deviation of the observed from the expected value. The summation extends over all observed values.

The degrees of freedom m in Eq. (10-13) equal the number of independent observations available for computing χ^2. In general, if $r - 1$ of the observations O are given, the remaining value of O can be deduced. In that case, there are $m = r - 1$ degrees of freedom, if the O values are not otherwise related to the E values. For example, if a coin is tossed n times and head shows k times, the number of tails is determined by subtracting k from n. If we are dealing with deviations, we can compute the deviation of the observed number of heads, $n/2 - k$. The deviation of the observed number of tails is thereby fixed at $k - n/2$. Hence, there is only one degree of freedom.

Let us suppose that we are concerned with the variance s^2 of a sample of size n drawn from a normal population with variance σ^2. A theorem of statistics states that the random variable given by the ratio of the sum of the squares of the deviations from the sample mean to the population variance has the χ^2 distribution with $n - 1$ degrees of freedom. Hence, for a sample from a normal distribution, by using Eq. (10-7), we can give χ^2 as

$$\chi^2 = \frac{(n - 1)s^2}{\sigma^2} \qquad (10\text{-}16)$$

For example, a plant management requires that the variance of weight of a product should not exceed 400. Ten samples of the product selected at random showed a variance of 800. Is quality control below standard? To answer, let us establish the hypothesis $H_o\colon \sigma^2 = 400$, with the alternative $H_a\colon \sigma^2 > 400$. Assume that the samples came from a normal population, and adopt a significance level of 5 percent. By Eq. (10-16),

$$\chi^2 = \frac{9 \times 800}{400} = 18$$

From Table 10-2, for $10 - 1 = 9$ degrees of freedom, and at a significance level of 5 percent, $\chi^2 = 16.92$. Hence, the statistic 18 lies in the critical region of values larger than 16.92. Thus, we must reject the null hypothesis. We can accept the alternative, that the sample came from a population with variance exceeding 400.

Consider now situations where an event E occurs k times in n trials.

You may wish to determine whether this could have been due to chance if the probability of E happening in one trial is p. The hypothesis can be subjected to a χ^2 test. For the purpose, Eq. (10-15) can be put in a convenient form:

$$\chi^2 = \frac{(k - np)^2}{np} + \frac{[n - k - n(1 - p)]^2}{n(1 - p)} = \frac{(k - np)^2}{npq} \tag{10-17}$$

where $q = 1 - p$. As pointed out previously for the tossing of a coin, there is only one degree of freedom.

Suppose, for example, that a manufacturer guarantees that not more than 10 percent of delivered items will be defective. A sample of 100 contains 12 defectives. Is this finding cause for rejection of the shipment? To answer, let us set up the hypothesis $H_o: p = 0.1$, with the alternative $H_a: p > 0.1$. Let us adopt a significance level of 5 percent. By Eq. (10-17),

$$\chi^2 = \frac{(12 - 100 \times 0.1)^2}{100 \times 0.1 \times 0.9} = 0.444$$

From Table 10-2, for one degree of freedom, and at a significance level of 5 percent, $\chi^2 = 3.84$. Hence, we cannot reject the hypothesis. There is insufficient evidence to conclude that $p > 0.1$.

In using the χ^2 test, you should not use expected values less than 5 in computing χ^2. Generally, it is desirable to combine two or more groups of observations so that the expected value exceeds 5.

10-7   *Student's t Test.* We observed in Art. 10-6 that when we know the variance of a population, we can use the χ^2 test to decide whether to reject hypotheses. When the population variance is not known and the sample size is so small that the sample variance is not a close approximation of the population variance, we cannot apply that test. We can, however, resort to the t test, which uses a sample mean and variance. The procedure for testing an hypothesis with the t test is as outlined in Art. 10-5.

That procedure starts with a null hypothesis $H_o: O = E$, the sample statistic O is not different from the population parameter E. Then, the method requires selection of a positive measure δ of the deviation of O from E, and δ is to have a known distribution $f(\delta)$. For the t test, $t = \delta$.

In this case, as in the χ^2 distribution, the density function contains two variables, t and degrees of freedom m. Thus,

$$f(t,m) = \frac{\Gamma[(m + 1)/2]}{\sqrt{m\pi}\ \Gamma(m/2)} \left(1 + \frac{t^2}{m}\right)^{-(m+1)/2} \qquad -\infty < t < \infty \tag{10-18}$$

Because of the complexity of this formula, probabilities of t are obtained from tables. Most books of statistics with tables contain t distribution tables. Table 10-3 is a small example of such tables. It gives the probability that each given value of t will be exceeded, for degrees of freedom up to 24.

The t distribution, for very large values of m, approaches the normal distribution with mean 0 and variance 1. Hence, Table 10-3 gives for $m = \infty$ values for that normal distribution. Generally, when $m > 24$, you can assume a normal distribution for t and use Table 10-1.

TABLE 10-3   Minimum Value of t for Significance at Various Levels

Degrees of freedom	Levels of significance			
	5%	2%	1%	0.1%
2	4.30	6.97	9.93	31.60
3	3.18	4.54	5.84	12.94
4	2.78	3.75	4.60	8.61
5	2.57	3.37	4.03	6.86
6	2.45	3.14	3.71	5.96
7	2.37	3.00	3.50	5.40
8	2.31	2.90	3.36	5.04
9	2.26	2.82	3.25	4.78
10	2.23	2.76	3.17	4.59
11	2.20	2.72	3.11	4.44
12	2.18	2.68	3.06	4.32
13	2.16	2.65	3.01	4.22
15	2.13	2.60	2.95	4.07
19	2.09	2.54	2.86	3.88
24	2.06	2.49	2.80	3.75
∞	1.96	2.33	2.58	3.29

One variate that has the distribution given by Eq. (10-18) is

$$t = \frac{|\bar{x} - \mu|}{\bar{\sigma}} \tag{10-19}$$

where $\bar{x}$ is the sample mean, μ the population mean, and $\bar{\sigma}$ the standard deviation, or standard error, of the means. Using Eq. (10-10), we can rewrite Eq. (10-19) in terms of the sample variance s^2 and standard deviation s:

$$t = \frac{|\bar{x} - \mu|}{\sqrt{s^2/n}} = \frac{|\bar{x} - \mu|\sqrt{n}}{s} \tag{10-20}$$

The degrees of freedom m in Eq. (10-18) equal the number of independent measurements that are available for computing s or s^2. When s or s^2 is determined from the n observations of a sample, $m = n - 1$. For s or s^2 determine the mean; and given the mean and $n - 1$ measurements, we can calculate the remaining one. So there are only $n - 1$ independent measurements for t in this case.

As an example of the use of the t test, let us go back to the packaging problem in Art. 10-4, the case of a manufacturer packaging 1-lb units by machine. This time, however, we are not given any information about the distribution of unit weights. Nine packages selected at random have a mean weight of 1.05 lb. The standard deviation of this sample, calculated from Eq. (10-6), is 0.05. Does the machine need adjustment?

To answer, let us establish the hypothesis $H_o: \mu = 1.00$, with the alternative $H_a: \mu > 1.00$. Let us adopt a significance level of 5 percent. By Eq. (10-20),

$$t = \frac{|1.05 - 1| \sqrt{9}}{0.05} = 3.00$$

From Table 10-3, for $m = 9 - 1 = 8$ degrees of freedom, and at a significance level of 5 percent, $t = 2.31$. Hence, the statistic 3.00 lies in the critical region of values larger than 2.31 and we must reject the hypothesis. We can accept the alternative, that the samples came from a population with mean weight greater than 1.00 lb. Therefore, the machine should be adjusted.

Sometimes you may encounter problems that require a comparison of the means of two sets of independent observations. This requires a test of the hypothesis $H_o: \mu_x = \mu_y$. For this purpose, use

$$t = \frac{\bar{x} - \bar{y}}{\bar{s} \sqrt{1/n_x + 1/n_y}} \qquad (10\text{-}21)$$

where $\bar{x}$ is the mean of a sample with n_x observations, $\bar{y}$ is the mean of a sample with n_y observations from an independent set, and $\bar{s}$, called the pooled estimate of standard deviation, is given by

$$\bar{s} = \sqrt{\frac{\Sigma e_x^2 + \Sigma e_y^2}{n_x + n_y - 2}} \qquad (10\text{-}22)$$

e_x and e_y are deviations from the means of the observations in each of the samples. The number of degrees of freedom $m = n_x + n_y - 2$.

Suppose, for example, that one laboratory tests 10 concrete specimens. It finds a mean strength of 3,000 psi. The sum of the squares of the deviations from this mean is 125,000. Another laboratory, making 17 tests, finds a mean strength of 3,080 psi.   The sum of the squares of the deviations from this mean is 75,000.   Is the difference of the means significant at the 5 percent level?

We start with the hypothesis that the means are equal.   The number of degrees of freedom is $10 + 17 - 2 = 25$.   Extrapolating in Table 10-3 for 25 degrees of freedom, and at a significance level of 5 percent, we obtain $t = 2.06$.   From Eq. (10-22), we get the pooled estimate of standard deviation

$$\bar{s} = \sqrt{\frac{125,000 + 75,000}{25}} = 89.4$$

Equation (10-21) then gives

$$t = \frac{3,080 - 3,000}{89.4 \sqrt{\frac{1}{10} + \frac{1}{17}}} = 2.24$$

Since the statistic $t = 2.24$ is in the critical region of values larger than 2.06, we reject the hypothesis.   We can conclude that the difference of the means is significant.

10-8   *Curve Fitting.* Statistics is often used to determine whether two or more variables are related and, if so, the relationship between them. The following illustrates some of the techniques for the case of two variables.

Engineers are accustomed to dealing with specific relations between variables, such as $y = f(x)$.   But you may encounter variables for which a relation cannot be given precisely; for example, the relation between height and weight of men or women.   Nevertheless, you may wish to give the relation in the form $y = f(x)$, or in an implicit form $f(x,y) = 0$, if possible.

To achieve this, the first step is to formulate an hypothesis as to the relation, or set up a mathematical model from theoretical considerations. Whether or not this is successful, the next step is to collect data, a sample, guided possibly by the conclusions from the first step.   The data should then be analyzed to see if it is in agreement with an hypothesis or if an hypothesis can be formulated from the observations.   The analysis often is aided by a plot of the data.   If the points plotted are clustered closely about a curve, the equation of the curve may be the relation sought.

The process of finding a curve that passes through or near a group of plotted points, indicating their general trend, is called curve fitting. The equation of the curve is called a **regression equation.**

An approximate method of fitting a curve to plotted points is to divide the data into several sets, find the arithmetic mean of the coordinates of each set (center of gravity), and draw the curve through these means (Fig. 10-1). For fitting a straight line, use two sets; for a second-degree curve, use three sets; for an nth-degree curve, use $n + 1$ sets.

If points are plotted on logarithmic coordinate paper, a straight line fitted to them represents $y = ax^n$, where n is the slope of the line and a is the intercept on the y axis.

If points are plotted on semilogarithmic paper (natural numbers along the x axis, $\log_{10}$ along the y axis), a straight line fitted to the points represents $y = a(10)^{kx}$, where k is the slope of the line and a is the intercept on the y axis.

The best-fitting curve of a given type is one that minimizes the sum of the squares of the deviations, or residuals, of the dependent variable from its values on the curve.

Suppose from theoretical considerations or from the pattern of plotted points that a straight line can be fitted to a set of n observations. Let the equation of the least-square line be $y = mx + b$, where m is its slope and b its intercept on the y axis. Then, m and b can be computed from simultaneous solution of the equations:

$$m\Sigma x_i + nb = \Sigma y_i$$
$$m\Sigma x_i{}^2 + b\Sigma x_i = \Sigma x_i y_i \tag{10-23}$$

For example, let us fit a straight line to the points $(1,3.4)$, $(2,3.6)$, $(3,4.6)$, and $(4,6.4)$ plotted in Fig. 10-2. Here, $n = 4$, $\Sigma x_i = 10$, $\Sigma x_i{}^2 = 30$, $\Sigma y_i = 18$, and $\Sigma x_i y_i = 50$. Substitution in Eqs. (10-23) yields

$$10m + 4b = 18$$
$$30m + 10b = 50$$

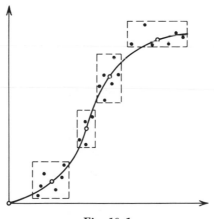

Fig. 10-1

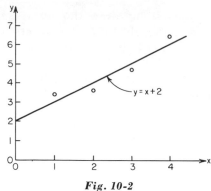

Fig. 10-2

The solution of these equations gives $m = 1$, $b = 2$. Hence, as shown in Fig. 10-2, the required line is $y = x + 2$.

To fit a second-degree least-squares curve to a group of points requires the solution of three simultaneous equations. The procedure, which can be extended to higher-degree equations, is as follows:

Let $y = a_1 + a_2x + a_3x^2$ be the equation of the curve. Let (x_i, y_i) with $i = 1, 2, \ldots, n$ be the observed data, or points. Also, let $v_i = a_1 + a_2x_i + a_3x_i^2 - y_i$ denote the residuals.

If the curve is to make Σv_i^2 a minimum, then the three equations to be solved are

$$\frac{\partial \Sigma v_i^2}{\partial a_k} = 2 \sum v_i \frac{\partial v_i}{\partial a_k} = 0 \qquad k = 1, 2, 3 \tag{10-24}$$

Since $\partial v_i/\partial a_1 = 1$, $\partial v_i/\partial a_2 = x_i$, and $\partial v_i/\partial a_3 = x_i^2$, the equations, when $a_1 + a_2x_i + a_3x_i^2 - y_i$ is substituted for v_i, become

$$\begin{aligned}
na_1 + a_2\Sigma x_i + a_3\Sigma x_i^2 &= \Sigma y_i \\
a_1\Sigma x_i + a_2\Sigma x_i^2 + a_3\Sigma x_i^3 &= \Sigma x_i y_i \\
a_1\Sigma x_i^2 + a_2\Sigma x_i^3 + a_3\Sigma x_i^4 &= \Sigma x_i^2 y_i
\end{aligned} \tag{10-25}$$

The summations are to be carried out from $i = 1$ to $i = n$.

10-9 *Correlation and Regression Coefficients.* In dealing with two variables, we can use as a measure of the variability of the data a parameter corresponding to variance, covariance.

Covariance is defined for two measures x_i and y_i by

$$\sigma_{xy}^2 = \frac{\Sigma(x_i - \bar{x})(y_i - \bar{y})}{n} \tag{10-26a}$$

where $\bar{x}$ is the mean of the set x_i, $\bar{y}$ is the mean of the set y_i, and n is the number of paired observations (x_i, y_i). This equation is equivalent to

$$\sigma_{xy}^2 = \frac{1}{n} \sum x_i y_i - \bar{x}\bar{y} \tag{10-26b}$$

The variables x and y may be related closely, vaguely, or not at all. To measure the degree of association between two variables, statisticians use a correlation coefficient.

Let $e_x = x_i - \bar{x}$ be the deviation of a measure x_i from the mean $\bar{x}$ of its set, and let $e_y = y_i - \bar{y}$ be the deviation of a measure y_i from the mean $\bar{y}$ of its group. The correlation coefficient for the two variables is then

$$r = \frac{\sigma_{xy}^2}{\sigma_x \sigma_y} = \frac{\Sigma e_x e_y}{n\sigma_x \sigma_y} = \frac{\Sigma x_i y_i - n\bar{x}\bar{y}}{n\sigma_x \sigma_y} \tag{10-27}$$

where n = number of paired observations (x_i, y_i)

σ_{xy}^2 = covariance of x and y

σ_x = standard deviation of x set

σ_y = standard deviation of y set

When x plotted against y yields points clustered closely about a straight line, correlation coefficient $r = \pm 1$ if the relation between x and y is perfect. If x and y are not related, $r = 0$.

If the line has a positive slope, r will be positive. If the line has a negative slope, r will be negative.

For point positions other than those about a straight line, the correlation coefficient may not be a reliable measure of relationship.

With data that can be correlated by a straight line, the least-squares line gives the best fit (Fig. 10-2), because it minimizes the sum of the squares of the deviations of the dependent variable. In terms of the correlation coefficient r and standard deviations σ_x and σ_y, we can write the equation of the least-squares line as

$$y_e = r\frac{\sigma_y}{\sigma_x}(x - \bar{x}) + \bar{y} \tag{10-28}$$

This is called the regression equation of the line. It can be used to estimate the probable value of y corresponding to a specific x.

When Eq. (10-28) is used for estimating y from x, the standard error of the estimate is

$$s_y = \sqrt{\frac{\Sigma(y_i - y_e)^2}{n - 2}} = \sigma_y \sqrt{1 - r^2} \tag{10-29}$$

where n is the number of measurements of y. The standard error represents the standard deviation of y from the least-squares line.

The regression coefficient m is the slope of the line.   From Eq. (10-28), the slope is

$$m = r \frac{\sigma_y}{\sigma_x} = \frac{\sigma_{xy}^2}{\sigma_x^2} \tag{10-30}$$

To illustrate the use of the preceding equations, let us try to evaluate the relation between shear strength and weld diameter of spot welds in steel, as observed in a laboratory.   Ten tests are made of spot welds in steel.   A plot of shear strength versus weld diameter (the independent variable, because it can be determined with greater precision) yields points apparently scattered about a straight line.   If the mean strength $\bar{y}$ is 975 lb, the mean diameter $\bar{x}$ is 224 mils, $\sigma_y = 182$ lb, $\sigma_x = 23$ mils, and the sum of the product of the pairs x_i and y_i, $\Sigma x_i y_i = 2{,}220{,}000$, determine the coefficients of correlation and regression, the regression equation, and the standard error of estimate.

By Eq. (10-27), the correlation coefficient is

$$r = \frac{\Sigma x_i y_i - n\bar{x}\bar{y}}{n\sigma_x\sigma_y} = \frac{2{,}220{,}000 - 10 \times 224 \times 975}{10 \times 23 \times 182} = 0.835$$

The regression coefficient is, by Eq. (10-30),

$$m = 0.835 \frac{182}{23} = 6.61$$

By Eq. (10-28), the regression equation is

$$y_e = 6.61(x - 224) + 975 = 6.61x - 507$$

And by Eq. (10-29), the standard error of estimate is

$$s_y = 182 \sqrt{1 - (0.835)^2} = 99.9$$

10-10   *Bibliography*

K. A. Brownlee, "Statistical Theory and Methodology in Science and Engineering," John Wiley & Sons, Inc., New York.

E. C. Bryant, "Statistical Analysis," McGraw-Hill Book Company, New York.

R. S. Burington and D. C. May, "Handbook of Probability and Statistics with Tables," McGraw-Hill Book Company, New York.

I. W. Burr, "Engineering Statistics and Quality Control," McGraw-Hill Book Company, New York.

W. G. Cochran, "Sampling Techniques," John Wiley & Sons, Inc., New York.

S. Ehrenfeld and S. B. Littauer, "Introduction to Statistical Analysis," McGraw-Hill Book Company, New York.

R. A. Fisher and F. Yates, "Statistical Tables for Biological, Agricultural and Medical Research," Oliver & Boyd Ltd., Edinburgh.

E. L. GRANT, "Statistical Quality Control," McGraw-Hill Book Company, New York.

I. GUTTMAN and S. S. WILKS, "Introductory Engineering Statistics," John Wiley & Sons, Inc., New York.

G. J. HAHN and S. S. SHAPIRO, "Statistical Models in Engineering," John Wiley & Sons, Inc., New York.

A. HALD, "Statistical Tables and Formulas," John Wiley & Sons, Inc., New York.

L. J. KAZMIER, "Statistical Analysis for Business and Economics," McGraw-Hill Book Company, New York.

E. S. PEARSON and H. O. HARTLEY, "Biometrika Tables for Statisticians," Cambridge University Press, London.

W. VOLK, "Applied Statistics for Engineers," McGraw-Hill Book Company, New York.

PROBLEMS

1. A building contractor keeping records of the costs of constructing office buildings one year recorded the following costs per square foot: two buildings from $17.50 to $22.50; four buildings from $22.50 to $27.50; nine buildings from $27.50 to $32.50; six buildings from $32.50 to $37.50; one building from $37.50 to $42.50.
    (a) What is the mean cost per square foot?
    (b) What is the median cost?
    (c) What is the 95th percentile?
    (d) What is the 9th decile?
    (e) What is the 3d quartile?
    (f) What is the mode?
2. Compare the arithmetic mean and geometric mean of 2, 4, 8, 16.
3. For the costs per square foot given in Prob. 1:
    (a) What is the range?
    (b) What is the semi-interquartile range, or quartile deviation?
    (c) What is the mean deviation?   (Use the median as base.)
    (d) What is the variance?   [Use Eq. (10-7).]
    (e) What is the standard deviation?
    (f) Check your solution to Prob. 3e with Eqs. (10-5) and (10-6).
4. Show that if $y = ax^n$, the variance of y equals $n^2 \bar{y}^2 \sigma^2(x)/\bar{x}^2$, where n and a are constants, x is the mean value of x, and $\bar{y} = a\bar{x}^n$.
5. Gas flow, as measured by a meter, is given by

$$Q = 1,000 \sqrt{\frac{HP}{W(t + 460)}}$$

where H = pressure differential, in. of water
    P = absolute pressure, psia
    W = molecular weight of gas
    t = temperature, °F

What is the variance of the flow if tests give $\bar{H} = 16$ with $\sigma(H) = 0.50$, $\bar{P} = 36$ with $\sigma(P) = 1.1$, $\bar{W} = 25$ with $\sigma(W) = 0.10$, and $\bar{t} = 69$ with $\sigma(t) = 0.30$?

Which variable contributes most to the variance of Q?   (*Hint:* Use the solution to Prob. 4.)

6. A machine is set to make wires 0.0625 in. in diameter.   Records indicate actual diameters have a normal distribution with standard deviation of 0.0025 in.

(a) What is the 95 percent confidence interval for these wires?

(b) What is the 99 percent confidence interval?

(c) What is the probability of a measurement in the range 0.0593 to 0.0650 in.?

(d) If 100 diameter measurements have a mean of 0.0610 with a standard deviation of 0.0050, should the wire machine be adjusted?   Assume a confidence level of 95 percent.

7. A brick manufacturer claims that the bricks that his plant produces have an ultimate compressive strength of at least 1,000 psi.   His records show the strengths to be normally distributed with a variance of 100.   A prospective purchaser has a laboratory test 25 bricks selected at random.   The laboratory reports a mean strength of 990 psi.   At a significance level of 5 percent, should the potential purchaser refuse to buy the bricks?

8. A company installs two types of lighting fixtures A and B in its factory. Initially, there are 30 percent of type A and 70 percent of type B.   After several years, records indicate that 189 lamps had burned out.   Of these, 42 were type A and 147 type B.   At a significance level of 5 percent, is there a difference between the two types?   (*Hint:* Use the χ^2 test.)

9. A sample containing 1,000 bolts included 14 defective ones.   Usually, 1 percent of such bolts are expected to be defective.   Is there cause for concern? (Assume a significance level of 5 percent and use the χ^2 test.)

10. A machine shop runs a comparison test on cutting machine A that it contemplates buying.   A random sample of 21 pieces cut by this machine has a variance of 0.030.   Long-time records for machine B now in the shop show a variance of 0.025.   Is there a difference in the performance of the machines at a significance level of 5 percent?   (*Hint:* Use the χ^2 test.)

11. A chemical is supposed to have a purity of 87.5 percent.   A daily report of 11 analyses indicates a mean purity of 87.1 percent and standard deviation of 0.80.   Is performance of the equipment acceptable at a significance level of 5 percent?   (*Hint:* Use the t test.)

12. Two laboratories tested five samples of the same material.   Laboratory A reported a mean of 30.0 and gave the sum of the squares of deviations from this mean as 0.672.   Laboratory B reported a mean of 30.8 and gave the sum of the squares of deviations from this mean as 0.112.   Should these results be considered different at the 5 percent level of significance?   (*Hint:* Use the t test.)

13. Given the following pairs of measurements, (1,7.1), (2,9.8), (3,13.4), and (4,15.7),

(a) Give the regression equation for the best-fitting straight line

(b) Compute the covariance of the variables

(c) Calculate the correlation coefficient

(d) Determine the regression coefficient

(e) Compute the standard error of the estimate of y from x

ANSWERS

1. (a) In each group of data, relate number of buildings to the group mean cost. Then, by Eq. (10-2), the mean cost is

$$\frac{2(\$20) + 4(\$25) + 9(\$30) + 6(\$35) + \$40}{2 + 4 + 9 + 6 + 1} = \$30$$

   (b) The median is $30. See Art. 10-1.
   (c) The 95th percentile is $40. See Art. 10-1.
   (d) The 9th decile is $35. See Art. 10-1.
   (e) The third quartile is $35. See Art. 10-1.
   (f) The mode is $30. See Art. 10-1.

2. By Eq. (10-2), the arithmetic mean is $\frac{1}{4}(2 + 4 + 8 + 16) = 7.5$. The geometric mean is $(2 \times 4 \times 8 \times 16)^{\frac{1}{4}} = 5.7$. (See Art. 10-1.) The arithmetic mean is larger.

3. (a) The range is $42.50 - \$17.50 = \$25.00$. See Art. 10-2.
   (b) The third quartile is $35 and the first quartile is $25. (See Art. 10-1.) Hence, the quartile deviation is $\frac{1}{2}(\$35 - \$25) = \$5$. See Art. 10-2.
   (c) The median is $30. The deviations from the median for each group are $20 - 30 = -10$, $25 - 30 = -5$, $30 - 30 = 0$, $35 - 30 = 5$, and $40 - 30 = 10$. Hence, the mean deviation is

$$\frac{2(\$10) + 4(\$5) + 0 + 6(\$5) + \$10}{22} = \$3.64$$

See Art. 10-2.
   (d) The mean is $30. The squares of the deviations from the mean for each group are $(20 - 30)^2 = 100, 25, 0, 25$, and 100. Hence, by Eq. (10-7), the variance is

$$\frac{2(100) + 4(25) + 0 + 6(25) + 100}{22 - 1} = 26.2$$

   (e) From the solution to Prob. 3d, the standard deviation is $\sqrt{26.2} = 5.12$. See Art. 10-2.
   (f) $\Sigma x_i^2 = 2(20)^2 + 4(25)^2 + 9(30)^2 + 6(35)^2 + (40)^2 = 20{,}350$. $(\Sigma x_i)^2 = 660^2 = 435{,}600$. By Eq. (10-5), $\Sigma e_i^2 = 20{,}350 - 435{,}600/22 = 550$. By Eq. (10-6), the standard deviation is $\sqrt{550/21} = 5.12$.

4. We can obtain $\sigma^2(y)$ from Eq. (10-8). For the purpose, first compute

$$\frac{\partial y}{\partial x} = \frac{\partial}{\partial x} ax^n = nax^{n-1}$$

By Eq. (10-8),

$$\sigma^2(y) = (na\bar{x}^{n-1})^2 \sigma^2(x) = n^2 a^2 \bar{x}^{2n-2} \sigma^2(x) \frac{\bar{x}^2}{\bar{x}^2} = n^2(a^2\bar{x}^{2n})\sigma^2(x)/\bar{x}^2$$

$$= n^2 \bar{y}^2 \sigma^2(x)/\bar{x}^2$$

5. We can obtain $\sigma^2(Q)$ from the sum of the variances of H, P, W, and t, using the solution to Prob. 4 with $n^2 = (\frac{1}{2})^2 = \frac{1}{4}$.

$$\bar{Q} = 1,000 \sqrt{\frac{16 \times 36}{25(69 + 460)}} = 209$$

Hence, from Prob. 4 on factoring out $\bar{Q}^2/4$, we get

$$\sigma^2(Q) = \frac{\bar{Q}^2}{4}\left[\frac{\sigma^2(H)}{\bar{H}^2} + \frac{\sigma^2(P)}{\bar{P}^2} + \frac{\sigma^2(W)}{\bar{W}^2} + \frac{\sigma^2(t)}{(\bar{t} + 460)^2}\right]$$
$$= \frac{43,600}{4}\left(\frac{0.25}{256} + \frac{1.21}{1,296} + \frac{0.01}{625} + \frac{0.09}{28,000}\right)$$
$$= 10,900 \times 10^{-4}(9.78 + 9.34 + 0.16 + 0.03) = 21.0$$

The biggest contribution to the variance of the flow comes from the pressure differential readings.

6. (a) From Table 10-1, $x - \bar{x} = \pm 1.96\sigma$ for 95 percent probability. Hence, the 95 percent confidence interval is $0.0625 \pm 1.96 \times 0.0025 = 0.0625 \pm 0.0049$. See Art. 10-4.

(b) From Table 10-1, $x - \bar{x} = \pm 2.58\sigma$ for 99 percent probability. Therefore, the 99 percent confidence interval is $0.0625 \pm 2.58 \times 0.0025 = 0.0625 \pm 0.0065$. See Art. 10-4.

(c) The standard deviate is $(0.0593 - 0.0625)/0.0025 = -1.28$ for a diameter of 0.0593. From Table 10-1, the probability is 39.97 percent of a standard deviate not smaller than -1.28. For a diameter of 0.0650, the standard deviate is $(0.0650 - 0.0625)/0.0025 = 1$. From Table 10-1, the probability is 34.13 percent of a standard deviate not larger than 1.00. Hence, the probability of a measurement between 0.0593 and 0.0650 is $0.3997 + 0.3413 = 0.7410$.

(d) Assume that H_o: $\mu = 0.0625$ is true and H_a: $\mu \neq 0.0625$ is an alternative. By Eq. (10-10), the standard deviation of the means is $\bar{s} = 0.0050/\sqrt{100} = 0.0005$. Hence, the standard deviate is

$$\frac{|\bar{x} - \mu|}{\bar{s}} = \frac{|0.0610 - 0.0625|}{0.0005} = 3.00$$

From Table 10-1, the probability is 95 percent that $|\bar{x} - \mu|/\bar{s} = 1.96$ will not be exceeded. Since the hypothesis gives a standard deviate greater than 1.96, the hypothesis should be rejected. The machine should be adjusted. See Arts. 10-3 through 10-5.

7. Assume that H_o: $\mu \geq 1,000$, with H_a: $\mu < 1,000$. Then, by Eq. (10-10),

$$\frac{\bar{x} - \mu}{\bar{s}} = \frac{\sqrt{n}\,(\bar{x} - \mu)}{\sigma} = \frac{\sqrt{25}\,(990 - 1,000)}{\sqrt{100}} = -5$$

From Table 10-1, the probability is $50 - 45 = 5$ percent of a standard deviate smaller than -1.64. Since the hypothesis gives a standard deviate smaller than

−1.64, the hypothesis can be rejected and the alternate accepted. Hence, the potential purchaser should refuse to buy the bricks. See Arts. 10-3 through 10-5.

8. Assume that $H_o: r_o = r_E$, where r_o is the ratio of the number of type A lamps that need replacement to the number of type B lamps that need replacement, and r_E is the expected value of this ratio. The alternative is $H_a: r_o < r_E$. If the two types performed equally well, the number of each type needing replacement would be proportional to the number installed, and r_o would be equal to $r_E = {}^{30}\!/_{70}$. Hence, we should expect that $0.30 \times 189 = 57$ type A lamps and 132 type B lamps should have been replaced. By Eq. (10-15),

$$\chi^2 = \frac{(42 - 57)^2}{57} + \frac{(147 - 132)^2}{132} = 3.95 + 1.70 = 5.65$$

Since the replacement total is given and only two measures comprise χ^2, there is only one degree of freedom. From Table 10-2, for one degree of freedom, there is only a 5 percent probability of $\chi^2 > 3.84$. The hypothesis gives a value of χ^2 larger than 3.84. Hence, we should reject the hypothesis in favor of the alternative, that type A performed better than type B. See Art. 10-6.

9. Assume that $H_o: O = E$ is true, where O is the observed number of defectives and $E = 0.01 \times 1,000 = 10$ is the expected number of defectives. The alternative is $H_a: 0 > E$. By Eq. (10-17),

$$\chi^2 = \frac{(14 - 10)^2}{10 \times 0.99} = 1.6$$

From Table 10-2, for one degree of freedom, there is only a 5 percent probability of $\chi^2 > 3.84$. Since the hypothesis gave $\chi^2 < 3.84$, it cannot be rejected. There is no cause for concern. See Art. 10-6.

10. Assume that $H_o: \sigma_A{}^2 = \sigma_B{}^2$ is true, with the alternative $H_a: \sigma_A{}^2 > \sigma_B{}^2$. By Eq. (10-16),

$$\chi^2 = \frac{(21 - 1)0.025}{0.030} = 16.67$$

From Table 10-2, there is only a 5 percent probability that χ^2 will exceed 31.40 (interpolated) for 20 degrees of freedom. Since the hypothesis gave $\chi^2 < 31.40$, it cannot be rejected. There is insufficient evidence to conclude that machine A performs differently from machine B. See Art. 10-6.

11. Assume that $H_o: \mu = 87.5$ is true, with the alternative $H_a: \mu < 87.5$. By Eq. (10-20),

$$t = \frac{\sqrt{11}\,|87.1 - 87.5|}{0.80} = 1.66$$

From Table 10-3, there is only a 5 percent probability that t will exceed 2.23 for 10 degrees of freedom. Since the hypothesis gave $t < 2.23$, we cannot reject it. There is insufficient evidence that the equipment is not performing satisfactorily. See Art. 10-7.

12. Assume that $H_o: \sigma_A = \sigma_B$ is true. The number of degrees of freedom is $5 + 5 - 2 = 8$. From Eq. (10-22), the pooled estimate of standard deviation is

$$\bar{s} = \sqrt{\frac{0.672 + 0.112}{8}} = 0.313$$

By Eq. (10-21),

$$t = \frac{30.8 - 30.0}{0.313 \sqrt{\tfrac{1}{5} + \tfrac{1}{5}}} = 4.04$$

From Table 10-3, there is only a 5 percent probability that 2.31 will be exceeded for eight degrees of freedom. Since the hypothesis gave $t > 2.31$, we should reject it. Hence, the results of the tests are not consistent. See Art. 10-7.

13. (a) Use Eq. (10-23), with $n = 4$, $\Sigma x_i = 10$, $\Sigma x_i^2 = 30$, $\Sigma y_i = 46.0$, and $\Sigma x_i y_i = 129.7$. The slope m and intercept b on the y axis are then given by

$$10m + 4b = 46.0$$
$$30m + 10b = 129.7$$

from which $m = 2.94$ and $b = 4.15$. Hence, the equation of the best-fitting line is $y = 2.94x + 4.15$.

(b) The means are $\bar{x} = (1/n)\Sigma x_i = 2.5$, and $\bar{y} = (1/n)\Sigma y_i = 11.5$. By Eq. (10-26b), the covariance is $129.7/4 - 2.5(11.5) = 3.68$.

(c) The variance of x_i, from Eqs. (10-3) and (10-5), is

$$\sigma^2(x) = \tfrac{1}{4}[30 - 2.5(10)] = 1.25$$

and the standard deviation $\sigma_x = 1.12$. The variance of y_i, with $\Sigma y_i^2 = 572.5$, is

$$\sigma^2(y) = \tfrac{1}{4}[572.5 - 11.5(46.0)] = 10.9$$

and the standard deviation is $\sigma_y = 3.30$. By Eq. (10-27) and the solution to Prob. 13b, the correlation coefficient is

$$r = \frac{3.68}{1.12 \times 3.30} = 0.994$$

(d) The regression coefficient is the slope of the line, $m = 2.94$. See Art. 10-9.

(e) As computed from $y_e = 2.94x_i + 4.15$ for $x_i = 1, 2, 3, 4$, $y_e = 7.09$, 10.03, 12.97, 15.91. The deviations of the corresponding values of y_i from these are 0, 0.2, 0.4, 0.2, and the sum of the squares of these deviations is 0.24. By Eq. (10-29), the standard error of the estimate is $\sqrt{0.24/(4 - 2)}$ $= 0.35$.

Index